VOLUME THREE HUNDRED AND FIFTY FIVE

INTERNATIONAL REVIEW OF CELL AND MOLECULAR BIOLOGY

Actin Cytoskeleton in Cancer Progression and Metastasis - Part A

INTERNATIONAL REVIEW OF CELL AND MOLECULAR BIOLOGY

VOLUME THREE HUNDRED AND FIFTY FIVE

INTERNATIONAL REVIEW OF CELL AND MOLECULAR BIOLOGY

Actin Cytoskeleton in Cancer Progression and Metastasis - Part A

Edited by

CLÉMENT THOMAS
Luxembourg Institute of Health,
Luxembourg City, Luxembourg

LORENZO GALLUZZI
Weill Cornell Medical College,
New York, NY, United States

ELSEVIER

ACADEMIC PRESS
An imprint of Elsevier

Academic Press is an imprint of Elsevier
50 Hampshire Street, 5th Floor, Cambridge, MA 02139, United States
525 B Street, Suite 1650, San Diego, CA 92101, United States
The Boulevard, Langford Lane, Kidlington, Oxford OX5 1GB, United Kingdom
125 London Wall, London, EC2Y 5AS, United Kingdom

First edition 2020

ISBN: 978-0-12-821281-3
ISSN: 1937-6448

For information on all Academic Press publications
visit our website at https://www.elsevier.com/books-and-journals

Publisher: Zoe Kruze
Acquisitions Editor: Ashlie M. Jackman
Editorial Project Manager: Leticia M. Lima
Production Project Manager: Denny Mansingh
Cover Designer: Victoria Pearson

Typeset by SPi Global, India

Contents

Contributors

Kyle M. Alpha
Department of Cell and Developmental Biology, SUNY Upstate Medical University, Syracuse, NY, United States

Martin Endara-Coll
Science Division, Biology Program, New York University Abu Dhabi (NYUAD), Abu Dhabi, United Arab Emirates; Department of Molecular Biosciences, The Wenner-Gren Institute, Stockholm University, Stockholm, Sweden

Jessica L. Henty-Ridilla
Department of Cell and Developmental Biology; Department of Biochemistry and Molecular Biology, State University of New York (SUNY) Upstate Medical University, Syracuse, NY, United States

Jessica Hotaling
Department of Cell and Developmental Biology, State University of New York (SUNY) Upstate Medical University, Syracuse, NY, United States

Raquel A.C. Machado
Cancer Cell Biology and Drug Discovery Group, Faculty of Science, Technology and Medicine, University of Luxembourg, Esch-sur-Alzette, Luxembourg

S. Raza Mahmood
Science Division, Biology Program, New York University Abu Dhabi (NYUAD), Abu Dhabi, United Arab Emirates

Ehsan Mohammadi
Faculty of Biomedical Engineering, Amirkabir University of Technology, Tehran, Iran

Piergiorgio Percipalle
Science Division, Biology Program, New York University Abu Dhabi (NYUAD), Abu Dhabi, United Arab Emirates; Department of Molecular Biosciences, The Wenner-Gren Institute, Stockholm University, Stockholm, Sweden

Morgan L. Pimm
Department of Cell and Developmental Biology, State University of New York (SUNY) Upstate Medical University, Syracuse, NY, United States

Elisabeth Schaffner-Reckinger
Cancer Cell Biology and Drug Discovery Group, Faculty of Science, Technology and Medicine, University of Luxembourg, Esch-sur-Alzette, Luxembourg

Mohammad Tafazzoli-Shadpour
Faculty of Biomedical Engineering, Amirkabir University of Technology, Tehran, Iran

Elham Torkashvand
Faculty of Biomedical Engineering, Amirkabir University of Technology, Tehran, Iran

Christopher E. Turner
Department of Cell and Developmental Biology, SUNY Upstate Medical University, Syracuse, NY, United States

Tomas Venit
Science Division, Biology Program, New York University Abu Dhabi (NYUAD), Abu Dhabi, United Arab Emirates

Weiyi Xu
Department of Cell and Developmental Biology, SUNY Upstate Medical University, Syracuse, NY, United States

Stefan Zahler
Department of Pharmacy, Ludwig-Maximilians-Universität, Munich, Germany

CHAPTER ONE

Paxillin family of focal adhesion adaptor proteins and regulation of cancer cell invasion

Kyle M. Alpha, Weiyi Xu, Christopher E. Turner*
Department of Cell and Developmental Biology, SUNY Upstate Medical University, Syracuse, NY, United States
*Corresponding author: e-mail address: turnerce@upstate.edu

Contents

Abstract

The paxillin family of proteins, including paxillin, Hic-5, and leupaxin, are focal adhesion adaptor/scaffolding proteins which localize to cell-matrix adhesions and are important in cell adhesion and migration of both normal and cancer cells. Historically, the role of these proteins in regulating the actin cytoskeleton through focal adhesion-mediated signaling has been well documented. However, studies in recent years have revealed additional functions in modulating the microtubule and intermediate filament cytoskeletons to affect diverse processes including cell polarization, vesicle trafficking and

International Review of Cell and Molecular Biology, Volume 355
ISSN 1937-6448
https://doi.org/10.1016/bs.ircmb.2020.05.003

mechanosignaling. Expression of paxillin family proteins in stromal cells is also important in regulating tumor cell migration and invasion through non-cell autonomous effects on the extracellular matrix. Both paxillin and Hic-5 can also influence gene expression through a variety of mechanisms, while their own expression is frequently dysregulated in various cancers. Accordingly, these proteins may serve as valuable targets for novel diagnostic and treatment approaches in cancer.

Abbreviations

CAF cancer-associated fibroblast
CDM cell-derived matrix
ECM extracellular matrix
EMT epithelial-mesenchymal transition
ERK extracellular signal-regulated kinase
FAK focal adhesion kinase
IF intermediate filament
MMP matrix metalloproteinase
MRTF myocardin-related transcription factor
MT microtubule
MTOC microtubule-organizing center
NSCLC non-small cell lung cancer
TGF-β transforming growth factor beta
YAP/TAZ Yes-associated protein/transcriptional coactivator with PDZ-binding motif

1. Introduction

Cell migration and invasion are finely regulated processes that are critical in many normal physiological processes including during embryonic development, wound repair and immune surveillance (Lauffenburger and Horwitz, 1996; Pollard and Borisy, 2003; Ridley et al., 2003). However, these dynamic cell movements are also crucial in cancer progression and metastasis (Hamidi and Ivaska, 2018). Metastasis is a complex process which includes invasion of primary tumor cells through the neighboring extracellular matrix (ECM) of the tissue stroma, intravasation into the bloodstream, circulatory transport, extravasation, and distant growth of tumor cells in various organs other than the primary site (Chaffer and Weinberg, 2011). Over 90% of cancer mortality is due to metastatic tumor growth, and therefore, understanding the mechanisms by which cells migrate and invade may prove useful in predicting and preventing lethal metastatic disease.

Much of our understanding of the mechanisms underlying cell motility has come from the study of adhesion-based translocation of cells on planar, or two-dimensional (2D) ECM environments (Ridley et al., 2003). However, 2D cell migration rarely occurs *in vivo*. Instead, most cells, including tumor cells, typically reside within, and migrate through, a complex three-dimensional (3D) ECM, where they may utilize both adhesion-based and adhesion-independent mechanisms of translocation (Friedl and Wolf, 2003; Hamidi and Ivaska, 2018; Harunaga and Yamada, 2011; van Helvert et al., 2018; Yamada and Sixt, 2019). Furthermore, there is now compelling evidence that the organization and composition of the 3D stromal ECM performs critical structural and signaling functions in facilitating tumor cell invasion and migration that involves both cell autonomous and non-cell autonomous signaling (Bonnans et al., 2014; Conklin et al., 2011; Provenzano et al., 2006; Valkenburg et al., 2018).

The paxillin family of cytoskeletal adaptor/scaffold proteins localize to sites of cell-ECM interaction called focal adhesions and have well-established structural and signaling roles in the regulation of the actin cytoskeleton and 2D cell migration (Deakin et al., 2012b; Turner, 2000a). In this chapter, we briefly review these findings and then focus on more recent advances in our understanding of how paxillin proteins contribute to tumor cell migration and invasion in 3D environments, discuss their emerging roles in the regulation of microtubule (MT)- and intermediate filament (IF)-based functions, and how their expression in cancer-associated fibroblasts (CAFs) impacts stroma ECM deposition and mechanosignaling to indirectly influence tumor progression.

2. The paxillin family

Paxillin was first described in 1990 as a 68 kDa, tyrosine-phosphorylated cytoskeletal protein that localizes to focal adhesions and interacts with another focal adhesion protein, vinculin (Brown et al., 1996; Turner, 2000a; Turner and Miller, 1994; Turner et al., 1990). Subsequently, two other proteins, Hic-5 (50 kDa; also called transforming growth factor beta 1 induced transcript 1 [TGFB1i1] and androgen receptor coactivator 55 kDa protein [ARA55]), and leupaxin (45 kDa), have been added to the paxillin family, each with similar domain structures, cellular localization, and evolutionarily conserved sequences (Fig. 1, domain maps & interactions) (Fujimoto et al., 1999; Jacob et al., 2016; Lipsky et al., 1998; Shibanuma et al., 1994; Thomas et al., 1999a,b). All three proteins possess an N-terminal domain

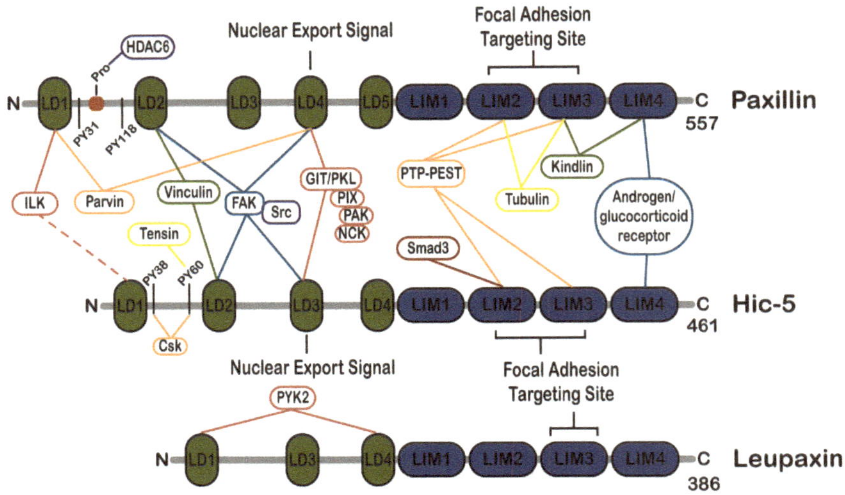

Fig. 1 Domain structure of paxillin family members and key binding partners. The three paxillin family members, paxillin, Hic-5 and leupaxin, exhibit extensive homology within the N-terminal leucine aspartate-rich LD motifs, the C-terminal LIM domains and key phosphorylation sites. These domains interact with shared, as well as distinct, binding partners in their role as adapter/scaffold proteins to regulate focal adhesion organization and signaling to control cytoskeletal remodeling and cell migration. A selection of key binding partners is indicated.

containing several short, amphipathic, helical leucine- and aspartate-rich LD motifs that participate in multiple protein-protein interactions and thereby facilitate their primary function as molecular adaptor/scaffold proteins (Brown and Turner, 2004; Brown et al., 1998; Tumbarello et al., 2002). The C-termini contain four highly conserved LIM domains, which are double-zinc-finger motifs involved in targeting paxillin proteins to focal adhesions and in binding additional structural and regulatory proteins (Brown et al., 1996; Smith et al., 2013). Signaling is accomplished via multiple phosphorylation sites that are distributed throughout the proteins' lengths (Bellis et al., 1995; Brown and Turner, 2004; López-Colomé et al., 2017; Turner, 2000a; Turner and Miller, 1994). Paxillin is expressed ubiquitously (Rashid et al., 2017; Turner et al., 1990), while Hic-5 is primarily expressed in smooth muscle tissues including the vasculature (Turner et al., 1991; Yuminamochi et al., 2003). Leupaxin expression is largely restricted to leukocytes (Lipsky et al., 1998), although it has also been observed in certain smooth muscle tissues and various cancers (Deakin et al., 2012b). Due to the lack of robust evidence regarding leupaxin functions in cancer, this review will focus primarily on paxillin and Hic-5.

The binding partners of, and pathways regulated by, paxillin and its family members have been studied extensively over the past thirty years and have been thoroughly discussed in several excellent reviews (Brown and Turner, 2004; Deakin and Turner, 2008; López-Colomé et al., 2017; Schaller, 2001; Turner, 2000a,b). Briefly, paxillin family members serve as molecular scaffolds/adapters to recruit downstream signaling proteins including kinases, such as focal adhesion kinase (FAK) (Brown et al., 1996; Thomas et al., 1999a,b; Turner and Miller, 1994) and extracellular signal-regulated kinase (ERK) (Ishibe et al., 2004; Ku and Meier, 2000; Subauste et al., 2004), the pseudokinase, integrin-linked kinase (ILK) (Nikolopoulos and Turner, 2001), and phosphatases, such as PTP-PEST (Brown and Turner, 2002; Côté et al., 1999; Jamieson et al., 2005). These phosphorylate or dephosphorylate the multiple serine, threonine, and tyrosine phosphorylation sites in paxillin, thereby regulating its interactions with other proteins (Bellis et al., 1995; Petit et al., 2000). Although they share many of the same binding partners (Fig. 1), there are a number of interactions that appear to be specific to individual paxillin family members. For example, paxillin binds kindlin, tubulin, and Crk (Brown and Turner, 2002; Theodosiou et al., 2016), Hic-5 binds tensin, Csk, and SMADS 3/7 (Goreczny et al., 2018; Shibanuma et al., 2004; Thomas et al., 1999a,b; Wang et al., 2008), and leupaxin binds the FAK-related Pyk2 in leukocytes (Lipsky et al., 1998).

Historically, the effects of paxillin family members on migration are primarily attributed to these phosphorylation changes, direct binding to actin-binding proteins including vinculin, talin and actopaxin/parvin (Humphries et al., 2007; Nikolopoulos and Turner, 2000; Petit et al., 2000; Theodosiou et al., 2016; Webb et al., 2004; Wood et al., 1994), and modulation of the activity of Rho GTPases family members, each of which ultimately impact actin cytoskeleton organization (Deakin and Turner, 2008; Deakin et al., 2012a; Etienne-Manneville and Hall, 2002; Tang et al., 2018). The Rho GTPases are a family of proteins that regulate actin cytoskeletal dynamics and include Cdc42, Rac1, and RhoA, among others (Sit and Manser, 2011). Their functions in 2D cell motility are well-understood (Ridley et al., 2003): Cdc42 is involved in cell polarization and filopodia formation (Faix and Rottner, 2006), Rac1 promotes the formation of broad lamellipodia (Raftopoulou and Hall, 2004), and RhoA promotes the formation of actin stress fibers and is important in the production of contractile forces (Lawson and Burridge, 2014). Both Cdc42 and Rac1 promote the formation of nascent adhesion structures, while RhoA promotes adhesion maturation at the cell's leading edge and disassembly at the cell rear (Ridley et al., 2003; Spiering and Hodgson, 2011). Through the recruitment of GTPase activating proteins (GAPs) and guanine-nucleotide exchange factors (GEFs) such as β-PIX, DOCK180, and p190 RhoGAP, paxillin family members help maintain the delicate balance between these GTPases that is required for optimal cell migration (Deakin and Turner, 2008; Tsubouchi et al., 2002; Turner et al., 1999; Vallés et al., 2004; Zaidel-Bar et al., 2005). In addition to these canonical regulatory networks, this chapter will emphasize novel ways in which paxillin family members regulate cytoskeletal dynamics and cell motility.

3. Fundamentals of cell migration

Migration and invasion are terms with varying connotations depending on the field in which they are used. In an *in vivo*, 3D context, migration typically refers to any form of directed cell movement through the interstitial ECM, which is the relatively porous matrix that surrounds cells throughout the body (Kramer et al., 2013). This may require some degree of ECM proteolysis, depending on the ECM environment and the extent of adhesion to other cells and to the matrix (Schaeffer et al., 2014). In contrast, invasion is a form of migration across some sort of tissue barrier, such as the basement membrane that separates epithelia from their surrounding stroma or surround the endothelial cells in blood vessels (Bravo-Cordero et al., 2012; Petrie and Yamada, 2016).

In both physiologic and tumor microenvironments, the ECM is composed of a network of matrix proteins and glycosaminoglycans that can act as both physical barriers to migration and as tracks along which cells can migrate (Mouw et al., 2014). The interstitial matrix consists of molecules including fibronectin, fibrous collagens, elastin, and proteoglycans, which form a 3D lattice (Frantz et al., 2010). In contrast, the basement membrane is composed primarily of laminins and collagen IV, which form a sheet-like layer between epithelial and stromal cells (Frantz et al., 2010). As an epithelial tumor forms within the normal tissue stroma, cancer cells can migrate outward through the adjacent tissue stroma as single cells or as collective groups (De Pascalis and Etienne-Manneville, 2017). Cells that invade collectively maintain their cell-cell adhesions, including cadherin-based adherens junctions, tight junctions, and desmosomes (Friedl and Alexander, 2011). Generally, mesenchymal or mesenchymal-like leader cells will use proteases to degrade and remodel the matrix, therefore forming paths for subsequent cells to follow (Friedl and Wolf, 2003). Through E-cadherin-dependent cell-cell adhesions, they pull follower cells out from the tumor (Padmanaban et al., 2019; Theveneau and Linker, 2017).

Single-cell migration in 3D can occur through two distinct phenotypes and mechanisms described as either amoeboid or mesenchymal (Condeelis and Segall, 2003; Sahai and Marshall, 2003; te Boekhorst and Friedl, 2016). Cells undergoing amoeboid migration are highly deformable and roughly spherical morphologically with leading lamella and few or no cell-matrix adhesions (Friedl and Wolf, 2003; Petrie and Yamada, 2012). Amoeboid migration requires RhoA-mediated cell contractility which drives cells to move via propulsive blebs and by forcing membrane protrusions through small, pre-existing gaps in the matrix (Friedl and Alexander, 2011; Sahai and Marshall, 2003). Therefore, amoeboid cells generally do not need to degrade the ECM in order to translocate (Lämmermann and Sixt, 2009). In contrast, cells undergoing 3D mesenchymal migration are elongated with small Rac-1-driven lamellipodia (Carragher et al., 2006; Lawson and Ridley, 2018; Sanz-Moreno et al., 2008). They form robust adhesions with the ECM and degrade small paths through it using proteolytic enzymes known as matrix metalloproteinases (MMPs), that are both secreted and membrane-bound (te Boekhorst and Friedl, 2016).

Remarkably, some cancer cells can switch frequently between amoeboid and mesenchymal modes of migration in a process known as plasticity, which is important in effectively invading through a variety of microenvironments (Petrie and Yamada, 2016). It has been hypothesized that this migratory plasticity is one of the primary methods by which certain cancer cell types are able to overcome therapies that target MMPs, which are required for effective mesenchymal migration (Friedl and Wolf, 2003). For a cell to transition from mesenchymal to amoeboid migration, it needs to disassemble its adhesions to the matrix and, reciprocally, an amoeboid cell must assemble and stabilize matrix adhesions to begin mesenchymal migration.

Since most migration *in vivo* occurs in 3D, various tools have been developed over the past several decades to study cell migration in 3D environments and to better understand the mechanisms that regulate these phenotypes (Cukierman et al., 2001). These methods range from simple 3D culture systems, such as *in vitro* collagen matrices, to highly sophisticated intravital imaging of labeled bioluminescent tumor cells (Condeelis and Weissleder, 2010; Shamir and Ewald, 2014). Many of the *in vitro* studies described below utilize 3D cell-derived matrices (CDMs), which are generated by culturing fibroblasts at high density for 1–2 weeks (Beacham et al., 2006). The fibroblasts secrete and remodel a matrix that is composed primarily of collagen and fibronectin fibers, similar to the matrix composition found at the interface between a tumor and surrounding normal stroma *in vivo* (Harunaga and Yamada, 2011). The ways by which members of the paxillin family of adapter proteins influence these various aspects of tumor cell invasion, migration, and stromal matrix remodeling will be discussed below.

4. Roles of paxillin and Hic-5 in regulating migration and invasion

4.1 Actin cytoskeleton-mediated effects

4.1.1 Cell plasticity and 3D migration phenotypes

MDA-MB-231 human breast cancer cells are highly invasive and metastatic *in vivo*. These cells exhibit plasticity in their 3D migration phenotypes and can invade by both mesenchymal and amoeboid modes, making them a useful tool for studying these processes (Deakin and Turner, 2011; Wolf et al., 2003). Paxillin and Hic-5 are both critical for optimal MDA-MB-231 cell migration and invasion in 3D (Fig. 2, amoeboid/mesenchymal migration) (Deakin and Turner, 2011; Gulvady et al., 2018). Furthermore, both proteins are required for MDA-MB-231 transendothelial migration *in vitro* and depletion of either protein significantly reduces the number of lung metastases in a mouse model of metastasis (Deakin and Turner, 2011). These data suggest that expression of both paxillin and Hic-5 is necessary for plasticity *in vivo*, which then enables effective lymphovascular invasion and the establishment of metastatic colonies.

Interestingly, while paxillin and Hic-5 co-ordinately regulate single cell 3D migration *in vitro*, they have dramatically opposing roles in dictating the migration phenotype utilized, which suggests that their relative endogenous expression levels, and/or their respective signaling activity, could play an important role in determining how tumor cells migrate *in vivo* (Deakin and Turner, 2011; Gulvady et al., 2018). Specifically, when MDA-MB-231 cells are seeded in a 3D matrix, acute knockdown of paxillin or overexpression of Hic-5 results in a hypermesenchymal phenotype, characterized by an exaggerated, elongated cell shape and robust, stable focal adhesions (Deakin and Turner, 2011). These robust adhesions take more time to assemble and disassemble, resulting in slower cell motility. Conversely, Hic-5 knockdown or paxillin overexpression results in cells exhibiting an amoeboid phenotype (Deakin and Turner, 2011; Gulvady et al., 2018).

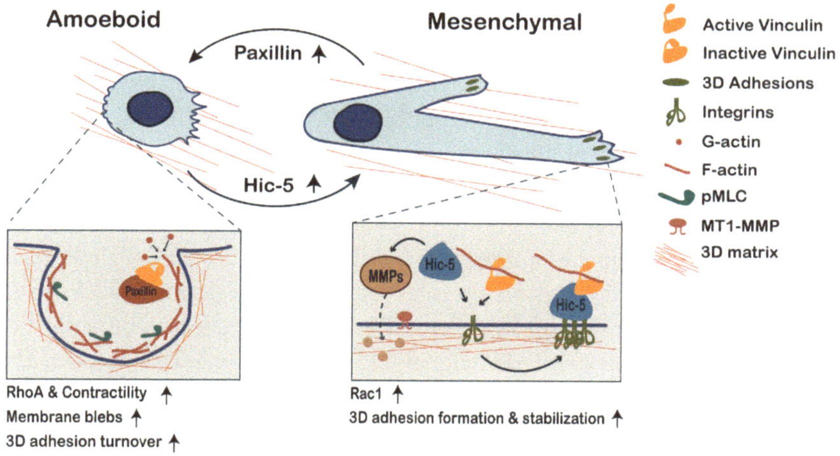

Fig. 2 Paxillin and Hic-5 play opposing roles in regulating tumor cell 3D migration phenotype. Tumor cells with high ratios of paxillin:Hic-5 expression (low Hic-5 levels, since paxillin expression is fairly consistent among tumor cells) more commonly exhibit amoeboid migration, which requires RhoA-mediated contractility. These cells have limited, or no 3D adhesions; any adhesions that form tend to disassemble rapidly. Paxillin may sequester inactive vinculin to promote adhesion disassembly. Amoeboid cells migrate by squeezing through existing spaces in the ECM via F-actin-rich plasma membrane blebs. Conversely, tumor cells with low paxillin:Hic-5 ratios (high Hic-5 levels) can exhibit plasticity and are more likely to exhibit mesenchymal migration, which is Rac1-mediated. These cells form relatively robust, stable 3D adhesions, and use MMPs to degrade the ECM. In these cells, Hic-5 binds active vinculin to promote 3D adhesion formation and stabilization. It is unclear precisely how paxillin versus Hic-5 "activity" is regulated to promote dynamic migration plasticity.

These stark differences in phenotypes are not recapitulated when the cells are plated on a 2D ECM, likely due in part to the fact that amoeboid movement is not observed in 2D cell culture. Instead, cells are well-spread with robust focal adhesions regardless of Hic-5 or paxillin expression levels, thus further emphasizing the importance of cellular context (Deakin and Turner, 2011; Gulvady et al., 2018).

Since paxillin and Hic-5 have highly homologous protein structures, including conserved LD motifs, LIM domains, and key phosphorylation sites, it is remarkable that they play such opposing roles in regulating 3D migration phenotypes (Deakin and Turner, 2011). Many factors could contribute to these differing roles, including differences in spatiotemporal localization, binding affinities for effectors, proteolysis, rates of autophagy of focal adhesions, regulation by kinases, and regulation of Rho GTPases (Cortesio et al., 2011; Deakin et al., 2012a; Nishiya et al., 2001; Sharifi et al., 2016, 2017). For example, tyrosine-phosphorylated paxillin binds the adapter protein Crk to promote migration in bladder carcinoma cells, while Hic-5 is unable to interact with Crk (Petit et al., 2000; Thomas et al., 1999a,b), but selectively interacts with the Src-inhibitory kinase, Csk (Thomas et al., 1999a,b). Importantly, the Rho GTPases play a key role in modulating 3D migration phenotypes: amoeboid migration requires elevated RhoA activity and mesenchymal migration requires Rac1 activity (Friedl et al., 2014). Numerous studies describe roles for both paxillin and Hic-5 in regulating Rho GTPase activity and their differing affinities for binding partners could affect the balance of this signaling. For example, it is known that paxillin and Hic-5 have different affinities for the GTPase activating protein GIT1 (Nishiya et al., 2002). Studies in 2D have shown that paxillin binds GIT1/2 through its LD4 motif, and the associated GIT-PIX-PAK-NCK protein complex regulates local Rac1 activity (Loo et al., 2004; Nayal et al., 2006; Turner et al., 1999; West et al., 2001). Other factors that may contribute to these opposing roles will be discussed in the following paragraphs.

Hic-5 knockdown in MDA-MB-231 cells promotes an amoeboid phenotype through increased activity of RhoA/ROCK and non-muscle myosin II (Deakin and Turner, 2011), which are critical for amoeboid motility due to their contribution to membrane blebbing, retraction, and contractility (Sahai and Marshall, 2003; Yoshida and Soldati, 2006). However, excessive RhoA and myosin activity, without balanced Rac1 activity, may result in dysregulation of blebbing and contractile dynamics that result in the decreases in migration velocity and invasion observed when MDA-MB-231 cells lacking Hic-5 are plated on 3D CDMs (Deakin and Turner, 2011; Ridley, 2015). It is important to note that Hic-5's effect on contractility has not always been consistent in the literature and varies between cell types. For example, it inhibits actomyosin contractility in smooth muscle and osteoblast-like cells by relocating from adhesion sites to actin stress fibers (Guignandon et al., 2006; Kim-Kaneyama et al., 2005), but also stimulates RhoA activity, stress fiber formation, and cell migration upon transforming growth factor beta (TGF-β)-induced epithelial-mesenchymal transition (Tumbarello and Turner, 2007). Therefore, Hic-5's role in modulating RhoA activity is likely cell- and context-dependent. It is also important to note that, despite paxillin's structural homology with Hic-5 and its role as one of the first proteins recruited to focal adhesions, paxillin is not sufficient to nucleate and stabilize 3D adhesions in MDA-MB-231 cells in the absence of Hic-5 (Deakin and Turner, 2011). Cells with an amoeboid morphology that cannot form adhesions and switch their 3D migration phenotype when the microenvironment favors mesenchymal motility are less able to effectively invade (Petrie and Yamada, 2012).

Conversely, the hypermesenchymal phenotype, caused by paxillin depletion in MDA-MB-231 cells, is associated with a significant increase Rac1 activity, which may impair migration persistence by promoting increased lateral protrusions that do not contribute to directed migration (Deakin and Turner, 2011). Paxillin knockdown also inhibits non-muscle myosin II activity, thereby impairing the actomyosin contractility crucial for amoeboid migration; contractility is important for releasing cell adhesions at the cell rear, so diminished contractility may promote a hypermesenchymal phenotype, partially through this failure of adhesion disassembly (Parsons et al., 2010). In a 2D ECM environment, paxillin has been shown to regulate only adhesion disassembly rates (Webb et al., 2004), but in 3D, paxillin depletion decreases both adhesion assembly and disassembly rates and increases the frequency of both short-lived, dynamic adhesions and stable, elongated, long-lasting adhesions (Deakin and Turner, 2011). This difference may be due in part to the fact that cell-matrix adhesions in 3D exhibit different functions, composition, and localization than those classically observed in 2D (Cukierman et al., 2001). With more stable, long-lasting adhesions and slower adhesion disassembly, cell motility is slowed because cells remain tethered more tightly to the matrix (Liu et al., 2015). These changes in adhesion dynamics are potentially due to significantly diminished levels of FAK activity following paxillin depletion (as measured by FAK Y397 phosphorylation) (Deakin and Turner, 2011), since FAK promotes both adhesion maturation and disassembly in 2D (Dumbauld et al., 2010; Webb et al., 2004). Interestingly, FAK Y397 phosphorylation is absent in the 3D adhesions formed by fibroblasts in CDMs (Cukierman et al., 2001) and FAK phosphorylation levels in epithelial cells are dictated by the compliance of the surrounding stromal matrix (Provenzano and Keely, 2011).

Consistent with the experimental manipulation of paxillin and Hic-5 levels described above, the endogenous Hic-5 expression level in tumor cells is indeed a reliable indicator of the cancer cells' migratory phenotype in 3D CDMs, and could therefore be used to predict how cells migrate in tumors *in vivo* (Gulvady et al., 2018). For example, in a panel of cancer cell lines derived from melanoma, fibrosarcoma, breast and pancreatic cancers, increased ratios of Hic-5-to-paxillin expression correlated with a more mesenchymal phenotype, as well as increased migration velocity, *in vitro* invasiveness, and plasticity (Gulvady et al., 2018). Additionally, ectopic overexpression of Hic-5 in the cells with low endogenous levels (e.g., A375MEA-3 and A375P melanoma cells) increased the plasticity and the percentage of cells exhibiting a mesenchymal phenotype and stimulated increased migration through 3D matrices. Interestingly, paxillin expression was relatively high and constant in the same panel of cells and thus was not a good indicator of their 3D phenotype (Gulvady et al., 2018). Nevertheless, this may indicate the absolute requirement for paxillin in tumor cells, possibly for functions unrelated to adhesion and motility. Therefore, Hic-5 expression is a primary determinant of cell morphology and migratory phenotype. Accordingly, Hic-5 levels could potentially be used to predict the predominant mode of motility in patient tumor samples and thereby guide treatment choices to best reduce cell invasion and metastasis. For example, cells from tumors with low Hic-5 expression may favor amoeboid migration and therefore be more resistant to MMP-targeting drugs such as andecaliximab, which has been used in clinical trials for treatment of several cancers (Winer et al., 2018).

Hic-5's role in promoting focal adhesion formation and a 3D mesenchymal phenotype depends on vinculin, another focal adhesion and actin-binding protein and direct binding partner of both Hic-5 and paxillin (Deakin et al., 2012a; Turner et al., 1990) (Fig. 2). Vinculin is well known for its role in mechanosensing/mechanotransduction and in 2D adhesion maturation and importantly, vinculin-null fibroblasts also exhibit a rounded, amoeboid morphology in 3D, which suggests that it may cooperate with Hic-5 to enable mesenchymal migration (Dumbauld et al., 2013; Thievessen et al., 2015). Indeed, experiments with vinculin mutants showed that ectopic expression of vinculin A50I, an "inactive" mutant, in MDA-MB-231 cells phenocopies Hic-5 knockdown, while expression of vinculin T12, a constitutively active vinculin mutant, promotes a mesenchymal phenotype (Gulvady et al., 2018). Importantly, co-expression of vinculin A50I with Hic-5 in cells with low endogenous Hic-5 expression (A375MEA-3 melanoma cells) blocks Hic-5 rescue of a mesenchymal phenotype (Gulvady et al., 2018). These data imply that both Hic-5 and active vinculin are required for driving a mesenchymal phenotype in 3D. Importantly, although Hic-5 does not interact with vinculin A50I at the plasma membrane, paxillin is known to do so and could therefore play a role in keeping vinculin, as well as talin, inactive at the membrane (Atherton et al., 2020;

Deakin et al., 2012a). This could partially explain paxillin's ability to promote adhesion disassembly and an amoeboid morphology and to antagonize the activity of Hic-5 (Fig. 2).

As previously described, Rac1 activity is important for 3D mesenchymal migration and Hic-5's role in mesenchymal migration may be partially mediated through Rac1 (Gulvady et al., 2018; Pankov et al., 2005; Petrie et al., 2012; Sanz-Moreno et al., 2008). Treatment of A375MEA-3 cells overexpressing Hic-5 with a Rac1 inhibitor prevents Hic-5-mediated induction of a mesenchymal phenotype, further connecting these proteins in mesenchymal migration (Gulvady et al., 2018). Additionally, Rac1 and RhoA activity both influence the binding preference of vinculin for Hic-5 versus paxillin in either 2D or 3D ECMs, suggesting that Rho GTPase activity and competition between Hic-5 and paxillin for vinculin binding may be a major mechanism regulating tumor cell migration and plasticity (Deakin et al., 2012a).

4.1.2 Invadopodia and matrix degradation

Cells undergoing mesenchymal migration degrade the ECM by several methods. Expression of membrane-bound membrane-type 1 MMP (MT1-MMP/MMP14) and of secreted MMP1, MMP9, MMP10, MMP11, and MMP13 is often upregulated in almost every cancer type and plays a key role in degrading the ECM, basement membranes, and vascular basal lamina to promote invasion and metastasis (Gobin et al., 2019). Hic-5 has been implicated in regulation of several of these proteins, which will be discussed in later sections. Briefly, Hic-5 is known to promote MT1-MMP localization to the membrane in both endothelial cells and fibroblasts (Dave et al., 2016; Petropoulos et al., 2016) and to promote MMP-9 expression in several cancer lines (Mori et al., 2019).

Cancer cells also utilize specialized adhesion structures known as invadopodia, which are actin-rich membrane protrusions that exhibit abundant MMP localization and activity, including MT1-MMP (Eddy et al., 2017; Linder, 2009; Murphy and Courtneidge, 2011). Although invadopodia have primarily been studied *in vitro,* intravital imaging and careful immunohistochemical staining provide evidence of their existence and importance for directed ECM degradation *in vivo*, where they have been observed in close proximity to areas of basement membrane degradation and at sites of tumor cell extravasation through the endothelium (Leong et al., 2014; Lohmer et al., 2014). Broadly speaking, individual invadopodia consist of a core of F-actin and actin regulatory and binding proteins, which are often surrounded by a ring of adhesion-associated proteins, including integrins, and various Rho GTPase family members (Eddy et al., 2017). The tyrosine kinase Src is particularly important in invadopodia formation (Chen et al., 1985), while cortactin, TKS4, and TKS5 play key roles in modulating their maturation and activity (Buschman et al., 2009; Oser et al., 2009; Sharma et al., 2013). In 2D, invadopodia (and related structures such as podosomes) are visualized as small, actin-rich puncta that colocalize with areas of matrix degradation. However, invadopodia can also self-assemble in 2D culture to form large superstructures known as rosettes which can degrade much larger areas of matrix (Linder, 2007). Both paxillin and Hic-5 localize to invadopodia and regulate their dynamics (Badowski et al., 2008; Gulvady et al., 2019; Petropoulos et al., 2016; Pignatelli et al., 2012b).

Paxillin has been shown to regulate invadopodia dynamics in Rous sarcoma virus (RSV)-transformed baby hamster kidney (BHK) cells and osteoclasts (Badowski et al., 2008). Tyrosine-phosphorylated paxillin is enriched at the inner rim of rosettes, and overexpression of a non-phosphorylatable paxillin mutant (Y31,118F) impaired disassembly of invadopodia actin cores (Badowski et al., 2008). This phenotype can be recapitulated using inhibitors of either calpain or ERK. Paxillin tyrosine phosphorylation activates Erk, which then activates calpain, a protease that promotes both adhesion and invadopodium disassembly through proteolysis (Boateng and Huttenlocher, 2012; Calle et al., 2006; Cortesio et al., 2011).

Although Hic-5 is not expressed in most epithelial cells, it promotes the formation of invadopodia following its upregulation during TGF-β-induced epithelial-mesenchymal transition (EMT) or ectopic overexpression in MCF10A normal mammary epithelial cells (Fig. 3A, invadopodia) (Pignatelli et al., 2012b). This is particularly relevant considering Hic-5's previously discussed role in promoting 3D mesenchymal migration, where cells must degrade the basement membrane in order to invade through the tumor stroma. Hic-5 tyrosine phosphorylation at Y38 and Y60 by Src is critical in this system, as overexpression of the Y38/Y60F non-phosphorylatable mutant or treatment with a Src inhibitor both prevent formation of invadopodia. Downstream signaling from Hic-5 requires RhoC activity for efficient invadopodia formation and p38 MAPK signaling (through Rac1) for matrix degradation, cell migration, and invasion (Fig. 3B) (Pignatelli et al., 2012b).

Hic-5 is also necessary for invadopodia rosette formation in NIH/3T3 Src-transformed fibroblasts (Gulvady et al., 2019). In this system, Hic-5

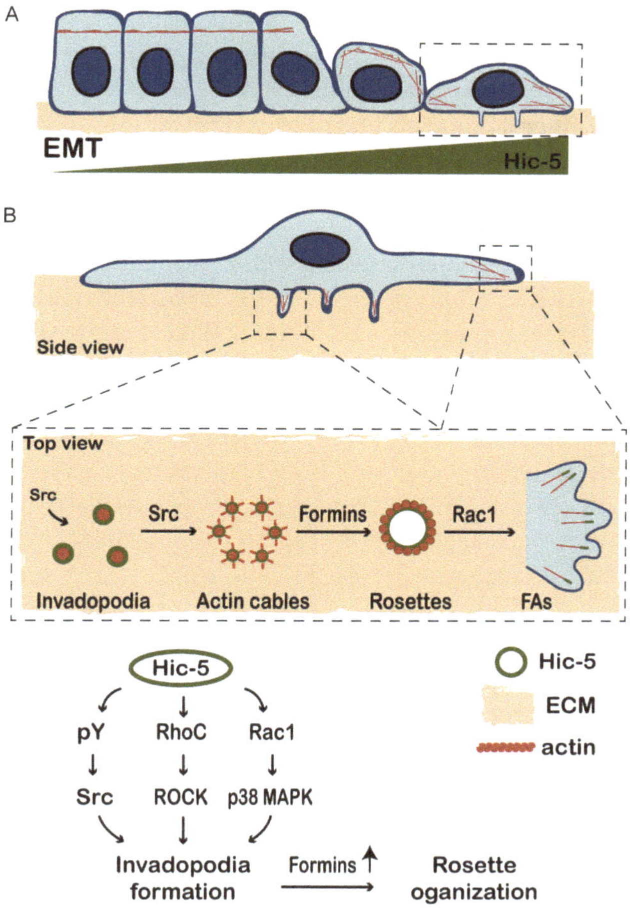

Fig. 3 Hic-5 contributes to the formation and dynamics of invadopodia and higher order rosette structures. (A) Hic-5 is upregulated during TGF-β-induced epithelial-mesenchymal transition (EMT) to promote RhoA/ROCK-mediated actin stress fiber ormation. (B) Hic-5 upregulation in normal mammary epithelial cells also promotes the formation of matrix-degrading invadopodia, via Src, Rac1 and RhoC, leading to increased cell invasion. Both paxillin and Hic-5 localize to, and regulate, the dynamics of individual invadopodia. Hic-5 also promotes the coalescence of numerous invadopodia into higher order rosettes in Src-transformed cells, involving formin-regulated actin assembly. Rapid rosette disassembly is Hic-5-dependent and is frequently associated with bursts of Rac1-driven lamellipodial extensions and the formation of new focal adhesions.

localization to invadopodia requires its LIM domains and complex rosette assembly requires its LD2 and LD3 motifs and phosphorylation at Y38/Y60. The expression of a Hic-5 mutant lacking its LD2 and LD3 domains (Hic-5 ΔLD2,3) or a non-phosphorylatable mutant (Hic-5 Y38,60F) does not prevent invadopodia formation but does alter their dynamics and impairs their organization into rosette structures. Furthermore, FAK kinase activity, Rac1 GTPase activity, and activity of the actin nucleating/polymerizing formin proteins are all required for this Hic-5 dependent invadopodia coalescence into rosettes (Gulvady et al., 2019). Pharmacologic inhibition of any of these proteins mimics the Hic-5 ΔLD2,3 phenotype. However, constitutively active FAK does not rescue rosette formation in cells expressing Hic-5 ΔLD2,3, implying that FAK acts upstream of Hic-5 to promote invadopodia coalescence into rosettes. Therefore, it is possible that FAK/Src-mediated phosphorylation of Hic-5 regulates its interaction with, or activation of, a Rac1 GEF, thereby promoting Rac1 activity that then induces formin-mediated actin assembly to interconnect invadopodia (Gulvady et al., 2019; Panzer et al., 2016) (Fig. 3).

4.2 Crosstalk with microtubules

To migrate effectively and persistently, cells must establish front-rear polarization and efficient anterograde vesicle trafficking to enable directed delivery of factors required for migration to the leading edge of the cell and for recycling of focal adhesion-associated integrins (Mellman and Nelson, 2008; Paul et al., 2015; Petrie et al., 2009; Wilson et al., 2018). The microtubule (MT) network is a major mediator of cell polarization through its role in coordinating cell-ECM adhesions, in positioning of the Golgi apparatus, and in facilitating vesicle trafficking (Garcin and Straube, 2019).

Microtubules are 25-nm wide tubes composed of protofilaments of tubulin subunits (Garcin and Straube, 2019). These subunits can be post-translationally modified by acetylation, phosphorylation, and tyrosination, among others, and these modifications affects MT dynamics (Janke and Bulinski, 2011). MT acetylation is commonly associated with longer-lived and more stable microtubules (Janke and Montagnac, 2017). Importantly, acetylated MTs are enriched at the Golgi apparatus and toward the leading edge in cells undergoing both 2D and 3D cell migration, suggesting that this post-translational modification may be important in establishing a polarized phenotype (Doyle et al., 2009; Ryan et al., 2012). One important regulator of MT acetylation and stability is the histone deacetylase 6 (HDAC6), which also deacetylates α-tubulin (Asthana et al., 2013). In fact, HDAC6 inhibitors have been successful in reducing tumor cell migration and invasion in pre-clinical studies and are currently being used in a number of clinical trials (Bian et al., 2018; Li et al., 2018).

It has long been known that paxillin is involved in establishing and maintaining mesenchymal cell front-rear polarity through its LD4 motif. Expression of a paxillin mutant lacking LD4 impairs cells' ability to orient their Golgi apparatus toward the leading edge and to migrate in a directed manner, but promotes the formation of random, non-directed protrusions, likely as a result of dysregulated Rac1 activity (West et al., 2001). However, recent work has more clearly established another mechanism for paxillin's crucial role in cell polarity.

Paxillin regulates MT acetylation by binding to and inhibiting HDAC6 (Fig. 4A, MT acetylation, polarity, & trafficking) (Deakin and Turner, 2014). Depletion of paxillin results in a significant decrease in MT

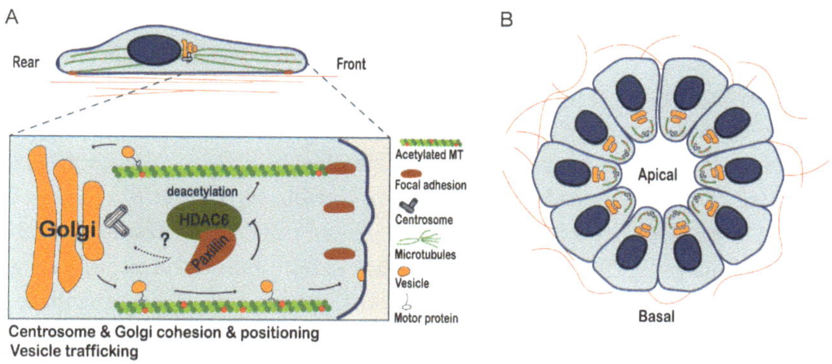

Fig. 4 Paxillin-dependent changes in microtubule acetylation and their effect on cell polarity, vesicle trafficking, and Golgi/centrosome cohesion. (A) Paxillin promotes front-rear polarity in mesenchymal cells through its regulation of microtubule (MT) acetylation. Paxillin binding to, and inhibition of the deacetylase HDAC6 promotes MT acetylation. Paxillin-dependent MT acetylation is also required for Golgi cohesion and positioning, as well as vesicle trafficking of pro-migratory factors, such as integrins from the endoplasmic reticulum via the Golgi apparatus to the plasma membrane. Microtubule targeting to focal adhesions promotes their turnover in a paxillin-dependent manner. Although primarily a focal adhesion protein, paxillin also localizes to the centrosome and is required for its cohesion and polarization during migration. (B) Paxillin also plays a key role in maintaining apical-basal polarity and lumen formation in normal epithelia through regulation of HDAC6 to control microtubule acetylation. Apically enriched acetylated microtubules facilitate basal-apical trafficking, cell shape and the integrity of cell-cell junctions.

acetylation but does not affect overall tubulin expression or distribution. The change in acetylation also reduces cell migration and importantly, can be reversed by treatment with the HDAC6-specific inhibitor tubacin or by HDAC6 knockdown (Deakin and Turner, 2014). Together with data demonstrating HDAC6 enrichment adjacent to focal adhesions and paxillin/HDAC6 binding in cells, these observations suggest that paxillin inhibition of HDAC6 functions to permit normal MT acetylation, thereby promoting cell motility, and is potentially linked to MT-mediated focal adhesion turnover (Efimov et al., 2008; Kaverina et al., 1999). However, since HDAC6 inhibitors can reduce tumor invasion *in vivo* as mentioned above, it is likely that a delicate balance of HDAC6 activity and microtubule acetylation is required for optimal cell migration and invasion.

Paxillin depletion and resulting MT deacetylation also affect the integrity and positioning of the Golgi apparatus and centrosomes (Deakin and Turner, 2014; Dubois et al., 2017). Paxillin knockdown causes fragmentation of the Golgi and prevents its efficient, polarized orientation toward the leading edge in 1D, 2D, and 3D systems (Dubois et al., 2017). This is likely due to MT-dependent effects, since MTs are known to tether Golgi stacks together and support their cohesion (Lippincott-Schwartz et al., 2000). Interestingly, paxillin is also required for centrosome polarization and cohesion—the maintenance of close proximity between centrioles in non-dividing cells (Fig. 4A) (Dubois et al., 2017). A small amount of paxillin localizes to the centrosome and paxillin depletion leads to significant centriole separation (Dubois et al., 2017). Centrioles are typically closely associated in normal cells during interphase (Gönczy, 2015). However, they are often aberrantly duplicated and separated in cancer cells and it is believed that these changes contribute to genomic instability due to chromosome segregation errors during mitosis (Pease and Tirnauer, 2011). Centriole cohesion also depends on MT integrity, suggesting that paxillin-mediated MT acetylation changes may be responsible for the separation phenotype previously observed following paxillin knockdown in MDA-MB-231 cells (Burakov et al., 2003; Dubois et al., 2017). Once again, these centrosome- and Golgi-related phenotypes can be rescued in paxillin-depleted cells by tubacin treatment or HDAC6 RNAi, confirming that they are mediated through paxillin-dependent effects on MT acetylation (Deakin and Turner, 2014; Dubois et al., 2017). However, the precise mechanism connecting paxillin, MT acetylation, and Golgi/centrosome cohesion and orientation has yet to be clearly elucidated.

Importantly, the Golgi apparatus and centrosomes are both microtubule-organizing centers (MTOCs) in the cell (Wu and Akhmanova, 2017).

Therefore, paxillin is required for the cohesion and effective polarization of each of the primary MTOCs in both normal and cancer cells. This mechanism is independent of paxillin localization to focal adhesions, since the observed phenotypes are evident even when cells were plated on poly-L-lysine-coated dishes, which prevents focal adhesion formation (Dubois et al., 2017). Thus, reduced paxillin expression in certain cancers could result in loss of cell polarization to impair effective cell migration/invasion, while a loss of centriole cohesion could promote carcinogenesis due to genomic instability.

Pharmacologic inhibition of FAK, an important signaling partner of paxillin, also prevents paxillin localization to the centrosome and mimics many of the phenotypes which occurred following paxillin depletion, suggesting that these proteins likely work together in regulating MT acetylation (Dubois et al., 2017). FAK stimulates MT acetylation through RhoA activation, so activated FAK could potentially modulate paxillin's regulation of HDAC6 activity (Palazzo et al., 2004).

Paxillin's regulation of MT acetylation also indirectly impacts anterograde vesicle trafficking (Fig. 4A) (Dubois et al., 2017). As mentioned previously, effective transport of pro-migratory components to the leading edge is crucial for optimal directed migration, and this transport often occurs along MTs (Naslavsky and Caplan, 2018). In paxillin-depleted cells, trafficking of GFP-tagged vesicular stomatitis virus G (VSVG)—a marker of anterograde trafficking—from the endoplasmic reticulum to the Golgi and from the Golgi to the plasma membrane are both impaired (Dubois et al., 2017). The mechanism by which this occurs remains to be elucidated, but it is possible that paxillin-dependent changes in MT acetylation affect the binding and/or activity of the molecular motors that traffic cargo along MTs or the activity of Rab GTPases, which function in vesicle trafficking and integrin recycling and therefore, focal adhesion dynamics (Garcin and Straube, 2019; Paul et al., 2015). Regardless of the mechanism, this result suggests an interesting focal adhesion-independent mechanism by which paxillin regulates cell migration.

Using a conditional paxillin knockout mouse, it has been shown that paxillin is also critical for the establishment of apical-basal polarity in normal mouse mammary glands *in vivo* and the formation of 3D polarized acini in a mammary epithelial cell *in vitro* model system, including the apical enrichment of acetylated MTs (Fig. 4B) (Xu et al., 2019). This newly described role for paxillin in normal epithelial cells could become dysregulated during carcinogenesis, as cancer cells often undergo an EMT and lose their apical-basal polarity prior to invasion (EMT will be discussed in a subsequent section). Indeed, 3D acini of mammary epithelial cells derived from the normal mammary gland of the paxillin knockout mice closely resemble the disorganized morphology and defects in cell-cell junction organization seen in tumor organoids generated using tumor fragments from conditional paxillin knockout MMTV-PyMT breast cancer mice (our unpublished results). Remarkably, pharmacologic inhibition of HDAC6 with tubacin rescued the defects in acini polarization (Xu et al., 2019), suggesting that paxillin regulation of MT acetylation not only affects front-rear polarization in single cells, but can also affect the polarization of multicellular units, potentially through its effects on intracellular trafficking and crosstalk with the actin cytoskeleton (Booth et al., 2014).

4.3 Crosstalk with intermediate filaments

Intermediate filaments (IFs) generally play an important role in determining cell morphology and response to mechanical stress due to their characteristic structural integrity (Leduc and Etienne-Manneville, 2015). IFs are built from several proteins and protein families, including cytokeratins, lamins, neurofilaments, desmin, and vimentin, and their expression varies widely by cell type and between normal and cancerous cells (Lowery et al., 2015). In fact, several IFs have been used clinically as markers for disease grade and tumor invasiveness for decades (Osborn and Weber, 1983). For example, vimentin expression is typically limited to mesenchymal cells, so its expression in epithelial cells may indicate a more malignant phenotype (Korsching et al., 2005). Furthermore, cytokeratins are used as markers for tumors of epithelial origin, and individual cytokeratins may be enriched in certain cell populations. For example, cells leading strands of collectively migrating cells in breast cancer are known to express cytokeratin 14 (Diepenbruck and Christofori, 2016). Different IFs have differing effects on cell migration (Leduc and Etienne-Manneville, 2015).

Vimentin is a type III IF expressed primarily in fibroblasts and cells that have undergone EMT (Lowery et al., 2015). Recent studies have shown a role for vimentin in promoting directional persistence of migrating cells, and vimentin filaments occasionally terminate at focal adhesions, suggesting some degree of crosstalk between the actin and IF cytoskeletons at these sites (Battaglia et al., 2018), a process that appears to be dependent on Hic-5 (Vohnoutka et al., 2019). Additionally, vimentin expression in fibroblasts protects against nuclear rupture and DNA damage as cells migrate, by forming a cage around the nucleus (Patteson et al., 2019). A similar mechanism could be important for protecting cancer cells migrating through tight spaces in the ECM, especially since DNA damage is an important part of carcinogenesis through its impact on genomic stability (Hanahan and Weinberg, 2011). Further evidence of focal adhesion/vimentin crosstalk and the role of Hic-5 will be discussed below.

4.4 Epithelial-mesenchymal transition

As briefly described above, epithelial-mesenchymal transition (EMT) is a key process in both normal development and in cancer cell invasion and metastasis, although its role in cancer is highly controversial. (Diepenbruck and Christofori, 2016; Ye and Weinberg, 2015). This controversy stems largely from the inability to monitor the transient and reversible EMT that occurs *in vivo* and evidence that EMT may not be required for metastasis in certain cancers (Fischer et al., 2015). Furthermore, immunohistochemical staining of tumors often show cells migrating collectively, which does not require a complete EMT; in fact, cells which have undergone a total EMT cannot invade collectively (Clark and Vignjevic, 2015). However, new tools have improved our understanding of EMT and its contribution to invasion in certain contexts, and a whole range of intermediate EMT states have been described (Pastushenko et al., 2018).

The process of EMT involves the disassembly of cell-cell junctions, the down-regulation of cell-cell junction components such as E-cadherin, and reorganization of the actin cytoskeleton to promote cell-matrix adhesion and increased migration (Brabletz et al., 2018). EMT also results in a switch from epithelial markers (including several cytokeratins) to mesenchymal markers (including vimentin) (Ye and Weinberg, 2015). Mesenchymal cells are much more motile than epithelial cells in both 2D and 3D culture systems (Aiello et al., 2018; Brabletz et al., 2018). TGF-β, a versatile cytokine which can, depending on the context, act as both an oncogene and a tumor suppressor, is a potent inducer of EMT and stimulates upregulation of Hic-5 expression which, as noted, is typically very low in epithelial cells (Fig. 3A) (Tumbarello and Turner, 2007). As mentioned above, Hic-5 is also known as transforming growth factor beta 1 induced transcript 1 (TGFB1i1) and participates in the non-canonical arm of TGF-β signaling, which primarily impacts RhoGTPase signaling and the actin cytoskeleton (Varney et al., 2016).

Following TGF-β stimulation, Hic-5 facilitates EMT through a downstream pathway involving RhoA and ROCK (Tumbarello and Turner, 2007).

Hic-5 RNAi after TGF-β stimulation results in reduced cell migration and the retention of epithelial morphology. Additionally, Hic-5 RNAi prevents TGF-β-induced RhoA activation, while ectopic Hic-5 overexpression promotes ROCK kinase activity and the formation of ROCK-dependent stress fibers (Tumbarello and Turner, 2007). Interestingly, inhibiting RhoA/ROCK activity by expressing dominant negative RhoA or by using a ROCK inhibitor prevents an increase in Hic-5 expression following TGF-β-induced EMT, suggesting that Hic-5 and RhoA/ROCK work in a positive feedback loop to promote EMT and maintain a mesenchymal phenotype. As discussed earlier, Hic-5 upregulation in response to TGF-β also promotes the formation of invadopodia that are important for the resulting mesenchymal invasion (Pignatelli et al., 2012a) (Fig. 3B).

5. Stroma-mediated effects on tumor invasion

Tumor cells are surrounded by an ECM which is deposited by resident fibroblasts and, along with local immune cells and the vasculature, makes up the tumor stroma (Quail and Joyce, 2013). While the idea that the microenvironment can affect cancer progression has been around since at least the proposition of the "seed-and-soil" hypothesis of metastasis, recent work has greatly increased out understanding of this field (Bissell et al., 2002; Humphrey et al., 2014; Kalluri, 2016; Langley and Fidler, 2011; Paget, 1889). In fact, multiple gene expression studies of the tumor stroma provide a more accurate prediction of patient outcomes than expression studies of the tumor epithelial cells themselves in several cancers, including colorectal and breast cancer (Calon et al., 2015; Finak et al., 2008; Todd et al., 2016). With these facts in mind, it is important to consider the roles of paxillin family members in stromal cells and how their expression in these cells can indirectly affect tumor cell migration and invasion.

5.1 Extracellular matrix deposition

During wound healing, resident tissue fibroblasts differentiate to become myofibroblasts, which participate in wound contraction and fibrotic changes (Bochaton-Piallat et al., 2016; Dabiri et al., 2008). In a similar manner, resident tumor fibroblasts are activated by ECM rigidity- and signaling-mediated changes to become smooth muscle actin-positive, highly contractile cancer-associated fibroblasts (CAFs), which generate and remodel the cancer ECM and play a significant role in tumor progression (Kalluri, 2016).

Crosstalk between CAFs and tumor cells can produce a feed-forward loop resulting in increased tumor cell invasion and metastasis through changes in the rigidity and molecular composition of the ECM and secretion of cytokines. For instance, more rigid and aligned matrices rich in fibrillar collagens and fibronectin generally promote cell invasion and, clinically, increasing tumor rigidity is associated with cancer progression, metastasis, and worse prognoses in breast and colorectal cancer, among others (Emon et al., 2018; Levental et al., 2009; Wei and Yang, 2016). Increased ECM rigidity can result from a number of factors, including increased expression of the collagen cross-linking protein lysyl oxidase (Omoto et al., 2018; Wei et al., 2017) or increased deposition of ECM fibers such as collagen and hyaluronan (Gkretsi and Stylianopoulos, 2018). The relative abundance of ECM proteins is also important for tumor stroma crosstalk. For example, increased stromal expression levels of fibronectin and tenascin C correlate with decreased overall patient survival in breast cancer (Ioachim et al., 2002). Increased collagen density and alignment, quantified by a "tumor-associated collagen signature" score, are also associated with negative patient outcomes (Provenzano et al., 2008).

Information about ECM rigidity and molecular composition is at least partially transmitted through focal adhesions in both tumor cells and CAFs. Multiple focal adhesion proteins can act as mechanosensors, including paxillin and Hic-5, as well as vinculin, talin, and integrins (Jansen et al., 2015). Tumor cells can then respond to these changes through complex signaling pathways involving activation of proteins such as Rho GTPases, MAPKs (mitogen-activated protein kinases), YAP/TAZ (Yes-associated protein/transcriptional coactivator with PDZ-binding motif), FAK, and MRTF (myocardin-related transcription factor) (Maller et al., 2013; Provenzano and Keely, 2011). Together, these pathways induce changes in cytoskeletal organization and gene expression that will be discussed in more detail below.

Hic-5 expression regulates myofibroblast function in non-cancerous wound healing and is also important in regulating CAF function in numerous cancers and pre-cancerous lesions (Fig. 5, CAFs & ECM deposition) (Dabiri et al., 2008; Du et al., 2019; Goreczny et al., 2017; Omoto et al., 2018; Varney et al., 2016). For example, Hic-5 deficiency in pancreatic fibroblasts (known as pancreatic stellate cells or PSCs) protects from cerulein-induced pancreatitis in mice (Chen et al., 2020). In humans, chronic pancreatitis predisposes patients to pancreatic cancer, and the PSCs of patients with pancreatitis exhibit increased Hic-5 expression.

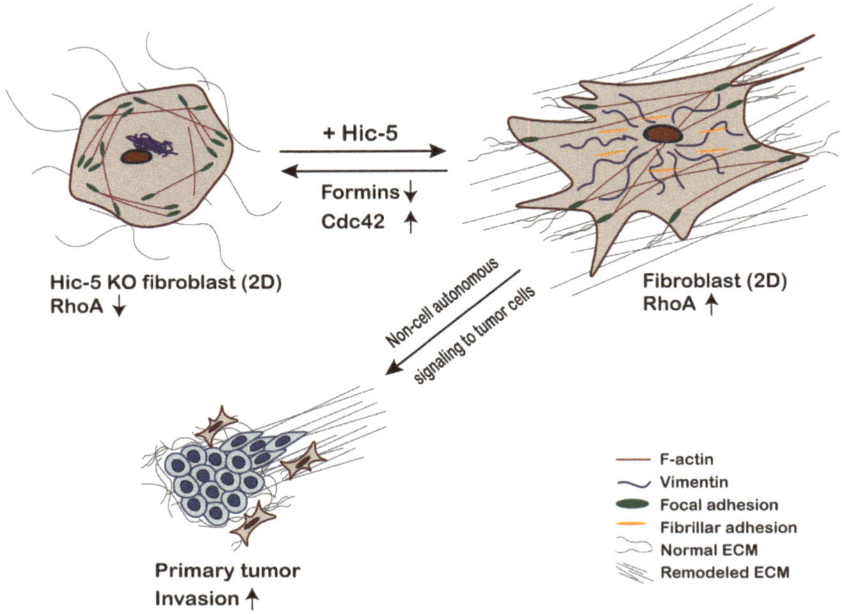

Fig. 5 Hic-5 expression in cancer-associated fibroblasts is required for effective extracellular matrix deposition and organization. Hic-5 expression promotes increased RhoA-mediated contractility in cancer-associated fibroblasts (CAFs), leading to a more "activated" myofibroblast-like phenotype. Hic-5 knockout CAFs are deficient in extracellular matrix (ECM) deposition and remodeling in 2D and 3D when compared to those expressing Hic-5; this is at least partially mediated by diminished fibronectin fiber assembly at fibrillar adhesions involving Src-regulated Hic-5 interactions with tensin. Furthermore, Hic-5 knockout CAFs exhibit a central F-actin "hole" and a collapsed vimentin intermediate filament cytoskeleton. Both of these phenotypes can be mimicked in Hic-5-expressing CAFs by pharmacologic formin inhibition or rescued in Hic-5 knockout CAFs by pharmacologic inhibition of the RhoGTPase Cdc42. Hic-5-dependent deposition and remodeling of the stromal ECM in vivo promotes non-cell autonomous signaling in the tumor cells to stimulate their invasion and dissemination.

If pancreatic cancer develops, Hic-5 expression in PSCs then promotes proliferation, survival, and invasion of pancreatic cancer cells *in vitro* when these cells are cultured together. Additionally, Hic-5 depletion in PSCs results in decreased MMP-9 expression in the fibroblasts (Qian et al., 2020). Furthermore, higher Hic-5 expression correlates with decreased post-operative survival. These data suggest that Hic-5-mediated fibrosis may play an important role in tissue stiffening and other pre-cancerous changes that may ultimately lead to cancer.

A recent study using patient-derived CAFs shows that Hic-5 expression is important in colorectal cancer (Omoto et al., 2018). Relative to normal fibroblasts from tumor-adjacent healthy tissue, colorectal CAFs have significantly higher Hic-5 expression. Hic-5 expression was induced in the normal fibroblasts by treatment with culture media that had been conditioned by CAFs or with cytokines including TGF-β, IL-1 β, and SDF-1/CXCL12. Colon cancer cell growth was shown to be inhibited in the presence of CAFs with low Hic-5 levels in *in vitro* assays, compared with cells cultured with wild-type CAFs (Omoto et al., 2018). Most interestingly, Hic-5 knockout mice did not develop any tumors in an azoxymethane-induced colorectal cancer model, in contrast to a >50% incidence in wild-type mice (Omoto et al., 2018). This was suggested to be partially due to Hic-5 stimulation of lysyl oxidase and collagen I expression. Lysyl oxidase is a secreted protein responsible for cross-linking collagen *in vivo*, and its expression correlates with increasing tumor stiffness, a negative prognostic marker in many cancers (Levental et al., 2009). Therefore, Hic-5 expression in human CAFs promotes a pro-tumorigenic stroma in colorectal cancer. Similar results were observed in oesophageal cancer CAFs (Du et al., 2019).

In the MMTV-PyMT mouse model of breast cancer, Hic-5 is surprisingly not expressed in the invasive tumor cells but its expression in the CAFs promotes tumor growth, decreases tumor latency, and enhances lung metastasis (Goreczny et al., 2017). Immunostaining of tumors from Hic-5 heterozygous and knockout mice show that collagen and fibronectin staining in the tumor stroma is decreased in the absence of Hic-5. This may be partially due to Hic-5's role in promoting fibrillar adhesion formation and maturation through its phosphorylation-dependent interaction with tensin-1 (Fig. 5) (Goreczny et al., 2018). These adhesions are important for assembly of fibronectin fibrils, which also affects collagen deposition and organization *in vivo*. Isolated CAFs from Hic-5 knockout mouse tumors are also less contractile than their Hic-5-expressing counterparts as a result of significantly reduced RhoA activity. In addition, the Hic-5 knockout CAFs produce thinner 3D CDMs with poorly aligned fibronectin and collagen fibers *in vitro* (Goreczny et al., 2017). This is a potentially clinically relevant phenotype since, as mentioned above, increasing density and alignment of tumor ECMs are inversely correlated with patient outcomes, including overall survival (Provenzano et al., 2008), Interestingly, MDA-MB-231 breast cancer cells seeded in CDMs generated by Hic-5-expressing CAFs are less likely to migrate with an amoeboid phenotype and have decreased directional persistence compared to those seeded in CDMs from

Hic-5 knockout CAFs (Goreczny et al., 2017). These changes in migration phenotype are likely one of the contributing factors to the decrease in lung metastases observed in Hic-5 knockout mice.

As noted above, there is growing evidence for crosstalk between the actin and IF cytoskeletons, which likely plays an important role in controling cell shape and motility (Burgstaller et al., 2010; Huber et al., 2015; Osmanagic-Myers et al., 2015).This, in turn, is likely important in myofibroblasts as they organize the surrounding stromal matrix. Evidence of a key role for Hic-5 in connecting vimentin organization with focal adhesions and, therefore, the actin cytoskeleton, was recently described. A collapse of the vimentin cytoskeleton was observed in Hic-5 knockout CAFs and normal lung fibroblasts (Vohnoutka et al., 2019). In contrast, depletion of paxillin did not cause a similar perturbation of the vimentin cytoskeleton. Hic-5 knockout CAFs also exhibit a reduction in centrally localized F-actin stress fibers (an actin "hole"), which could be directly linked to this vimentin phenotype as both are rescued in knockout CAFs by inhibition of the Rho GTPase Cdc42, and phenocopied by pharmacological inhibition of formins in Hic-5 expressing CAFs (Vohnoutka et al., 2019). Together, Hic-5-mediated regulation of Rho GTPase and formin activity may co-ordinately affect both the actin and vimentin cytoskeletons to allow disease-relevant changes in fibroblast phenotype. For instance, low Hic-5 and a collapsed vimentin cytoskeleton may be necessary for a pro-migratory fibroblast phenotype, while high Hic-5 and robust actin and vimentin cytoskeletons are necessary for a highly contractile phenotype, where cells deposit and remodel the surrounding ECM (Fig. 5). Although these observations were made in CAFs, it is interesting to speculate whether a similar phenotype may occur in cancer cells depending on the level of Hic-5 expression and, if so, whether a collapsed vimentin network can still protect against nuclear rupture or facilitate cancer cell amoeboid motility.

5.2 Tumor neovascularization

The tumor stroma also includes blood vessels which support the growing tumor by providing oxygen and nutrients and removing carbon dioxide and waste (Nishida et al., 2006). From early in tumorigenesis, tumor cells promote neovascularization of the tumor stroma by secreting factors which activate an "angiogenic switch," including vascular endothelial growth factor-A (VEGF-A) and fibroblast growth factor (FGF) (Hanahan and Weinberg, 2011). As tumor progression proceeds, this neovascularization is modulated by both tumor and stromal cells. The process of angiogenesis involves activation of endothelial cells by pro-angiogenic factors, production of proteases to degrade the matrix, cell migration/invasion and proliferation, tube formation, synthesis of new basement membrane, and recruitment of supporting cells including smooth muscle cells and pericytes (Rajabi and Mousa, 2017). The degree of neovascularization varies by cancer type, but most solid tumors are highly vascularized with aberrant, convoluted vessels that support the tumors' high metabolic requirements. The success of anti-angiogenic drugs such as bevacizumab and cabozantinib in treating a variety of cancers highlights the importance of neovascularization for tumor growth and survival (Rajabi and Mousa, 2017).

Hic-5 is highly expressed in endothelial cells and vascular smooth muscle cells (Kim-Kim-Kaneyama et al., 2011, 2012). Depletion of Hic-5 reduces endothelial sprouting and lumen formation in a 3D *in vitro* assay, suggesting Hic-5 may play a role in regulating vascularization (Dave et al., 2016). Additionally, inducing sprouting with pro-angiogenic factors including VEGF, FGF, and sphingosine 1-phosphate also increases protein complex formation between Hic-5, FAK, and the matrix metalloproteinase MT1-MMP, which is a membrane-anchored protease critical for endothelial cell invasion during sprouting (Dave et al., 2016). The interaction with MT1-MMP is mediated by Hic-5's LIM2 and LIM3 domains. While Hic-5 knockdown does not alter MT1-MMP activity, it does reduce its localization at the membrane during endothelial sprouting, which results in impaired invasion.

Hic-5 also plays a role in vascular remodeling after injury due to its mechanosensitive properties (Kim-Kaneyama et al., 2008, 2011). When undisturbed, the femoral arteries of Hic-5-null mice appear similar to those of wild-type mice (Kim-Kaneyama et al., 2008). However, after surgical injury by a wire, the arteries in Hic-5-null mice have delayed recovery and higher numbers of apoptotic cells in the vascular walls. When mechanically stretched *in* vitro, smooth muscle cells derived from the Hic-5-null mice exhibit an increased cytoplasmic localization of vinculin, another mechanosensing focal adhesion protein that also binds actin (Atherton et al., 2020; Humphrey et al., 2014; Kim-Kaneyama et al., 2005). This provides further evidence that stable focal adhesions help protect against stretch-induced apoptosis and that Hic-5 is required for stable integration of vinculin and other focal adhesion proteins, as previously discussed in the context of mesenchymal tumor cell morphology. This idea is well-supported by other studies and is an important consideration in cells in the vascular wall, which are constantly experiencing mechanical changes (Katsumi et al., 2005; Kim-Kaneyama et al., 2005). Although these studies don't examine vascular injury in a cancer-related context, the processes of normal vascular wound healing and angiogenesis bear many similarities, suggesting that Hic-5 may be an interesting target to pursue when investigating tumor vascularization (Dabiri et al., 2008; Kareva et al., 2016; Kim-Kaneyama et al., 2012).

Although less well-studied, paxillin also plays a role in remodeling the vasculature. The angiogenic cytokine angiopoietin-1 requires paxillin expression for effective promotion of endothelial cell polarization, migration, and sprouting (Boscher et al., 2019). Angiopoietin-1 binding induces PAK2-mediated paxillin phosphorylation which is necessary for activation of the Rho GTPase Cdc42 at the leading edge and recruitment of the polarity protein Par3. Paxillin also mediates endothelial permeability through control of Rho GTPase signaling, which may be important in allowing tumor cell intravasation and extravasation (Birukova et al., 2009; García-Román and Zentella-Dehesa, 2013; Gawlak et al., 2014).

5.3 Immune cell function

A healthy and intact immune system is a critical defence measure against cancer development (Gonzalez et al., 2018). Accordingly, tumors often demonstrate highly dysregulated immune environments with suppression of normal immune pathways that would destroy the tumor, and activation of inflammatory conditions that promote tumor progression (Pandya et al., 2016). While an in-depth analysis of immune dysregulation is beyond the scope of this review, it is important to note that paxillin has been shown to modulate immune cell signaling, and that a healthy and properly-regulated immune response can impair tumor cell migration and invasion (Gauthier et al., 2017;Herreros et al., 2000 ; Robertson and Ostergaard, 2011).

In a mouse orthotopic colon cancer model, down-regulation of paxillin expression in macrophages indirectly inhibited tumor growth through effects on the immune system (Zhang et al., 2018). Macrophages are phagocytic immune cells that play roles in both the innate and adaptive immune response, and the M2 class of tumor-associated macrophages (TAMs) are anti-inflammatory, promote tumor growth and progression, and correlate with negative patient prognoses (Aras and Zaidi, 2017). Paxillin expression is upregulated during M2 macrophage activation (Zhang et al., 2018). Furthermore, paxillin depletion inhibits M2 polarization and reduces macrophage invasion and proliferation *in vivo*, consistent with its role in epithelial cell polarization (Dubois et al., 2017; Xu et al., 2019). Co-injection of colon cancer cells with either wild-type or paxillin-depleted macrophages in a nude mouse model results in significantly decreased tumor volume in mice with paxillin-depleted macrophages, suggesting that paxillin can modulate the immune response in this circumstance to promote tumor growth (Zhang et al., 2018).

Multiple studies have addressed a role for paxillin in T-cells, an important class of cytotoxic cells in the adaptive immune system that are responsible for killing epithelial tumor cells (Thommen and Schumacher, 2018). Cytotoxic T-cells interface with their target cells at immunologic synapses, which are supramolecular clusters of several transmembrane and membrane-associated proteins. When one of these immunologic synapse proteins, CD103 (the α subunit of integrin $\alpha_E\beta_7$), binds to a tumor-specific antigen, paxillin is phosphorylated and binds to the CD103 cytoplasmic tail at the immunologic synapse (Gauthier et al., 2017). If paxillin expression is knocked down by shRNA or its phosphorylation is inhibited by the Src kinase inhibitor saracatinib, T-cell spreading and effector activities are impaired *in vitro*, including their abilities to lyse tumor cells and release the anti-tumor cytokine INFγ (Gauthier et al., 2017).

Interestingly, another role for paxillin in T-cells is related to its ability to regulate the positioning of the MTOC, as discussed earlier in a different context. When a T-cell binds to a target cell, paxillin is recruited to both the immunologic synapse and the MTOC (Herreros et al., 2000; Robertson and Ostergaard, 2011). Importantly, paxillin contributes to reorientation of the MTOC toward the target cell. This is a crucial step in T-cell activation because it allows for directed delivery of cytokines to the plasma membrane for secretion to enhance target cell killing (Martín-Cófreces et al., 2008). These and other studies suggest that paxillin expression in T-cells is important for effective anti-tumor immunity, potentially through HDAC6-mediated changes in MT acetylation and MTOC polarization.

While paxillin is expressed in most circulating and tissue-resident immune cells, Hic-5 exhibits only slightly enhanced RNA expression in basophils and minimal expression in any other immune cells, suggesting it has little direct impact on immune function (paxillin and Hic-5 expression data from proteinatlas.org, v19.3) (Thul et al., 2017). Similarly, although leupaxin expression was first identified in leukocytes, there a lack of subsequent research indicating an important role for this protein in cancer-related immunity (Lipsky et al., 1998).

6. Regulation of gene expression

Acute and long-term changes in gene expression are also important in cancer progression and metastasis. Expression of paxillin family members can be repressed through the action of microRNA, and both paxillin and Hic-5 can act in the nucleus to regulate gene transcription, similar to several other focal adhesion proteins including FAK and zyxin (Kadrmas and Beckerle, 2004; Kasai et al., 2003; Kleinschmidt and Schlaepfer, 2017; Shibanuma et al., 2003).

6.1 Regulation of paxillin expression by miRNA

Paxillin expression can be post-transcriptionally regulated by miRNAs during cancer invasion and metastasis. miRNAs are a class of 19–22 nt single strand non-coding RNAs that typically act by binding to target mRNAs, thus inducing the degradation of the mRNAs or inhibition of their translation (Bartel, 2009; Kusenda et al., 2006). miRNAs are highly involved in tumor progression processes such as tumor cell adhesion, migration, invasion, and proliferation (Hayes et al., 2014; Peng and Croce, 2016). miR-218 negatively regulates paxillin translation (Wu et al., 2010). Accordingly, low miR-218 levels in non-small cell lung cancer correlate with high paxillin expression and, furthermore, with decreased patient survival and increased relapse rates (Wu et al., 2010). A similar miR-218-dependent, post-transcriptional regulation of paxillin has been observed in oral cavity squamous cell carcinoma (Wu et al., 2014b). miR-137 inhibits paxillin translation in colorectal cancer, so low miR-137 levels result in high paxillin expression and reduced survival, (Chen et al., 2013). However, miR-137-mediated paxillin inhibition in lung cancer cells results in reduced proliferation and migration, suggesting that the result of paxillin down-regulation is again, context-specific (Bi et al., 2014). Several other studies in cancer cell lines have shown that paxillin expression can be directly inhibited by other miRNAs, thus impairing the ability of cancer cells to migrate and invade (Li et al., 2015; Matsuyama et al., 2016; Qin et al., 2015; Tao et al., 2016). For example, miR-145 interacts with paxillin mRNA in human colorectal cancer cells to suppress its translation (Qin et al., 2015), and miR-27b inhibits paxillin expression and integrin-mediated cell adhesion (Matsuyama et al., 2016).

In addition to direct binding to paxillin mRNA, miRNAs can also regulate paxillin through indirect mechanisms. For example, the miR-200 family/ZEB1 axis potentiates signaling through integrin β1 in response to binding extracellular collagen I, which promotes FAK-dependent paxillin phosphorylation and thereby increases lung cancer cell invasion and metastasis (Ungewiss et al., 2016). In ovarian cancer, miR-708 inhibits the GTPase activating protein Rap1B, which is a regulator of integrin-based adhesion signaling (Hattori and Minato, 2003). This inhibition results in reduced paxillin phosphorylation at focal adhesions and decreased lung metastases (Lin et al., 2015).

6.2 Regulation of gene expression by Hic-5 and paxillin

As mentioned above, Hic-5 is also known as androgen receptor coactivator 55 kDa protein (ARA55) (Fujimoto et al., 1999). In addition to its localization at focal adhesions, Hic-5 can also localize to the nucleus as a coactivator for the androgen receptor transcription factor (Heitzer and DeFranco, 2006a; Shibanuma et al., 2003), and studies have shown that it can regulate expression of numerous genes by promoting the formation of a transcriptional complex with the transcriptional coactivator p300 and transcription factors Sp1 and SMAD3 in a variety of cell types (Heitzer and DeFranco, 2006b; Shibanuma et al., 2012; Yang et al., 2000). One particularly relevant study examined Hic-5 responsive gene expression changes in response to glucocorticoid treatment in U2OS osteosarcoma cells (Chodankar et al., 2014). Four of the top 15 gene ontology categories represented in the Hic-5-responsive group of genes were related to cell migration and adhesion, suggesting that Hic-5 may regulate migration indirectly through transcriptional changes.

Hic-5 also plays an important role in transducing mechanical and actin cytoskeleton-related signals to regulate gene expression through pathways including the actin-MRTF-SRF pathway (Varney et al., 2016). MRTFs bind to, and are sequestered by monomeric G-actin, but incorporation of G-actin monomers into F-actin fibers release MRTFs, allowing them to translocate to the nucleus and interact with the transcription factor serum response factor (SRF) (Olson and Nordheim, 2010). Therefore, actin polymerization into stress fibers promotes MRTF-SRF-dependent transcription of genes including several genes involved in contractility, such as smooth muscle actin and calponin (Crider et al., 2011). In fibroblasts, MRTF-SRF signaling promotes Hic-5 expression (Wang et al., 2011), but Hic-5 is also required for TGF-β-dependent nuclear localization of MRTF-A and promotion of smooth muscle actin expression (Varney et al., 2016).

These data indicate that Hic-5 promotes fibroblast differentiation via a TGF-β-dependent feed-forward loop that regulates expression of actin cytoskeleton- and contractility-related genes.

Hic-5 may also play a role in other mechanosensitive transcription pathways involving the YAP/TAZ-TEAD (transcriptional enhancer factor-domain) axis. YAP/TAZ nuclear localization and activation correlate with several Hic-5-related phenotypes, including EMT and ECM rigidity-dependent fibroblast activation, although a direct role for Hic-5 in YAP/TAZ signaling has not been reported (Calvo et al., 2013; Shao et al., 2014). Furthermore, YAP/TAZ activity is modulated by several pathways, including a mechanosensitive regulation similar to MRTF-SRF signaling: high ECM rigidity and/or cellular contractility promotes release of YAP/TAZ from cytoplasmic sequestration and allows nuclear translocation (Totaro et al., 2018). YAP/TAZ activity is elevated in many cancers, and results in increased aggressiveness and cell proliferation, as well as induction of expression of several actin cytoskeleton- and focal adhesion-related genes, including formins, integrins, and RhoGEFs like the Dock proteins (Pocaterra et al., 2020; Totaro et al., 2018). Therefore, a role for Hic-5 in mediating YAP/TAZ transcription of proteins relevant to the actin cytoskeleton and cancer cell migration would not be surprising.

Although paxillin contains a nuclear export sequence and localizes most strongly to focal adhesions, several studies have demonstrated that it can also be shuttled into the nucleus to regulate androgen- and MAPK signaling pathway gene transcription (Kasai et al., 2003; Ma and Hammes, 2018). For example, RNAseq analysis in prostate cancer cells produced a panel of over 1000 paxillin-dependent, androgen-responsive genes (Ma et al., 2019). Paxillin expression in these cells increased transcription of pro-proliferative genes, including the CyclinD/Rb/E2F pathway, while it decreased transcription of pro-apoptotic genes, including CASP1 and TNSF10.

Although much of the signaling through paxillin family members occurs directly at focal adhesions, their respective roles in modulating gene expression is clearly also important to consider when studying their impact on tumor progression.

7. Paxillin family member mutations and expression in human cancers

Although somatic mutations of paxillin family members in cancer are generally considered rare, recent evidence indicates that they may play a role in some cancers (Deakin et al., 2012b). A study of lung cancer tissues showed that paxillin had a 9.5% somatic mutation rate, with the most common mutation (A127T) causing increased proliferation *in vitro* (possibly through binding with the cell survival protein Bcl-2) and increased invasion in a mouse xenograft model (Jagadeeswaran et al., 2008). Even without mutations, expression levels of paxillin family members are often highly dysregulated in a variety of cancers (Fig. 6, table of expression levels) (Deakin et al., 2012b; Rhodes et al., 2004). However, the lack of a consistent

Cancer Type	Paxillin		Hic-5		Leupaxin	
	↑	↓	↑	↓	↑	↓
Bladder Cancer				2		
Brain & CNS Cancer	5		4			
Breast Cancer		2		1	1	1
Cervical Cancer	1					
Head & Neck Cancer	1				1	
Kidney Cancer				1		
Leukemia		9				
Liver Cancer			1			
Lung Cancer				3		2
Lymphoma	4		8			
Melanoma	1					
Oesophageal Cancer	2		1			
Other Cancer			1			
Ovarian Cancer				1		
Pancreatic Cancer			2			
Prostate Cancer				1		
Sarcoma			1			

Fig. 6 Expression of paxillin family members varies among cancers. Numbers in paxillin, Hic-5, and leupaxin columns indicate how many studies show increased or decreased mRNA expression of that gene in tumor relative to normal tissue (thresholds for inclusion: $P < 0.0001$, fold change ≥ 2, top 10% gene rank). Darker cell color indicates higher gene rank percentile for the studies in that cell. Data were obtained from the Oncomine cancer expression database (Thermo Fisher Scientific Inc., www.oncomine.org).

increase or decrease in expression—even among cancers of a specific organ—suggests that their role as either oncogenes or tumor suppressors is context-specific and may vary based on tumor type, stage, and other factors. For example, the opposing roles for paxillin and Hic-5 in regulating 3D migratory phenotype suggest that changes in their expression in response to changes in the surrounding ECM environment may promote more effective migration and, therefore, cancer progression (Deakin and Turner, 2011; Gulvady et al., 2018). Additionally, many of the studies which describe up- or down-regulation of paxillin family members use large-scale genomic, proteomic, or transcriptomic analyses and do not always specify whether gene expression is measured in the tumor cells, stromal cells, or both (López-Colomé et al., 2017). While numerous cell-based studies have focused on the modulation of paxillin family member expression in a variety of tumor cell types, few have carefully studied the role of these proteins using patient data. Here, we will briefly present data from studies describing paxillin's role in gliomas to provide one example of the many contexts in which paxillin family members can affect cancer cell invasion and metastasis.

Paxillin clearly functions as an oncogene in gliomas, and its expression is inversely correlated with patient outcomes (Chen et al., 2016; Sun et al., 2017). RNA sequencing data and immunohistochemical staining both demonstrate increased paxillin expression in glioma tissue relative to normal adjacent tissue, and paxillin expression positively correlates with tumor grade and metastasis (Sun et al., 2017). In a multivariate analysis, where tumors were stratified as either high- or low paxillin expressors, higher-than-median paxillin expression correlated with decreased overall survival with a roughly twofold greater risk of mortality within the study observation period. Gene set enrichment analysis using glioma samples showed that paxillin expression significantly affected gene sets related to cell adhesion, proliferation, and migration. These results were validated using *in vitro* assays of glioma cell lines (Sun et al., 2017). Ectopic expression of paxillin increased proliferation in clonogenic assays and invasion in transwell assays, while paxillin knockdown decreased both. Similar results have been observed in many other cancers as described throughout this review, suggesting that paxillin family proteins may be useful as both biomarkers and potential treatment targets in cancer.

8. Therapeutic potential

Several current cancer therapeutics work by targeting pathways that involve paxillin and/or Hic-5. Therefore, a better understanding of the role of these proteins may lead to the development of novel therapeutics and diagnostics as well as more effective, targeted application of those already employed in the clinic. Here, we will briefly discuss several examples of chemotherapeutics that are already used clinically and whose mechanism of action is somehow impacted by paxillin or Hic-5 signaling as well as potential future directions in targeting these proteins for treatment

Since paxillin family members have no intrinsic enzymatic activity and belong to an emerging family of proteins containing intrinsically disordered regions (Wright and Dyson, 2014), they are difficult to target directly with small molecule inhibitors (Ambadipudi and Zweckstetter, 2016; Neerathilingam et al., 2016). Intrinsically disordered proteins (IDPs) are proteins which do not fold into consistent, defined, stable structures, but instead fluctuate among a range of 3D conformations (Wright and Dyson, 2014). Post-translational modifications can bias these proteins toward a particular set of conformations, allowing the same protein to favor different interactions in different signaling environments. The interactions between an IDP and a binding partner can often be targeted to prevent signaling, especially if the IDP also contains highly ordered regions, such as the paxillin and Hic-5 LD motifs. This approach has been used to understand the function of the paxillin LD motifs by studying their interactions with FAK (Neerathilingam et al., 2016).

The non-receptor tyrosine kinase, FAK is an important signaling partner of both paxillin and Hic-5, and its activation promotes several pro-tumorigenic pathways, including cell survival, migration, and angiogenesis (Lee et al., 2015; Sulzmaier et al., 2014). Therefore, many small molecule inhibitors have been designed to inhibit its kinase activity and have shown some promise in pre-clinical studies as combinatorial therapies (Kanteti et al., 2016; Lee et al., 2015). However, it is also possible to inhibit FAK signaling by impairing its interaction with paxillin (Deramaudt et al., 2014). Expression of a FAK mutant that cannot bind paxillin results in reduced phosphorylation of FAK and its target substrates, including paxillin, leading to altered adhesion dynamics, reduced cell adhesion, and impaired cell migration. Taking this idea one step further, synthetic antibody inhibitors have been designed against paxillin's LD2 and LD4 motifs, which are important for its binding to FAK (Nocula-Lugowska et al., 2015; Tumbarello et al., 2002). These synthetic antibodies can act as competitive inhibitors to compete with the focal adhesion targeting domain of FAK and reduce its binding to paxillin (Nocula-Lugowska et al., 2015). The functional outcomes of this inhibition have not yet been studied in detail but based on results seen with FAK mutants deficient in paxillin binding, targeting the paxillin/FAK interaction in this manner may 1 day be an effective therapeutic approach.

Advances in gene therapy may also soon allow modulation of paxillin and Hic-5 expression *in vivo* by specifically targeting tumor cells with RNAi or CRISPR technologies. For example, siRNAs to paxillin have been be packaged and delivered using carbonate apatite nanocarriers, which accumulate in tumors following intravenous administration in mice (Ashaie et al., 2019). This treatment resulted in significantly decreased tumor volume and weight at all time points in a mouse model of breast cancer relative to those treated with empty carbonate apatite nanocarriers. While CRISPR is currently being used only in clinical trials for gene editing of autologous T-cells (which currently does not involve targeting paxillin or Hic-5), development of delivery vehicles such as these carbonate apatite nanocarriers, polymeric nanoparticles, and liposomes could eventually allow delivery of CRISPR/Cas9 encoding DNA and RNA to tumor cells for *in vivo* modulation of paxillin expression and its downstream signaling partners (Zhan et al., 2019).

In non-small cell lung cancer (NSCLC), cisplatin is a commonly used treatment that interferes with DNA replication, thereby killing fast-growing cells (Wu et al., 2014a). However, cisplatin resistance is a common problem. Interestingly, when paxillin is overexpressed in lung cancer cell lines, the percentage of apoptotic cells following cisplatin treatment is decreased, and increased doses of cisplatin are required for the same inhibitory effect on cell viability (Wu et al., 2014a). Paxillin has previously been shown to suppress apoptosis through FAK-mediated signaling (Zouq et al., 2009), and this cisplatin-related effect was shown to be mediated by phosphorylation of paxillin at Y31 and Y118, which activates ERK signaling and thereby increases expression of the anti-apoptotic protein Bcl-2 (Wu et al., 2014a). However, cisplatin resistance could be partially overcome in a mouse model by combining cisplatin with an ERK inhibitor. In human NSCLC samples, paxillin phosphorylation positively correlates with ERK phosphorylation and Bcl-2 expression and patients with positive staining for these markers had poorer responses to cisplatin-based chemotherapy than patients with negative staining (Salgia et al., 1999; Wu et al., 2014a). Therefore, ERK inhibitors may serve as a valuable tool to prevent paxillin-mediated signaling and overcome cisplatin resistance in NSCLC.

In addition to targeting downstream kinases, other paxillin-regulated proteins such as MTs are commonly targeted in cancer (Jordan and Wilson, 2004). Two major classes of chemotherapeutic agents—taxanes and vinca alkaloids—bind tubulin and inhibit mitosis by either over-stabilizing microtubules or preventing their assembly, respectively, thereby preventing the assembly of a functional mitotic spindle (Calligaris et al., 2010). Since paxillin also regulates MT stability via acetylation, it would be interesting to explore whether the MT-stabilizing taxanes are more effective in tumors with higher paxillin expression.

Finally, Hic-5 expression may be used as a marker to guide treatment decisions in certain situations. As discussed above, it's role in ECM deposition and remodeling has now been described in several contexts, including human colorectal cancer and mouse breast cancer. In aggressive human breast cancer subtypes, including Her2-positive, basal-like, and grade three tumors, increased Hic-5 expression correlates with reduced patient survival, suggesting that assessment of tumor stroma Hic-5 levels could be used to guide therapy choices (Goreczny et al., 2018).

9. Conclusion

The paxillin family of proteins are critical players in cancer cell morphology, migration, and invasion. In addition to their well-studied effects on the actin cytoskeleton, their roles in crosstalk with the MT and IF cytoskeletons to modulate cell polarity, trafficking, and ECM remodeling are exciting emerging avenues of investigation. Despite the importance of this crosstalk with other cytoskeletons, the actin cytoskeleton remains a key partner of paxillin, Hic-5, and other focal adhesion proteins in regulating cell migration through dynamic changes in cell morphology and the formation of matrix-degrading structures such as invadopodia. Due to their dysregulation in a wide variety of cancers, targeting paxillin family proteins and their downstream signaling partners may serve as valuable opportunities for diagnostic evaluation and therapeutic intervention in the future.

Funding

Work in the author's lab was funded by grants from the National Institutes of Health (R35 GM131709, RO1 GM047607, RO1 CA163296) and the Carol M. Baldwin Breast Cancer Research Fund of CNY. We thank the other members of the Turner lab for helpful discussions and apologize to those investigators whose studies we were unable to include in this chapter.

References

Aiello, N.M., Maddipati, R., Norgard, R.J., Balli, D., Li, J., Yuan, S., Yamazoe, T., Black, T., Sahmoud, A., Furth, E.E., Bar-Sagi, D., Stanger, B.Z., 2018. EMT subtype influences epithelial plasticity and mode of cell migration. Dev. Cell 45, 681–695.

Ambadipudi, S., Zweckstetter, M., 2016. Targeting intrinsically disordered proteins in rational drug discovery. Expert Opin. Drug Discov. 11, 65–77.
Aras, S., Zaidi, M.R., 2017. TAMeless traitors: macrophages in cancer progression and metastasis. Br. J. Cancer 117, 1583–1591.
Ashaie, M.A., Islam, R.A., Kamaruzman, N.I., Ibnat, N., Tha, K.K., Chowdhury, E.H., 2019. Targeting cell adhesion molecules via carbonate apatite-mediated delivery of specific siRNAs to breast cancer cells in vitro and in vivo. Pharmaceutics 11, 309.
Asthana, J., Kapoor, S., Mohan, R., Panda, D., 2013. Inhibition of HDAC6 deacetylase activity increases its binding with microtubules and suppresses microtubule dynamic instability in MCF-7 cells. J. Biol. Chem. 288, 22516–22526.
Atherton, P., Lausecker, F., Carisey, A., Gilmore, A., Critchley, D., Barsukov, I., Ballestrem, C., 2020. Relief of talin autoinhibition triggers a force-independent association with vinculin. J. Cell Biol. 219, e201903134.
Badowski, C., Pawlak, G., Grichine, A., Chabadel, A., Oddou, C., Jurdic, P., Pfaff, M., Albiges-Rizo, C., Block, M.R., 2008. Paxillin phosphorylation controls invadopodia/podosomes spatiotemporal organization. Mol. Biol. Cell 19, 633–645.
Bartel, D.P., 2009. microRNAs: target recognition and regulatory functions. Cell 136, 215–233.
Battaglia, R.A., Delic, S., Herrmann, H., Snider, N.T., 2018. Vimentin on the move: new developments in cell migration. F1000Research 7, 1796.
Beacham, D.A., Amatangelo, M.D., Cukierman, E., 2006. Preparation of extracellular matrices produced by cultured and primary fibroblasts. Curr. Protoc. Cell Biol. 33, 10.9.1–10.9.21.
Bellis, S.L., Miller, J.T., Turner, C.E., 1995. Characterization of tyrosine phosphorylation of paxillin in vitro by focal adhesion kinase. J. Biol. Chem. 270, 17437–17441.
Bi, Y., Han, Y., Bi, H., Gao, F., Wang, X., 2014. miR-137 impairs the proliferative and migratory capacity of human non-small cell lung cancer cells by targeting paxillin. Hum. Cell 27, 95–102.
Bian, X., Liang, Z., Feng, A., Salgado, E., Shim, H., 2018. HDAC inhibitor suppresses proliferation and invasion of breast cancer cells through regulation of miR-200c targeting CRKL. Biochem. Pharmacol. 147, 30–37.
Birukova, A.A., Cokic, I., Moldobaeva, N., Birukov, K.G., 2009. Paxillin is involved in the differential regulation of endothelial barrier by HGF and VEGF. Am. J. Respir. Cell Mol. Biol. 40, 99–107.
Bissell, M.J., Radisky, D.C., Rizki, A., Weaver, V.M., Petersen, O.W., 2002. The organizing principle: microenvironmental influences in the normal and malignant breast. Differentiation 70, 537–546.
Boateng, L.R., Huttenlocher, A., 2012. Spatiotemporal regulation of Src and its substrates at invadosomes. Eur. J. Cell Biol. 91, 878–888.
Bochaton-Piallat, M.-L., Gabbiani, G., Hinz, B., 2016. The myofibroblast in wound healing and fibrosis: answered and unanswered questions. F1000Research 5, 752.
Bonnans, C., Chou, J., Werb, Z., 2014. Remodelling the extracellular matrix in development and disease. Nat. Rev. Mol. Cell Biol. 15, 786–801.
Booth, A.J.R., Blanchard, G.B., Adams, R.J., Röper, K., 2014. A dynamic microtubule cytoskeleton directs medial actomyosin function during tube formation. Dev. Cell 29, 562–576.
Boscher, C., Gaonac'h-Lovejoy, V., Delisle, C., Gratton, J.-P., 2019. Polarization and sprouting of endothelial cells by angiopoietin-1 require PAK2 and paxillin-dependent Cdc42 activation. Mol. Biol. Cell 30, 2227–2239.
Brabletz, T., Kalluri, R., Nieto, M.A., Weinberg, R.A., 2018. EMT in cancer. Nat. Rev. Cancer 18, 128–134.

Bravo-Cordero, J.J., Hodgson, L., Condeelis, J., 2012. Directed cell invasion and migration during metastasis. Curr. Opin. Cell Biol. 24, 277–283.
Brown, M.C., Turner, C.E., 2002. Roles for the tubulin- and PTP–PEST-binding paxillin LIM domains in cell adhesion and motility. Int. J. Biochem. Cell Biol. 34, 855–863.
Brown, M.C., Turner, C.E., 2004. Paxillin: adapting to change. Physiol. Rev. 84, 1315–1339.
Brown, M.C., Perrotta, J.A., Turner, C.E., 1996. Identification of LIM3 as the principal determinant of paxillin focal adhesion localization and characterization of a novel motif on paxillin directing vinculin and focal adhesion kinase binding. J. Cell Biol. 135, 1109–1123.
Brown, M.C., Curtis, M.S., Turner, C.E., 1998. Paxillin LD motifs may define a new family of protein recognition domains. Nat. Struct. Biol. 5, 677–678.
Burakov, A., Nadezhdina, E., Slepchenko, B., Rodionov, V., 2003. Centrosome positioning in interphase cells. J. Cell Biol. 162, 963–969.
Burgstaller, G., Gregor, M., Winter, L., Wiche, G., 2010. Keeping the vimentin network under control: cell-matrix adhesion-associated plectin 1f affects cell shape and polarity of fibroblasts. Mol. Biol. Cell 21, 3362–3375.
Buschman, M.D., Bromann, P.A., Cejudo-Martin, P., Wen, F., Pass, I., Courtneidge, S.A., 2009. The novel adaptor protein Tks4 (SH3PXD2B) is required for functional podosome formation. Mol. Biol. Cell 20, 1302–1311.
Calle, Y., Carragher, N.O., Thrasher, A.J., Jones, G.E., 2006. Inhibition of calpain stabilises podosomes and impairs dendritic cell motility. J. Cell Sci. 119, 2375–2385.
Calligaris, D., Verdier-Pinard, P., Devred, F., Villard, C., Braguer, D., Lafitte, D., 2010. Microtubule targeting agents: from biophysics to proteomics. Cell. Mol. Life Sci. 67, 1089–1104.
Calon, A., Lonardo, E., Berenguer-Llergo, A., Espinet, E., Hernando-Momblona, X., Iglesias, M., Sevillano, M., Palomo-Ponce, S., Tauriello, D.V.F., Byrom, D., Cortina, C., Morral, C., Barceló, C., Tosi, S., Riera, A., Attolini, C.S.O., Rossell, D., Sancho, E., Batlle, E., 2015. Stromal gene expression defines poor-prognosis subtypes in colorectal cancer. Nat. Genet. 47, 320–329.
Calvo, F., Ege, N., Grande-Garcia, A., Hooper, S., Jenkins, R.P., Chaudhry, S.I., Harrington, K., Williamson, P., Moeendarbary, E.E., Charras, G., Sahai, E., 2013. Mechanotransduction and YAP-dependent matrix remodelling is required for the generation and maintenance of cancer-associated fibroblasts. Nat. Cell Biol. 15, 637–646.
Carragher, N.O., Walker, S.M., Scott Carragher, L.A., Harris, F., Sawyer, T.K., Brunton, V.G., Ozanne, B.W., Frame, M.C., 2006. Calpain 2 and Src dependence distinguishes mesenchymal and amoeboid modes of tumour cell invasion: a link to integrin function. Oncogene 25, 5726–5740.
Chaffer, C.L., Weinberg, R.A., 2011. A perspective on cancer cell metastasis. Science (80-) 331, 1559–1564.
Chen, W.T., Chen, J.M., Parsons, S.J., Parsons, J.T., 1985. Local degradation of fibronectin at sites of expression of the transforming gene product pp60src. Nature 316, 156–158.
Chen, D.L., Wang, D.S., Wu, W.J., Zeng, Z.L., Luo, H.Y., Qiu, M.Z., Ren, C., Zhang, D.S., Wang, Z.Q., Wang, F.H., Li, Y.H., Kang, T.B., Xu, R.H., 2013. Overexpression of paxillin induced by miR-137 suppression promotes tumor progression and metastasis in colorectal cancer. Carcinogenesis 34, 803–811.
Chen, B., Xia, L., Xu, C., Xiao, F., Wang, Y., 2016. Paxillin functions as an oncogene in human gliomas by promoting cell migration and invasion. Onco. Targets. Ther. 9, 6935–6943.
Chen, H., Tan, P., Qian, B., Du, Y., Wang, A., Shi, H., Huang, Z., Huang, S., Liang, T., Fu, W., 2020. Hic-5 deficiency protects cerulein-induced chronic pancreatitis via down-regulation of the NF-κB (p65)/IL-6 signalling pathway. J. Cell. Mol. Med. 24, 1488–1503.

Chodankar, R., Wu, D.Y., Schiller, B.J., Yamamoto, K.R., Stallcup, M.R., 2014. Hic-5 is a transcription coregulator that acts before and/or after glucocorticoid receptor genome occupancy in a gene-selective manner. Proc. Natl. Acad. Sci. U. S. A. 111, 4007–4012.
Clark, A.G., Vignjevic, D.M., 2015. Modes of cancer cell invasion and the role of the microenvironment. Curr. Opin. Cell Biol. 36, 13–22.
Condeelis, J., Segall, J.E., 2003. Intravital imaging of cell movement in tumours. Nat. Rev. Cancer 3, 921–930.
Condeelis, J., Weissleder, R., 2010. In vivo imaging in cancer. Cold Spring Harb. Perspect. Biol. 2, a003848.
Conklin, M.W., Eickhoff, J.C., Riching, K.M., Pehlke, C.A., Eliceiri, K.W., Provenzano, P.P., Friedl, A., Keely, P.J., 2011. Aligned collagen is a prognostic signature for survival in human breast carcinoma. Am. J. Pathol. 178, 1221–1232.
Cortesio, C.L., Boateng, L.R., Piazza, T.M., Bennin, D.A., Huttenlocher, A., 2011. Calpain-mediated proteolysis of paxillin negatively regulates focal adhesion dynamics and cell migration. J. Biol. Chem. 286, 9998–10006.
Côté, J.F., Turner, C.E., Tremblay, M.L., 1999. Intact LIM 3 and LIM 4 domains of paxillin are required for the association to a novel polyproline region (Pro 2) of protein-tyrosine phosphatase-PEST. J. Biol. Chem. 274, 20550–20560.
Crider, B.J., Risinger, G.M., Haaksma, C.J., Howard, E.W., Tomasek, J.J., 2011. Myocardin-related transcription factors A and B are key regulators of TGF-B1-induced fibroblast to myofibroblast differentiation. J. Invest. Dermatol. 131, 2378–2385.
Cukierman, E., Pankov, R., Stevens, D.R., Yamada, K.M., 2001. Taking cell-matrix adhesions to the third dimension. Science (80-) 294, 1708–1712.
Dabiri, G., Tumbarello, D.A., Turner, C.E., Van De Water, L., 2008. Hic-5 promotes the hypertrophic scar myofibroblast phenotype by regulating the TGF-β1 autocrine loop. J. Invest. Dermatol. 128, 2518–2525.
Dave, J.M., Abbey, C.A., Duran, C.L., Seo, H., Johnson, G.A., Bayless, K.J., 2016. Hic-5 mediates the initiation of endothelial sprouting by regulating a key surface metalloproteinase. J. Cell Sci. 129, 743–756.
De Pascalis, C., Etienne-Manneville, S., 2017. Single and collective cell migration: the mechanics of adhesions. Mol. Biol. Cell 28, 1833–1846.
Deakin, N.O., Turner, C.E., 2008. Paxillin comes of age. J. Cell Sci. 121, 2435–2444.
Deakin, N.O., Turner, C.E., 2011. Distinct roles for paxillin and Hic-5 in regulating breast cancer cell morphology, invasion, and metastasis. Mol. Biol. Cell 22, 327–341.
Deakin, N.O., Turner, C.E., 2014. Paxillin inhibits HDAC6 to regulate microtubule acetylation, Golgi structure, and polarized migration. J. Cell Biol. 206, 395–413.
Deakin, N.O., Ballestrem, C., Turner, C.E., 2012a. Paxillin and Hic-5 interaction with vinculin is differentially regulated by Rac1 and RhoA. PLoS One 7, e37990.
Deakin, N.O., Pignatelli, J., Turner, C.E., 2012b. Diverse roles for the paxillin family of proteins in cancer. Genes Cancer 3, 362–370.
Deramaudt, T.B.T.B., Dujardin, D., Noulet, F., Martin, S., Vauchelles, R., Takeda, K., Ronde, P., Rondé, P., 2014. Altering FAK-paxillin interactions reduces adhesion, migration and invasion processes. PLoS One 9, e92059.
Diepenbruck, M., Christofori, G., 2016. Epithelial-mesenchymal transition (EMT) and metastasis: yes, no, maybe? Curr. Opin. Cell Biol. 43, 7–13.
Doyle, A.D., Wang, F.W., Matsumoto, K., Yamada, K.M., 2009. One-dimensional topography underlies three-dimensional fibrillar cell migration. J. Cell Biol. 184, 481–490.
Du, X., Xu, Q., Pan, D., Xu, D., Niu, B., Hong, W., Zhang, R., Li, X., Chen, S., 2019. Hic-5 in cancer-associated fibroblasts contributes to esophageal squamous cell carcinoma progression. Cell Death Dis. 10, 873.
Dubois, F., Alpha, K., Turner, C.E., 2017. Paxillin regulates cell polarization and anterograde vesicle trafficking during cell migration. Mol. Biol. Cell 28, 3815–3831.

Dumbauld, D.W., Shin, H., Gallant, N.D., Michael, K.E., Radhakrishna, H., García, A.J., 2010. Contractility modulates cell adhesion strengthening through focal adhesion kinase and assembly of vinculin-containing focal adhesions. J. Cell. Physiol. 223, 746–756.

Dumbauld, D.W., Lee, T.T., Singh, A., Scrimgeour, J., Gersbach, C.A., Zamir, E.A., Fu, J., Chen, C.S., Curtis, J.E., Craig, S.W., García, A.J., 2013. How vinculin regulates force transmission. Proc. Natl. Acad. Sci. U. S. A. 110, 9788–9793.

Eddy, R.J., Weidmann, M.D., Sharma, V.P., Condeelis, J.S., 2017. Tumor cell invadopodia: invasive protrusions that orchestrate metastasis. Trends Cell Biol. 27, 595–607.

Efimov, A., Schiefermeier, N., Grigoriev, I., Brown, M.C., Turner, C.E., Small, J.V., Kaverina, I., 2008. Paxillin-dependent stimulation of microtubule catastrophes at focal adhesion sites. J. Cell Sci. 121, 196–204.

Emon, B., Bauer, J., Jain, Y., Jung, B., Saif, T., 2018. Biophysics of tumor microenvironment and cancer metastasis—a mini review. Comput. Struct. Biotechnol. J. 16, 279–287.

Etienne-Manneville, S., Hall, A., 2002. Rho GTPases in cell biology. Nature 420, 629–635.

Faix, J., Rottner, K., 2006. The making of filopodia. Curr. Opin. Cell Biol. 18, 18–25.

Finak, G., Bertos, N., Pepin, F., Sadekova, S., Souleimanova, M., Zhao, H., Chen, H., Omeroglu, G., Meterissian, S., Omeroglu, A., Hallett, M., Park, M., 2008. Stromal gene expression predicts clinical outcome in breast cancer. Nat. Med. 14, 518–527.

Fischer, K.R., Durrans, A., Lee, S., Sheng, J., Li, F., Wong, S.T.C., Choi, H., El Rayes, T., Ryu, S., Troeger, J., Schwabe, R.F., Vahdat, L.T., Altorki, N.K., Mittal, V., Gao, D., 2015. Epithelial-to-mesenchymal transition is not required for lung metastasis but contributes to chemoresistance. Nature 527, 472–476.

Frantz, C., Stewart, K.M., Weaver, V.M., 2010. The extracellular matrix at a glance. J. Cell Sci. 123, 4195–4200.

Friedl, P., Alexander, S., 2011. Cancer invasion and the microenvironment: plasticity and reciprocity. Cell 147, 992–1009.

Friedl, P., Wolf, K., 2003. Tumour-cell invasion and migration: diversity and escape mechanisms. Nat. Rev. Cancer 3, 362–374.

Friedl, P., Wolf, K., Zegers, M.M., 2014. Rho-directed forces in collective migration. Nat. Cell Biol. 16, 208–210.

Fujimoto, N., Yeh, S., Kang, H.Y., Inui, S., Chang, H.C., Mizokami, A., Chang, C., 1999. Cloning and characterization of androgen receptor coactivator, ARA55, in human prostate. J. Biol. Chem. 274, 8316–8321.

García-Román, J., Zentella-Dehesa, A., 2013. Vascular permeability changes involved in tumor metastasis. Cancer Lett. 335, 259–269.

Garcin, C., Straube, A., 2019. Microtubules in cell migration. Essays Biochem. 63, 509–520.

Gauthier, L., Corgnac, S., Boutet, M., Gros, G., Validire, P., Bismuth, G., Mami-Chouaib, F., 2017. Paxillin binding to the cytoplasmic domain of CD103 promotes cell adhesion and effector functions for CD8+ resident memory T cells in tumors. Cancer Res. 77, 7072–7082.

Gawlak, G., Tian, Y., O'Donnell, J.J., Tian, X., Birukova, A.A., Birukov, K.G., 2014. Paxillin mediates stretch-induced Rho signaling and endothelial permeability via assembly of paxillin-p42/44MAPK-GEF-H1 complex. FASEB J. 28, 3249–3260.

Gkretsi, V., Stylianopoulos, T., 2018. Cell adhesion and matrix stiffness: coordinating cancer cell invasion and metastasis. Front. Oncol. 8, 145.

Gobin, E., Bagwell, K., Wagner, J., Mysona, D., Sandirasegarane, S., Smith, N., Bai, S., Sharma, A., Schleifer, R., She, J.X., 2019. A pan-cancer perspective of matrix metalloproteases (MMP) gene expression profile and their diagnostic/prognostic potential. BMC Cancer 19, 581.

Gönczy, P., 2015. Centrosomes and cancer: revisiting a long-standing relationship. Nat. Rev. Cancer 15, 639–652.

Gonzalez, H., Hagerling, C., Werb, Z., 2018. Roles of the immune system in cancer: from tumor initiation to metastatic progression. Genes Dev. 32, 1267–1284.
Goreczny, G.J., Ouderkirk-Pecone, J.L., Olson, E.C., Krendel, M., Turner, C.E., 2017. Hic-5 remodeling of the stromal matrix promotes breast tumor progression. Oncogene 36, 2693–2703.
Goreczny, G.J., Forsythe, I.J., Turner, C.E., 2018. Hic-5 regulates fibrillar adhesion formation to control tumor extracellular matrix remodeling through interaction with tensin1. Oncogene 37, 1699–1713.
Guignandon, A., Boutahar, N., Rattner, A., Vico, L., Lafage-Proust, M.H., 2006. Cyclic strain promotes shuttling of PYK2/Hic-5 complex from focal contacts in osteoblast-like cells. Biochem. Biophys. Res. Commun. 343, 407–414.
Gulvady, A.C., Dubois, F., Deakin, N.O., Goreczny, G.J., Turner, C.E., 2018. Hic-5 expression is a major indicator of cancer cell morphology, migration, and plasticity in three-dimensional matrices. Mol. Biol. Cell 29, 1704–1717.
Gulvady, A.C., Forsythe, I.J., Turner, C.E., 2019. Hic-5 regulates Src-induced invadopodia rosette formation and organization. Mol. Biol. Cell 30, 1298–1313.
Hamidi, H., Ivaska, J., 2018. Every step of the way: integrins in cancer progression and metastasis. Nat. Rev. Cancer 18, 533–548.
Hanahan, D., Weinberg, R.A., 2011. Hallmarks of cancer: the next generation. Cell 144, 646–674.
Harunaga, J.S., Yamada, K.M., 2011. Cell-matrix adhesions in 3D. Matrix Biol. 30, 363–368.
Hattori, M., Minato, N., 2003. Rap1 GTPase: functions, regulation, and malignancy. J. Biochem. 134, 479–484.
Hayes, J., Peruzzi, P.P., Lawler, S., 2014. MicroRNAs in cancer: biomarkers, functions and therapy. Trends Mol. Med. 20, 460–469.
Heitzer, M.D., DeFranco, D.B., 2006a. Hic-5, an adaptor-like nuclear receptor coactivator. Nucl. Recept. Signal. 4, e019.
Heitzer, M.D., DeFranco, D.B., 2006b. Mechanism of action of Hic-5/androgen receptor activator 55, a LIM domain-containing nuclear receptor coactivator. Mol. Endocrinol. 20, 56–64.
Herreros, L., Rodríguez-Fernández, J.L., Brown, M.C., Alonso-Lebrero, J.L., Cabañas, C., Sánchez-Madrid, F., Longo, N., Turner, C.E., Sánchez-Mateos, P., 2000. Paxillin localizes to the lymphocyte microtubule organizing center and associates with the microtubule cytoskeleton. J. Biol. Chem. 275, 26436–26440.
Huber, F., Boire, A., López, M.P., Koenderink, G.H., 2015. Cytoskeletal crosstalk: when three different personalities team up. Curr. Opin. Cell Biol. 32, 39–47.
Humphrey, J.D., Dufresne, E.R., Schwartz, M.A., 2014. Mechanotransduction and extracellular matrix homeostasis. Nat. Rev. Mol. Cell Biol. 15, 802–812.
Humphries, J.D., Wang, P., Streuli, C., Geiger, B., Humphries, M.J., Ballestrem, C., 2007. Vinculin controls focal adhesion formation by direct interactions with talin and actin. J. Cell Biol. 179, 1043–1057.
Ioachim, E., Charchanti, A., Briasoulis, E., Karavasilis, V., Tsanou, H., Arvanitis, D., Agnantis, N., Pavlidis, N., 2002. Immunohistochemical expression of extracellular matrix components tenascin, fibronectin, collagen type IV and laminin in breast cancer: their prognostic value and role in tumour invasion and progression. Eur. J. Cancer 38, 2362–2370.
Ishibe, S., Joly, D., Liu, Z.X., Cantley, L.G., 2004. Paxillin serves as an ERK-regulated scaffold for coordinating FAK and Rac activation in epithelial morphogenesis. Mol. Cell 16, 257–267.
Jacob, A.E., Turner, C.E., Amack, J.D., 2016. Evolution and expression of paxillin genes in teleost fish. PLoS One 11, e0165266.

Jagadeeswaran, R., Surawska, H., Krishnaswamy, S., Janamanchi, V., Mackinnon, A.C., Seiwert, T.Y., Loganathan, S., Kanteti, R., Reichman, T., Nallasura, V., Schwartz, S., Faoro, L., Wang, Y.C., Girard, L., Tretiakova, M.S., Ahmed, S., Zumba, O., Soulii, L., Bindokas, V.P., Szeto, L.L., Gordon, G.J., Bueno, R., Sugarbaker, D., Lingen, M.W., Sattler, M., Krausz, T., Vigneswaran, W., Natarajan, V., Minna, J., Vokes, E.E., Ferguson, M.K., Husain, A.N., Salgia, R., 2008. Paxillin is a target for somatic mutations in lung cancer: implications for cell growth and invasion. Cancer Res. 68, 132–142.

Jamieson, J.S., Tumbarello, D.A., Hallé, M., Brown, M.C., Tremblay, M.L., Turner, C.E., 2005. Paxillin is essential for PTP-PEST-dependent regulation of cell spreading and motility: a role for paxillin kinase linker. J. Cell Sci. 118, 5835–5847.

Janke, C., Bulinski, J.C., 2011. Post-translational regulation of the microtubule cytoskeleton: mechanisms and functions. Nat. Rev. Mol. Cell Biol. 12, 773–786.

Janke, C., Montagnac, G., 2017. Causes and consequences of microtubule acetylation. Curr. Biol. 27, R1287–R1292.

Jansen, K.A., Donato, D.M., Balcioglu, H.E., Schmidt, T., Danen, E.H.J., Koenderink, G.H., 2015. A guide to mechanobiology: where biology and physics meet. Biochim. Biophys. Acta Mol. Cell Res. 1853, 3043–3052.

Jordan, M.A., Wilson, L., 2004. Microtubules as a target for anticancer drugs. Nat. Rev. Cancer 4, 253–265.

Kadrmas, J.L., Beckerle, M.C., 2004. The LIM domain: from the cytoskeleton to the nucleus. Nat. Rev. Mol. Cell Biol. 5, 920–931.

Kalluri, R., 2016. The biology and function of fibroblasts in cancer. Nat. Rev. Cancer 16, 582–598.

Kanteti, R., Batra, S.K., Lennon, F.E., Salgia, R., 2016. FAK and paxillin, two potential targets in pancreatic cancer. Oncotarget 7, 31586–31601.

Kareva, I., Abou-Slaybi, A., Dodd, O., Dashevsky, O., Klement, G.L., 2016. Normal wound healing and tumor angiogenesis as a game of competitive inhibition. PLoS One 11, e0166655.

Kasai, M., Guerrero-Santoro, J., Friedman, R., Leman, E.S., Getzenberg, R.H., DeFranco, D.B., 2003. The group 3 LIM domain protein paxillin potentiates androgen receptor transactivation in prostate cancer cell lines. Cancer Res. 63, 4927–4935.

Katsumi, A., Naoe, T., Matsushita, T., Kaibuchi, K., Schwartz, M.A., 2005. Integrin activation and matrix binding mediate cellular responses to mechanical stretch. J. Biol. Chem. 280, 16546–16549.

Kaverina, I., Krylyshkina, O., Small, J.V., 1999. Microtubule targeting of substrate contacts promotes their relaxation and dissociation. J. Cell Biol. 146, 1033–1043.

Kim-Kaneyama, J., Suzuki, W., Ichikawa, K., Ohki, T., Kohno, Y., Sata, M., Nose, K., Shibanuma, M., 2005. Uni-axial stretching regulates intracellular localization of Hic-5 expressed in smooth-muscle cells in vivo. J. Cell Sci. 118, 937–949.

Kim-Kaneyama, J.r., Wachi, N., Sata, M., Enomoto, S., Fukabori, K., Koh, K., Shibanuma, M., Nose, K., 2008. Hic-5, an adaptor protein expressed in vascular smooth muscle cells, modulates the arterial response to injury in vivo. Biochem. Biophys. Res. Commun. 376, 682–687.

Kim-Kaneyama, J.R., Takeda, N., Sasai, A., Miyazaki, A., Sata, M., Hirabayashi, T., Shibanuma, M., Yamada, G., Nose, K., 2011. Hic-5 deficiency enhances mechanosensitive apoptosis and modulates vascular remodeling. J. Mol. Cell. Cardiol. 50, 77–86.

Kim-Kaneyama, J.r., Lei, X.F., Arita, S., Miyauchi, A., Miyazaki, T., Miyazaki, A., 2012. Hydrogen peroxide-inducible clone 5 (Hic-5) as a potential therapeutic target for vascular and other disorders. J. Atheroscler. Thromb. 19, 601–609.

Kleinschmidt, E.G., Schlaepfer, D.D., 2017. Focal adhesion kinase signaling in unexpected places. Curr. Opin. Cell Biol. 45, 24–30.

Korsching, E., Packeisen, J., Liedtke, C., Hungermann, D., Wülfing, P., van Diest, P.J., Brandt, B., Boecker, W., Buerger, H., 2005. The origin of vimentin expression in invasive breast cancer: epithelial-mesenchymal transition, myoepithelial histogenesis or histogenesis from progenitor cells with bilinear differentiation potential? J. Pathol. 206, 451–457.

Kramer, N., Walzl, A., Unger, C., Rosner, M., Krupitza, G., Hengstschläger, M., Dolznig, H., 2013. In vitro cell migration and invasion assays. Mutat. Res. Rev. Mutat. Res. 752, 10–24.

Ku, H., Meier, K.E., 2000. Phosphorylation of paxillin via the ERK mitogen-activated protein kinase cascade in EL4 thymoma cells. J. Biol. Chem. 275, 11333–11340.

Kusenda, B., Mraz, M., Mayer, J., Pospisilova, S., 2006. microRNA biogenesis, functionality and cancer relevance. Biomed. Pap. 150, 205–215.

Lämmermann, T., Sixt, M., 2009. Mechanical modes of "amoeboid" cell migration. Curr. Opin. Cell Biol. 21, 636–644.

Langley, R.R., Fidler, I.J., 2011. The seed and soil hypothesis revisited—the role of tumor-stroma interactions in metastasis to different organs. Int. J. Cancer 128, 2527–2535.

Lauffenburger, D.A., Horwitz, A.F., 1996. Cell migration: a physically integrated molecular process. Cell 84, 359–369.

Lawson, C.D., Burridge, K., 2014. The on-off relationship of Rho and Rac during integrin-mediated adhesion and cell migration. Small GTPases 5, e27958.

Lawson, C.D., Ridley, A.J., 2018. Rho GTPase signaling complexes in cell migration and invasion. J. Cell Biol. 217, 447–457.

Leduc, C., Etienne-Manneville, S., 2015. Intermediate filaments in cell migration and invasion: the unusual suspects. Curr. Opin. Cell Biol. 32, 102–112.

Lee, B.Y., Timpson, P., Horvath, L.G., Daly, R.J., 2015. FAK signaling in human cancer as a target for therapeutics. Pharmacol. Ther. 146, 132–149.

Leong, H.S., Robertson, A.E., Stoletov, K., Leith, S.J., Chin, C.A., Chien, A.E., Hague, M.N., Ablack, A., Carmine-Simmen, K., McPherson, V.A., Postenka, C.O., Turley, E.A., Courtneidge, S.A., Chambers, A.F., Lewis, J.D., 2014. Invadopodia are required for cancer cell extravasation and are a therapeutic target for metastasis. Cell Rep. 8, 1558–1570.

Levental, K.R., Yu, H., Kass, L., Lakins, J.N., Egeblad, M., Erler, J.T., Fong, S.F.T., Csiszar, K., Giaccia, A., Weninger, W., Yamauchi, M., Gasser, D.L., Weaver, V.M., 2009. Matrix crosslinking forces tumor progression by enhancing integrin signaling. Cell 139, 891–906.

Li, D., Li, Z., Xiong, J., Gong, B., Zhang, G., Cao, C., Jie, Z., Liu, Y., Cao, Y., Yan, Y., Xiong, H., Qiu, L., Yang, M., Chen, H., Jiang, S., Yang, X., Chen, H., 2015. microRNA-212 functions as an epigenetic-silenced tumor suppressor involving in tumor metastasis and invasion of gastric cancer through down-regulating PXN expression. Am. J. Cancer Res. 5, 2980–2997.

Li, T., Zhang, C., Hassan, S., Liu, X., Song, F., Chen, K., Zhang, W., Yang, J., 2018. Histone deacetylase 6 in cancer. J. Hematol. Oncol. 11, 111.

Lin, K.T., Yeh, Y.M., Chuang, C.M., Yang, S.Y., Chang, J.W., Sun, S.P., Wang, Y.S., Chao, K.C., Wang, L.H., 2015. Glucocorticoids mediate induction of microRNA-708 to suppress ovarian cancer metastasis through targeting Rap1B. Nat. Commun. 6, 5917.

Linder, S., 2007. The matrix corroded: podosomes and invadopodia in extracellular matrix degradation. Trends Cell Biol. 17, 107–117.

Linder, S., 2009. Invadosomes at a glance. J. Cell Sci. 122, 3009–3013.

Lippincott-Schwartz, J., Roberts, T.H., Hirschberg, K., 2000. Secretory protein trafficking and organelle dynamics in living cells. Annu. Rev. Cell Dev. Biol. 16, 557–589.
Lipsky, B.P., Beals, C.R., Staunton, D.E., 1998. Leupaxin is a novel LIM domain protein that forms a complex with PYK2. J. Biol. Chem. 273, 11709–11713.
Liu, Y.J., Le Berre, M., Lautenschlaeger, F., Maiuri, P., Callan-Jones, A., Heuzé, M., Takaki, T., Voituriez, R., Piel, M., 2015. Confinement and low adhesion induce fast amoeboid migration of slow mesenchymal cells. Cell 160, 659–672.
Lohmer, L.L., Kelley, L.C., Hagedorn, E.J., Sherwood, D.R., 2014. Invadopodia and basement membrane invasion in vivo. Cell Adh. Migr. 8, 246–255.
Loo, T.-H., Ng, Y.-W., Lim, L., Manser, E., 2004. GIT1 activates p21-activated kinase through a mechanism independent of p21 binding. Mol. Cell. Biol. 24, 3849–3859.
López-Colomé, A.M., Lee-Rivera, I., Benavides-Hidalgo, R., López, E., 2017. Paxillin: a crossroad in pathological cell migration. J. Hematol. Oncol. 10, 50.
Lowery, J., Kuczmarski, E.R., Herrmann, H., Goldman, R.D., 2015. Intermediate filaments play a pivotal role in regulating cell architecture and function. J. Biol. Chem. 290, 17145–17153.
Ma, X., Hammes, S.R., 2018. Paxillin actions in the nucleus. Steroids 133, 87–92.
Ma, X., Biswas, A., Hammes, S.R., 2019. Paxillin regulated genomic networks in prostate cancer. Steroids 151, 108463.
Maller, O., DuFort, C.C., Weaver, V.M., 2013. YAP forces fibroblasts to feel the tension. Nat. Cell Biol. 15, 570–572.
Martín-Cófreces, N.B., Robles-Valero, J., Cabrero, J.R., Mittelbrunn, M., Gordón-Alonso, M., Sung, C.H., Alarcón, B., Vázquez, J., Sánchez-Madrid, F., 2008. MTOC translocation modulates IS formation and controls sustained T cell signaling. J. Cell Biol. 182, 951–962.
Matsuyama, R., Okuzaki, D., Okada, M., Oneyama, C., 2016. microRNA-27b suppresses tumor progression by regulating ARFGEF1 and focal adhesion signaling. Cancer Sci. 107, 28–35.
Mellman, I., Nelson, W.J., 2008. Coordinated protein sorting, targeting and distribution in polarized cells. Nat. Rev. Mol. Cell Biol. 9, 833–845.
Mori, K., Uchida, T., Yoshie, T., Mizote, Y., Ishikawa, F., Katsuyama, M., Shibanuma, M., 2019. A mitochondrial ROS pathway controls matrix metalloproteinase 9 levels and invasive properties in RAS-activated cancer cells. FEBS J. 286, 459–478.
Mouw, J.K., Ou, G., Weaver, V.M., 2014. Extracellular matrix assembly: a multiscale deconstruction. Nat. Rev. Mol. Cell Biol. 15, 771–785.
Murphy, D.A., Courtneidge, S.A., 2011. The "ins" and "outs" of podosomes and invadopodia: characteristics, formation and function. Nat. Rev. Mol. Cell Biol. 12, 413–426.
Naslavsky, N., Caplan, S., 2018. The enigmatic endosome—sorting the ins and outs of endocytic trafficking. J. Cell Sci. 131, jcs216499.
Nayal, A., Webb, D.J., Brown, C.M., Schaefer, E.M., Vicente-Manzanares, M., Horwitz, A.R., 2006. Paxillin phosphorylation at Ser273 localizes a GIT1-PIX-PAK complex and regulates adhesion and protrusion dynamics. J. Cell Biol. 173, 587–599.
Neerathilingam, M., Bairy, S.G., Mysore, S., 2016. Deciphering mode of action of functionally important regions in the intrinsically disordered paxillin (residues 1-313) using its interaction with FAT (focal adhesion targeting domain of focal adhesion kinase). PLoS One 11, e0150153.
Nikolopoulos, S.N., Turner, C.E., 2000. Actopaxin, a new focal adhesion protein that binds paxillin LD motifs and actin and regulates cell adhesion. J. Cell Biol. 151, 1435–1447.
Nikolopoulos, S.N., Turner, C.E., 2001. Integrin-linked kinase (ILK) binding to paxillin LD1 motif regulates ILK localization to focal adhesions. J. Biol. Chem. 276, 23499–23505.

Nishida, N., Yano, H., Nishida, T., Kamura, T., Kojiro, M., 2006. Angiogenesis in cancer. Vasc. Health Risk Manag. 2, 213–219.

Nishiya, N., Tachibana, K., Shibanuma, M., Mashimo, J., Nose, K., 2001. Hic-5-reduced cell spreading on fibronectin: competitive effects between paxillin and Hic-5 through interaction with focal adhesion kinase. Mol. Cell. Biol. 21, 5332–5345.

Nishiya, N., Shirai, T., Suzuki, W., Nose, K., 2002. Hic-5 interacts with GIT1 with a different binding mode from paxillin. J. Biochem. 132, 279–289.

Nocula-Lugowska, M., Lugowski, M., Salgia, R., Kossiakoff, A.A., 2015. Engineering synthetic antibody inhibitors specific for LD2 or LD4 motifs of paxillin. J. Mol. Biol. 427, 2532–2547.

Olson, E.N., Nordheim, A., 2010. Linking actin dynamics and gene transcription to drive cellular motile functions. Nat. Rev. Mol. Cell Biol. 11, 353–365.

Omoto, T., Kim-Kaneyama, J., Lei, X.-F., Orimo, A., Ohnishi, K., Yoshihara, K., Miyauchi, A., Li, S., Gao, L., Umemoto, T., Tanaka, J., Nakahara, K., Takeya, M., Ishida, F., Kudo, S., Haraguchi, S., Miyazaki, T., Miyazaki, A., 2018. The impact of stromal Hic-5 on the tumorigenesis of colorectal cancer through lysyl oxidase induction and stromal remodeling. Oncogene 37, 1205–1219.

Osborn, M., Weber, K., 1983. Tumor diagnosis by intermediate filament typing: a novel tool for surgical pathology. Lab. Investig. 48, 372–394.

Oser, M., Yamaguchi, H., Mader, C.C., Bravo-Cordero, J.J., Arias, M., Chen, X., DesMarais, V., Van Rheenen, J., Koleske, A.J., Condeelis, J., 2009. Cortactin regulates cofilin and N-WASp activities to control the stages of invadopodium assembly and maturation. J. Cell Biol. 186, 571–587.

Osmanagic-Myers, S., Rus, S., Wolfram, M., Brunner, D., Goldmann, W.H., Bonakdar, N., Fischer, I., Reipert, S., Zuzuarregui, A., Walko, G., Wiche, G., 2015. Plectin reinforces vascular integrity by mediating crosstalk between the vimentin and the actin networks. J. Cell Sci. 128, 4138–4150.

Padmanaban, V., Krol, I., Suhail, Y., Szczerba, B.M., Aceto, N., Bader, J.S., Ewald, A.J., 2019. E-cadherin is required for metastasis in multiple models of breast cancer. Nature 573, 439–444.

Paget, S., 1889. The distribution of secondary growths in cancer of the breast. Lancet 133, 571–573.

Palazzo, A.F., Eng, C.H., Schlaepfer, D.D., Marcantonio, E.E., Gundersen, G.G., 2004. Localized stabilization of microtubules by integrin- and FAK-facilitated Rho signaling. Science (80-) 303, 836–839.

Pandya, P.H., Murray, M.E., Pollok, K.E., Renbarger, J.L., 2016. The immune system in cancer pathogenesis: potential therapeutic approaches. J. Immunol. Res. 2016, 4273943.

Pankov, R., Endo, Y., Even-Ram, S., Araki, M., Clark, K., Cukierman, E., Matsumoto, K., Yamada, K.M., 2005. A Rac switch regulates random versus directionally persistent cell migration. J. Cell Biol. 170, 793–802.

Panzer, L., Trü be, L., Klose, M., Joosten, B., Slotman, J., Cambi, A., Linder, S., 2016. The formins FHOD1 and INF2 regulate inter- and intra-structural contractility of podosomes. J. Cell Sci. 129, 298–313.

Parsons, J.T., Horwitz, A.R., Schwartz, M.A., 2010. Cell adhesion: integrating cytoskeletal dynamics and cellular tension. Nat. Rev. Mol. Cell Biol. 11, 633–643.

Pastushenko, I., Brisebarre, A., Sifrim, A., Fioramonti, M., Revenco, T., Boumahdi, S., Van Keymeulen, A., Brown, D., Moers, V., Lemaire, S., De Clercq, S., Minguijón, E., Balsat, C., Sokolow, Y., Dubois, C., De Cock, F., Scozzaro, S., Sopena, F., Lanas, A., D'Haene, N., Salmon, I., Marine, J.-C., Voet, T., Sotiropoulou, P.A., Blanpain, C., 2018. Identification of the tumour transition states occurring during EMT. Nature 556, 463–468.

Patteson, A.E., Vahabikashi, A., Pogoda, K., Adam, S.A., Mandal, K., Kittisopikul, M., Sivagurunathan, S., Goldman, A., Goldman, R.D., Janmey, P.A., 2019. Vimentin protects cells against nuclear rupture and DNA damage during migration. J. Cell Biol. 218, 4079–4092.

Paul, N.R., Jacquemet, G., Caswell, P.T., 2015. Endocytic trafficking of integrins in cell migration. Curr. Biol. 25, R1092–R1105.

Pease, J.C., Tirnauer, J.S., 2011. Mitotic spindle misorientation in cancer—out of alignment and into the fire. J. Cell Sci. 124, 1007–1016.

Peng, Y., Croce, C.M., 2016. The role of microRNAs in human cancer. Signal Transduct. Target. Ther. 1, 15004.

Petit, V., Boyer, B., Lentz, D., Turner, C.E., Thiery, J.P., Valles, A.M., 2000. Phosphorylation of tyrosine residues 31 and 118 on paxillin regulates cell migration through an association with CRK in NBT-II cells. J. Cell Biol. 148, 957–969.

Petrie, R.J., Yamada, K.M., 2012. At the leading edge of three-dimensional cell migration. J. Cell Sci. 125, 5917–5926.

Petrie, R.J., Yamada, K.M., 2016. Multiple mechanisms of 3D migration: the origins of plasticity. Curr. Opin. Cell Biol. 42, 7–12.

Petrie, R.J., Doyle, A.D., Yamada, K.M., 2009. Random versus directionally persistent cell migration. Nat. Rev. Mol. Cell Biol. 10, 538–549.

Petrie, R.J., Gavara, N., Chadwick, R.S., Yamada, K.M., 2012. Nonpolarized signaling reveals two distinct modes of 3D cell migration. J. Cell Biol. 197, 439–455.

Petropoulos, C., Oddou, C., Emadali, A., Hiriart-Bryant, E., Boyault, C., Faurobert, E., Pol, S.V., Kim-Kaneyama, J.R., Kraut, A., Coute, Y., Block, M.R., Albiges-Rizo, C., Destaing, O., 2016. Roles of paxillin family members in adhesion and ECM degradation coupling at invadosomes. J. Cell Biol. 213, 585–599.

Pignatelli, J., LaLonde, S.E., LaLonde, D.P., Clarke, D., Turner, C.E., 2012a. Actopaxin (α-parvin) phosphorylation is required for matrix degradation and cancer cell invasion. J. Biol. Chem. 287, 37309–37320.

Pignatelli, J., Tumbarello, D.A., Schmidt, R.P., Turner, C.E., 2012b. Hic-5 promotes invadopodia formation and invasion during TGF-β-induced epithelial-mesenchymal transition. J. Cell Biol. 197, 421–437.

Pocaterra, A., Romani, P., Dupont, S., 2020. YAP/TAZ functions and their regulation at a glance. J. Cell Sci. 133, jcs230425.

Pollard, T.D., Borisy, G.G., 2003. Cellular motility driven by assembly and disassembly of actin filaments. Cell 112, 453–465.

Provenzano, P.P., Keely, P.J., 2011. Mechanical signaling through the cytoskeleton regulates cell proliferation by coordinated focal adhesion and Rho GTPase signaling. J. Cell Sci. 124, 1195–1205.

Provenzano, P.P., Eliceiri, K.W., Campbell, J.M., Inman, D.R., White, J.G., Keely, P.J., 2006. Collagen reorganization at the tumor-stromal interface facilitates local invasion. BMC Med. 4, 38.

Provenzano, P.P., Inman, D.R., Eliceiri, K.W., Knittel, J.G., Yan, L., Rueden, C.T., White, J.G., Keely, P.J., 2008. Collagen density promotes mammary tumor initiation and progression. BMC Med. 6, 1–15.

Qian, B., Wei, L., Yang, Z., He, Q., Chen, H., Wang, A., Yang, D., Li, Q., Li, J., Zheng, S., Fu, W., 2020. Hic-5 in pancreatic stellate cells affects proliferation, apoptosis, migration, invasion of pancreatic cancer cells and postoperative survival time of pancreatic cancer. Biomed. Pharmacother. 121, 109355.

Qin, J., Wang, F., Jiang, H., Xu, J., Jiang, Y., Wang, Z., 2015. microRNA-145 suppresses cell migration and invasion by targeting paxillin in human colorectal cancer cells. Int. J. Clin. Exp. Pathol. 8, 1328–1340.

Quail, D.F., Joyce, J.A., 2013. Microenvironmental regulation of tumor progression and metastasis. Nat. Med. 19, 1423–1437.
Raftopoulou, M., Hall, A., 2004. Cell migration: Rho GTPases lead the way. Dev. Biol. 265, 23–32.
Rajabi, M., Mousa, S., 2017. The role of angiogenesis in cancer treatment. Biomedicines 5, 34.
Rashid, M., Belmont, J., Carpenter, D., Turner, C.E., Olson, E.C., 2017. Neural-specific deletion of the focal adhesion adaptor protein paxillin slows migration speed and delays cortical layer formation. Development 144, 4002–4014.
Rhodes, D.R., Yu, J., Shanker, K., Deshpande, N., Varambally, R., Ghosh, D., Barrette, T., Pandey, A., Chinnaiyan, A.M., 2004. ONCOMINE: a cancer microarray database and integrated data-mining platform. Neoplasia 6, 1–6.
Ridley, A.J., 2015. Rho GTPase signalling in cell migration. Curr. Opin. Cell Biol. 36, 103–112.
Ridley, A.J., Schwartz, M.A., Burridge, K., Firtel, R.A., Ginsberg, M.H., Borisy, G., Parsons, J.T., Horwitz, A.R., 2003. Cell migration: integrating signals from front to back. Science (80-) 302, 1704–1709.
Robertson, L.K., Ostergaard, H.L., 2011. Paxillin associates with the microtubule cytoskeleton and the immunological synapse of CTL through its leucine-aspartic acid domains and contributes to microtubule organizing center reorientation. J. Immunol. 187, 5824–5833.
Ryan, S.D., Bhanot, K., Ferrier, A., De Repentigny, Y., Chu, A., Blais, A., Kothary, R., 2012. Microtubule stability, Golgi organization, and transport flux require dystonin-a2-MAP1B interaction. J. Cell Biol. 196, 727–742.
Sahai, E., Marshall, C.J., 2003. Differing modes for tumour cell invasion have distinct requirements for Rho/ROCK signalling and extracellular proteolysis. Nat. Cell Biol. 5, 711–719.
Salgia, R., Li, J.-L., Ewaniuk, D.S., Wang, Y.-B., Sattler, M., Chen, W.-C., Richards, W., Pisick, E., Shapiro, G.I., Rollins, B.J., Chen, L.B., Griffin, J.D., Sugarbaker, D.J., 1999. Expression of the focal adhesion protein paxillin in lung cancer and its relation to cell motility. Oncogene 18, 67–77.
Sanz-Moreno, V., Gadea, G., Ahn, J., Paterson, H., Marra, P., Pinner, S., Sahai, E., Marshall, C.J., 2008. Rac activation and inactivation control plasticity of tumor cell movement. Cell 135, 510–523.
Schaeffer, D., Somarelli, J.A., Hanna, G., Palmer, G.M., Garcia-Blanco, M.A., 2014. Cellular migration and invasion uncoupled: increased migration is not an inexorable consequence of epithelial-to-mesenchymal transition. Mol. Cell. Biol. 34, 3486–3499.
Schaller, M.D., 2001. Paxillin: a focal adhesion-associated adaptor protein. Oncogene 20, 6459–6472.
Shamir, E.R., Ewald, A.J., 2014. Three-dimensional organotypic culture: experimental models of mammalian biology and disease. Nat. Rev. Mol. Cell Biol. 15, 647–664.
Shao, D.D., Xue, W., Krall, E.B., Bhutkar, A., Piccioni, F., Wang, X., Schinzel, A.C., Sood, S., Rosenbluh, J., Kim, J.W., Zwang, Y., Roberts, T.M., Root, D.E., Jacks, T., Hahn, W.C., 2014. KRAS and YAP1 converge to regulate EMT and tumor survival. Cell 158, 171–184.
Sharifi, M.N., Mowers, E.E., Drake, L.E., Collier, C., Chen, H., Zamora, M., Mui, S., Macleod, K.F., 2016. Autophagy promotes focal adhesion disassembly and cell motility of metastatic tumor cells through the direct interaction of paxillin with LC3. Cell Rep. 15, 1660–1672.
Sharifi, M.N., Mowers, E.E., Macleod, K.F., 2017. Autophagic degradation of focal adhesions underlies metastatic cancer dissemination. Mol. Cell. Oncol. 4, 1–3.

Sharma, V.P., Eddy, R., Entenberg, D., Kai, M., Gertler, F.B., Condeelis, J., 2013. Tks5 and SHIP2 regulate invadopodium maturation, but not initiation, in breast carcinoma cells. Curr. Biol. 23, 2079–2089.
Shibanuma, M., Mashimo, J.I., Kuroki, T., Nose, K., 1994. Characterization of the TGFβ1-inducible hic-5 gene that encodes a putative novel zinc finger protein and its possible involvement in cellular senescence. J. Biol. Chem. 269, 26767–26774.
Shibanuma, M., Kim-Kaneyama, J., Ishino, K., Sakamoto, N., Hishiki, T., Yamaguchi, K., Mori, K., Mashimo, J., Nose, K., 2003. Hic-5 communicates between focal adhesions and the nucleus through oxidant-sensitive nuclear export signal. Mol. Biol. Cell 14, 1158–1171.
Shibanuma, M., Kim-Kaneyama, J.R., Sato, S., Nose, K., 2004. A LIM protein, Hic-5, functions as a potential coactivator for Sp1. J. Cell. Biochem. 91, 633–645.
Shibanuma, M., Mori, K., Nose, K., 2012. Hic-5: a mobile molecular scaffold regulating the anchorage dependence of cell growth. Int. J. Cell Biol. 2012, 426138.
Sit, S.-T., Manser, E., 2011. Rho GTPases and their role in organizing the actin cytoskeleton. J. Cell Sci. 124, 679–683.
Smith, M.A., Blankman, E., Deakin, N.O., Hoffman, L.M., Jensen, C.C., Turner, C.E., Beckerle, M.C., 2013. LIM domains target actin regulators paxillin and zyxin to sites of stress fiber strain. PLoS One 8, e69378.
Spiering, D., Hodgson, L., 2011. Dynamics of the rho-family small GTPases in actin regulation and motility. Cell Adh. Migr. 5, 170–180.
Subauste, M.C., Pertz, O., Adamson, E.D., Turner, C.E., Junger, S., Hahn, K.M., 2004. Vinculin modulation of paxillin-FAK interactions regulates ERK to control survival and motility. J. Cell Biol. 165, 371–381.
Sulzmaier, F.J., Jean, C., Schlaepfer, D.D., 2014. FAK in cancer: mechanistic findings and clinical applications. Nat. Rev. Cancer 14, 598–610.
Sun, L.-H., Yang, F.-Q., Zhang, C.-B., Wu, Y.-P., Liang, J.-S., Jin, S., Wang, Z., Wang, H.-J., Bao, Z.-S., Yang, Z.-X., Jiang, T., 2017. Overexpression of paxillin correlates with tumor progression and predicts poor survival in glioblastoma. CNS Neurosci. Ther. 23, 69–75.
Tang, K., Boudreau, C.G., Brown, C.M., Khadra, A., 2018. Paxillin phosphorylation at serine 273 and its effects on Rac, Rho and adhesion dynamics. PLoS Comput. Biol. 14, 1–29.
Tao, W.Y., Wang, C.Y., Sun, Y.H., Su, Y.H., Pang, D., Zhang, G.Q., 2016. microRNA-34c suppresses breast cancer migration and invasion by targeting GIT1. J. Cancer 7, 1653–1662.
te Boekhorst, V., Friedl, P., 2016. Plasticity of cancer cell invasion: mechanisms and implications for therapy. In: Advances in Cancer Research, first ed. Elsevier Inc.
Theodosiou, M., Widmaier, M., Böttcher, R.T., Rognoni, E., Veelders, M., Bharadwaj, M., Lambacher, A., Austen, K., Müller, D.J., Zent, R., Fässler, R., 2016. Kindlin-2 cooperates with talin to activate integrins and induces cell spreading by directly binding paxillin. Elife 5, e10130.
Theveneau, E., Linker, C., 2017. Leaders in collective migration: are front cells really endowed with a particular set of skills? F1000Research 6, 1899.
Thievessen, I., Fakhri, N., Steinwachs, J., Kraus, V., McIsaac, R.S., Gao, L., Chen, B.C., Baird, M.A., Davidson, M.W., Betzig, E., Oldenbourg, R., Waterman, C.M., Fabry, B., 2015. Vinculin is required for cell polarization, migration, and extracellular matrix remodeling in 3D collagen. FASEB J. 29, 4555–4567.
Thomas, J.W., Cooley, M.A., Broome, J.M., Salgia, R., Griffin, J.D., Lombardo, C.R., Schaller, M.D., 1999a. The role of focal adhesion kinase binding in the regulation of tyrosine phosphorylation of paxillin. J. Biol. Chem. 274, 36684–36692.

Thomas, S.M., Hagel, M., Turner, C.E., 1999b. Characterization of a focal adhesion protein, Hic-5, that shares extensive homology with paxillin. J. Cell Sci. 112, 181–190.
Thommen, D.S., Schumacher, T.N., 2018. T cell dysfunction in cancer. Cancer Cell 33, 547–562.
Thul, P.J., Akesson, L., Wiking, M., Mahdessian, D., Geladaki, A., Ait Blal, H., Alm, T., Asplund, A., Björk, L., Breckels, L.M., Bäckström, A., Danielsson, F., Fagerberg, L., Fall, J., Gatto, L., Gnann, C., Hober, S., Hjelmare, M., Johansson, F., Lee, S., Lindskog, C., Mulder, J., Mulvey, C.M., Nilsson, P., Oksvold, P., Rockberg, J., Schutten, R., Schwenk, J.M., Sivertsson, A., Sjöstedt, E., Skogs, M., Stadler, C., Sullivan, D.P., Tegel, H., Winsnes, C., Zhang, C., Zwahlen, M., Mardinoglu, A., Pontén, F., Von Feilitzen, K., Lilley, K.S., Uhlén, M., Lundberg, E., 2017. A subcellular map of the human proteome. Science 356, eaal3321.
Todd, J.R., Ryall, K.A., Vyse, S., Wong, J.P., Natrajan, R.C., Yuan, Y., Tan, A.C., Huang, P.H., 2016. Systematic analysis of tumour cell-extracellular matrix adhesion identifies independent prognostic factors in breast cancer. Oncotarget 7, 62939–62953.
Totaro, A., Panciera, T., Piccolo, S., 2018. YAP/TAZ upstream signals and downstream responses. Nat. Cell Biol. 20, 888–899.
Tsubouchi, A., Sakakura, J., Yagi, R., Mazaki, Y., Schaefer, E., Yano, H., Sabe, H., 2002. Localized suppression of RhoA activity by Tyr31/118-phosphorylated paxillin in cell adhesion and migration. J. Cell Biol. 159, 673–683.
Tumbarello, D.A., Turner, C.E., 2007. Hic-5 contributes to epithelial-mesenchymal transformation through a RhoA/ROCK-dependent pathway. J. Cell. Physiol. 211, 736–747.
Tumbarello, D.A., Brown, M.C., Turner, C.E., 2002. The paxillin LD motifs. FEBS Lett. 513, 114–118.
Turner, C.E., 2000a. Paxillin and focal adhesion signalling. Nat. Cell Biol. 2, E231–E236.
Turner, C.E., 2000b. Paxillin interactions. J. Cell Sci. 113, 4139–4140.
Turner, C.E., Miller, J.T., 1994. Primary sequence of paxillin contains putative SH2 and SH3 domain binding motifs and multiple LIM domains: identification of a vinculin and pp125 (Fak)-binding region. J. Cell Sci. 107, 1583–1591.
Turner, C.E., Glenney, J.R., Burridge, K., 1990. Paxillin: a new vinculin-binding protein present in focal adhesions. J. Cell Biol. 111, 1059–1068.
Turner, C.E., Kramarcy, N., Sealock, R., Burridge, K., 1991. Localization of paxillin, a focal adhesion protein, to smooth muscle dense plaques, and the myotendinous and neuromuscular junctions of skeletal muscle. Exp. Cell Res. 192, 651–655.
Turner, C.E., Brown, M.C., Perrotta, J.A., Riedy, M.C., Nikolopoulos, S.N., McDonald, A.R., Bagrodia, S., Thomas, S., Leventhal, P.S., 1999. Paxillin LD4 motif binds PAK and PIX through a novel 95-kD ankyrin repeat, ARF-GAP protein: a role in cytoskeletal remodeling. J. Cell Biol. 145, 851–863.
Ungewiss, C., Rizvi, Z.H., Roybal, J.D., Peng, D.H., Gold, K.A., Shin, D.-H., Creighton, C.J., Gibbons, D.L., 2016. The microRNA-200/Zeb1 axis regulates ECM-dependent β1-integrin/FAK signaling, cancer cell invasion and metastasis through CRKL. Sci. Rep. 6, 18652.
Valkenburg, K.C., de Groot, A.E., Pienta, K.J., 2018. Targeting the tumour stroma to improve cancer therapy. Nat. Rev. Clin. Oncol. 15, 366–381.
Vallés, A.M., Beuvin, M., Boyer, B., 2004. Activation of Rac1 by paxillin-Crk-DOCK180 signaling complex is antagonized by Rap1 in migrating NBT-II cells. J. Biol. Chem. 279, 44490–44496.
van Helvert, S., Storm, C., Friedl, P., 2018. Mechanoreciprocity in cell migration. Nat. Cell Biol. 20, 8–20.
Varney, S.D., Betts, C.B., Zheng, R., Wu, L., Hinz, B., Zhou, J., Van De Water, L., 2016. Hic-5 is required for myofibroblast differentiation by regulating mechanically dependent MRTF-A nuclear accumulation. J. Cell Sci. 129, 774–787.

Vohnoutka, R.B., Gulvady, A.C., Goreczny, G., Alpha, K., Handelman, S.K., Sexton, J.Z., Turner, C.E., 2019. The focal adhesion scaffold protein Hic-5 regulates vimentin organization in fibroblasts. Mol. Biol. Cell 30, 3037–3056.

Wang, H., Song, K., Krebs, T.L., Yang, J., Danielpour, D., 2008. Smad7 is inactivated through a direct physical interaction with the LIM protein Hic-5/ARA55. Oncogene 27, 6791–6805.

Wang, X., Hu, G., Betts, C., Harmon, E.Y., Keller, R.S., Van De Water, L., Zhou, J., 2011. Transforming growth factor-β1-induced transcript 1 protein, a novel marker for smooth muscle contractile phenotype, is regulated by serum response factor/myocardin protein. J. Biol. Chem. 286, 41589–41599.

Webb, D.J., Donais, K., Whitmore, L.A., Thomas, S.M., Turner, C.E., Parsons, J.T., Horwitz, A.F., 2004. FAK-Src signalling through paxillin, ERK and MLCK regulates adhesion disassembly. Nat. Cell Biol. 6, 154–161.

Wei, S.C., Yang, J., 2016. Forcing through tumor metastasis: the interplay between tissue rigidity and epithelial-mesenchymal transition. Trends Cell Biol. 26, 111–120.

Wei, B., Zhou, X., Liang, C., Zheng, X., Lei, P., Fang, J., Han, X., Wang, L., Qi, C., Wei, H., 2017. Human colorectal cancer progression correlates with LOX-induced ECM stiffening. Int. J. Biol. Sci. 13, 1450–1457.

West, K.A., Zhang, H., Brown, M.L.C., Nikolopoulos, S.N., Riedy, M.C., Horwitz, A.F., Turner, C.E., 2001. The LD4 motif of paxillin regulates cell spreading and motility through an interaction with paxillin kinase linker (PKL). J. Cell Biol. 154, 161–176.

Wilson, B.J., Allen, J.L., Caswell, P.T., 2018. Vesicle trafficking pathways that direct cell migration in 3D matrices and in vivo. Traffic 19, 899–909.

Winer, A., Adams, S., Mignatti, P., 2018. Matrix metalloproteinase inhibitors in cancer therapy: turning past failures into future successes. Mol. Cancer Ther. 17, 1147–1155.

Wolf, K., Mazo, I., Leung, H., Engelke, K., Von Andrian, U.H., Deryugina, E.I., Strongin, A.Y., Bröcker, E.B., Friedl, P., 2003. Compensation mechanism in tumor cell migration: mesenchymal-amoeboid transition after blocking of pericellular proteolysis. J. Cell Biol. 160, 267–277.

Wood, C.K., Turner, C.E., Jackson, P., Critchley, D.R., 1994. Characterisation of the paxillin-binding site and the C-terminal focal adhesion targeting sequence in vinculin. J. Cell Sci. 107, 709–717.

Wright, P.E., Dyson, H.J., 2014. Intrinsically disordered proteins in cellular signalling and regulation. Nat. Rev. Mol. Cell Biol. 16, 18–29.

Wu, J., Akhmanova, A., 2017. Microtubule-organizing centers. Annu. Rev. Cell Dev. Biol. 33, 51–75.

Wu, D.W., Cheng, Y.W., Wang, J., Chen, C.Y., Lee, H., 2010. Paxillin predicts survival and relapse in non-small cell lung cancer by microRNA-218 targeting. Cancer Res. 70, 10392–10401.

Wu, D.-W., Wu, T.C., Wu, J.Y., Cheng, Y.-W., Chen, Y.C., Lee, M.C., Chen, C.Y., Lee, H., 2014a. Phosphorylation of paxillin confers cisplatin resistance in non-small cell lung cancer via activating ERK-mediated Bcl-2 expression. Oncogene 33, 4385–4395.

Wu, D.W., Chuang, C.Y., Lin, W.L., Sung, W.W., Cheng, Y.W., Lee, H., 2014b. Paxillin promotes tumor progression and predicts survival and relapse in oral cavity squamous cell carcinoma by microRNA-218 targeting. Carcinogenesis 35, 1823–1829.

Xu, W., Gulvady, A.C., Goreczny, G.J., Olson, E.C., Turner, C.E., 2019. Paxillin-dependent regulation of apical-basal polarity in mammary gland morphogenesis. Development 146, dev174367.

Yamada, K.M., Sixt, M., 2019. Mechanisms of 3D cell migration. Nat. Rev. Mol. Cell Biol. 20, 738–752.

Yang, L., Guerrero, J., Hong, H., DeFranco, D.B., Stallcup, M.R., 2000. Interaction of the τ2 transcriptional activation domain of glucocorticoid receptor with a novel steroid receptor coactivator, Hic-5, which localizes to both focal adhesions and the nuclear matrix. Mol. Biol. Cell 11, 2007–2018.

Ye, X., Weinberg, R.A., 2015. Epithelial-mesenchymal plasticity: a central regulator of cancer progression. Trends Cell Biol. 25, 675–686.

Yoshida, K., Soldati, T., 2006. Dissection of amoeboid movement into two mechanically distinct modes. J. Cell Sci. 119, 3833–3844.

Yuminamochi, T., Yatomi, Y., Osada, M., Ohmori, T., Ishii, Y., Nakazawa, K., Hosogaya, S., Ozaki, Y., 2003. Expression of the LIM proteins paxillin and Hic-5 in human tissues. J. Histochem. Cytochem. 51, 513–521.

Zaidel-Bar, R., Kam, Z., Geiger, B., 2005. Polarized downregulation of the paxillin-p130CAS-Rac1 pathway induced by shear flow. J. Cell Sci. 118, 3997–4007.

Zhan, T., Rindtorff, N., Betge, J., Ebert, M.P., Boutros, M., 2019. CRISPR/Cas9 for cancer research and therapy. Semin. Cancer Biol. 55, 106–119.

Zhang, L.L., Zhang, L.F., Shi, Y.B., 2018. Down-regulated paxillin suppresses cell proliferation and invasion by inhibiting M2 macrophage polarization in colon cancer. Biol. Chem. 399, 1285–1295.

Zouq, N.K., Keeble, J.A., Lindsay, J., Valentijn, A.J., Zhang, L., Mills, D., Turner, C.E., Streuli, C.H., Gilmore, A.P., 2009. FAK engages multiple pathways to maintain survival of fibroblasts and epithelia: differential roles for paxillin and p130Cas. J. Cell Sci. 122, 357–367.

CHAPTER TWO

Nuclear actin in cancer biology

Stefan Zahler*
Department of Pharmacy, Ludwig-Maximilians-Universität, Munich, Germany
*Corresponding author: e-mail address: stefan.zahler@cup.uni-muenchen.de

Contents

Abstract

The presence of actin in the nucleus has been a matter of debate for many years. In recent years many important roles of actin in the nucleus (transcriptional regulation, chromatin remodeling, DNA repair, cell division, maintenance of nuclear architecture) have been identified, and the precise control of nuclear actin levels has been demonstrated. The vital importance of the actin driven processes in the cell make it highly likely that dysregulation of nuclear actin dynamics and structure can be linked to tumor induction and -progression. In this chapter I summarize our current knowledge about nuclear actin in the cancer context.

1. Historical perspective

The presence of actin, or at least an actin like protein in the nuclear fraction of calf thymus has been described as early as 1963 (Ohnishi et al., 1963). However, actin is the most abundant protein in the cytoplasm, and tightly attached to the nuclear envelope. It was, therefore, a matter of debate, whether nuclear actin really exists, or if it is only an artifact caused by cytoplasmic contamination. In the 1970s many independent groups were able to reproduce the detection of actin in cell nuclei of organisms ranging from slime molds (Jockusch et al., 1970) to mammalian cells (Douvas et al., 1975; Lestourgeon et al., 1975). To avoid contamination due to the conventional isolation procedure, hand-isolated nuclei from Xenopus eggs were

International Review of Cell and Molecular Biology, Volume 355
ISSN 1937-6448
https://doi.org/10.1016/bs.ircmb.2020.04.001

established as reliable model, and confirmed the existence of nuclear actin (Clark and Merriam, 1977). As to the functional role of nuclear actin, its shape (F- versus G-actin) and concentration, however, the debate continued (Goldstein et al., 1977).

2. There and back again: How actin shuttles to and from the nucleus

In initial papers a nuclear concentration of actin similar to the cytosolic levels was reported, leading to the conclusion that actin can shuttle through the nuclear envelope relatively easily, and that nuclear actin levels are just the consequence of a passive equilibrium distribution (Goldstein et al., 1977). However, the size limit for passive diffusion through nuclear pores is around 40 kDa, and actin with its 42 kDa seems to be just above the threshold of diffusive transport. Later it was discovered that the polymerization state of nuclear actin dramatically changes under stressful conditions (e.g., high concentrations of DMSO or heat shock), and that this phenomenon of formation of actin (or actin-cofilin) rods is accompanied by a transient redistribution of cytoplasmic actin into the nucleus (Fukui, 1978; Iida et al., 1986). This of course requires active regulation of actin shuttling into and from the nucleus. In 1998 two putative nuclear export sequences (NES) were identified in actin (Wada et al., 1998). Mutational disruption of both NESs inhibited nuclear export of actin in rat cells, as did treatment with leptomycin B (LMB), an inhibitor of exportin 1 (XPO1 or CRM1). Interestingly, the nuclear accumulation of actin went along with decreased proliferative capacity (Wada et al., 1998). In a more recent paper (Stuven et al., 2003), this pathway could not be corroborated, but instead a co-transport of actin and profilin via exportin 6 (Exp6) was suggested. It could well be that the respective major pathway depends on the cell type and species used. Interestingly, Exp6 is not expressed in the giant nuclei of *Xenopus* oocytes (100,000 times larger than nuclei of somatic cells), causing a very high level of nuclear actin in these cells, which is needed for mechanical stabilization (Bohnsack et al., 2006). The relevant point here is that nuclear actin levels are regulated at all: this clearly indicates functional implications of actin in the nucleus. A further interesting question is that if actin has to be pumped from the nucleus, how does it get there in the first place? It could of course be that some residual cytoplasmic actin remains in the nucleus after reconstitution of the nuclear envelope in the telophase of the cell cycle, or that some actin leaks into the nucleus passively through nuclear pores, establishing an equilibrium sooner or later. Since actin itself lacks a nuclear localization sequence (NLS), only co-transport with an actin binding protein (ABP) is feasible as an active pathway. Indeed, it was demonstrated that actin is co-transported with cofilin into the nucleus of mast cells upon ATP depletion or treatment with latrunculin B, an actin depolymerizing natural compound (Pendleton et al., 2003). Life cell microscopy studies using overexpression of GFP-actin have revealed that actin is actively shuttled into the nucleus by importin 9 (Ipo9) together with cofilin (Dopie et al., 2012). In this study the relevance of Crm1 for nuclear export could not be reproduced, leaving Exp6 as the main exit route of actin from the nucleus. From the kinetic data in this study it can be concluded that nuclear actin exists in a monomeric (about 75%), as well as in a polymeric form, and is to some part tightly integrated into protein complexes. Similar proportions have also been previously described by a different group (McDonald et al., 2006). A key determinant of actin shuttling seems to be the availability of G-actin, which at least in part explains the increase of nuclear actin levels upon treatment of cells with actin depolymerizing compounds (Gegenfurtner et al., 2018; McDonald et al., 2006; Pendleton et al., 2003). Interestingly, nuclear actin levels seem to vary considerably between cells of the same type (Skarp et al., 2013). This could be due to different cell cycle stages or to DNA replication stress, and is coordinated with the shuttling of the actin cofactors IQGAP1 and Rac1 (Johnson et al., 2013).

3. Now you see it, now you don't: The many faces of nuclear actin

Our concepts about reality are often based on visual impressions, and, thus, the acceptance of the existence of actin in the nucleus has been largely enhanced by the visualization of nuclear actin. This issue, however, is far from trivial, and there have been some excellent publications on the visualization of nuclear actin (Belin et al., 2013; Melak et al., 2017). The tools used for actin visualization in the nucleus can be divided into five groups: (1) *Phalloidin*, a small molecular natural compound, which binds to F-actin, and which can be specifically labeled with fluorophores without loosing its affinity to F-actin. Staining fixed and permeabilized cells with phalloidin is the classical approach for visualizing F-actin. Since this did not work in many cases for visualizing nuclear actin, the existence of canonical F-actin fibers in the nucleus was often questioned. We now know that the successful use of phalloidin requires specific fixation and staining procedures and depends on the experimental system used (Melak et al., 2017). Shortcomings of phalloidin are its poor membrane permeability, making it only applicable in fixed and permeabilized cells, and precludes its use for dynamic studies in living cells, as well as its binding mode. Since phalloidin binds close to a site where also actin binding proteins like, e.g., cofilin bind, competition may occur (Mcgough et al., 1997), and F-actin filaments decorated with cofilin might become invisible to phalloidin labeling. This might in part explain the missing phalloidin staining of nuclear actin structures like actin (or actin-cofilin) rods (Nishida et al., 1987). A different small molecule with a fluorescent tag, SiR actin, which is based on jasplakinolide, has recently been developed (Lukinavicius et al., 2014). The advantage of this probe is its enhanced membrane permeability, which allows dynamic studies. The issues about competitive binding or artifacts due to its intrinsic F-actin stabilizing effects, however, persist. (2) *Overexpression of fluorescent fusion proteins of actin.* Though this approach has been widely used for studying cytoplasmic actin dynamics, it is not really appropriate for nuclear actin for several reasons: by overexpression the levels of actin shift, which might artificially enhance nuclear actin levels, and the bulky GFP tag might exclude GFP-actin from specific kinds of filaments (Doyle and Botstein, 1996). (3) *Fluorescence tagged fusion proteins displaying actin binding domains (ABDs).* To circumvent the problems of overexpressing actin, probes with domains from proteins naturally interacting with actin have been developed. These comprise LifeAct (Riedl et al., 2008), UtrCH (Burkel et al., 2007), F-tractin (Belin et al., 2014), and Actin-Chromobody (Plessner et al., 2015). All of them are valuable tools, but still have issues concerning cell behavior depending on expression levels, and relatively high fluorescence background, which makes nuclear studies difficult (Melak et al., 2017). (4) *Fluorescence tagged fusion proteins with nuclear localization and export sequences.* These are the latest generation of nuclear actin probes, and reduce the problem of cytoplasmic background by confining the signal to the nucleus due to an NIS encoded in the probes. Since these constructs still could shift the trafficking of actin to and from the nucleus, an even more sophisticated probe with combined NLS and NES was constructed to maintain nuclear shuttling (Plessner et al., 2015). Expression levels seem to be quite important in all genetically encoded probes, and working with stably expressing cell lines might give the best results (Melak et al., 2017). (5) *Antibodies against special forms of actin.* Conventional antibodies against actin (which are hard to generate due to the highly conserved nature of the actin molecule) are of little use for microscopy applications, since they bind to G-actin, which is so abundant in all cells, that no structural details can be seen against the background. Antibodies raised against specific conformations of actin, like, e.g., 2G2 (Gonsior et al., 1999) or 1C7 (Schoenenberger et al., 2005), however, were able to specifically stain structures also in the nucleus after mild fixation. These studies have led to the hypothesis that actin might exist in different conformations, depending on the functional context and interacting proteins (e.g., profilin, which causes a more "open" configuration, or cofilin, which causes a compaction of the actin monomer), or different dimerized or even non-canonical polymerized forms (Jockusch et al., 2006). Along these lines, a crystal structure for nuclear actin (N-actin) in complex with Arp4 has been published, showing a more twisted conformation than conventional G-actin (Cao et al., 2016).

A further level of complexity is added by the fact that actin is the target of more than 140 post translational modifications (PTMs), like actetylation, oxidation, phosphorylation or SUMOylation (Varland et al., 2019). It is highly likely that these modifications specifically alter actin conformation, and, thus, its binding to other proteins and its functions. Furthermore, binding of actin probes, like phalloidin or antibodies might be affected.

Keeping in mind that all the approaches mentioned above are susceptible to artifacts, we can hypothesize that actin in the nuclear compartment could occur in the following shapes:

- *Canonical G-actin monomers*
- *Actin with a divergent conformation due to binding to some interacting protein*
- *Canonical F-actin*
- *F-actin decorated with, e.g., cofilin or other proteins precluding phalloidin binding*
- *Non-canonical actin polymers*

Since all these forms of actin have been described in specific cell types and under specific functional conditions (e.g., cellular stress or cell spreading), it lies close at hand to speculate, that they do not occur all at the same time, and that specific functions might be ascribed to them (Table 1).

4. Nuclear actin: What it's good for

The tight regulation of nuclear actin levels clearly indicates its vital functional roles. To date, several of these (sometimes tightly interwoven) functions have been undisputedly revealed:

- Transcriptional regulation
- Chromatin organization and movement
- DNA repair
- Nuclear structure and stability

Table 1 Shape and functions of nuclear actin.

Shape of nuclear actin	Functional context
Canonical G-actin monomers	Monomer pool for subsequent polymerization or for integration into protein complexes, regulation of MKL1/MRTF activity (Vartiainen et al., 2007)
Actin with a divergent conformation due to binding to some interacting protein	Part, e.g., of chromatin remodeling complexes, like INO80, BAF or Swr1 (Knoll et al., 2018)
Canonical F-actin	Mechanical stabilization of large nuclei (Feric and Brangwynne, 2013), generation of nuclear protrusions in early G1 phase (Baarlink et al., 2017)
F-actin decorated with, e.g., cofilin or other proteins precluding phalloidin binding	Response to cellular stresses (Fukui, 1978; Nishida et al., 1987)
Non-canonical actin polymers	Function and physiological relevance not known

The first three aspects are closely related to the fact that nuclear actin is an integral part of several chromatin remodeling complexes, and interacts with all three forms of RNA polymerases. This will be reviewed elsewhere in this issue (see chapter "Nuclear actin and myosin in chromatin regulation and maintenance of genome integrity" by Percipalle).

One additional and very intriguing mode of transcriptional regulation by actin is its interaction with transcription factors or coactivators of mechanosensitive signaling pathways, like MRTF (MAL) or YAP. The functional logic behind this is quite convincing: whenever cells experience mechanical stress or have to move, they will remodel their actin cytoskeleton and shift the balance of G- and F-actin. This change translates into transcriptional regulation by altering the nuclear access of MRTF or YAP. MRTF is bound to G-actin, which masks the NLS of MRTF (Mouilleron et al., 2011). If the G-actin pool is reduced by polymerization into F-actin, MRTF is set free and enters the nucleus, where it interacts with SRF to induce transcription of downstream genes (Posern and Treisman, 2006). The same regulation of MRTF activity by G-actin seems to occur in the nucleus (Vartiainen et al., 2007): high levels of nuclear G-actin bind

MRTF and keep it from binding to SRF, functional states of the cell, which cause formation of nuclear F-actin liberate MRTF and induce SRF binding and activation. For YAP, the situation is similar, but a bit more complicated: the AMOT protein binds to YAP and causes its cytoplasmic localization. F-actin competes with YAP for binding of AMOT. Consequently, when F-actin is formed, YAP is released from AMOT and is able to translocate into the nucleus (Mana-Capelli et al., 2014). It has been shown very recently that in the nucleus YAP is inactivated by the SWI/SNF complex. Upon formation of nuclear F-actin, SWI/SNF is bound to F-actin fibers, thus releasing YAP for acting as transcriptional regulator (Chang et al., 2018).

The mechanically stabilizing role of nuclear actin completely relies on F-actin. In the enormously large oocytes of *Xenopus laevis*, a well structured network of F-actin can be observed, which is based on the high levels of nuclear actin in this cell type due to the lack of Exp6 and the resulting absence of active export from the nucleus (Bohnsack et al., 2006; Stuven et al., 2003). In smaller cells, which are not threatened by nuclear collapse due to gravitation, F-actin still seems to play a role for re-shaping the nucleus in dependence of the cell cycle (Baarlink et al., 2017). The cell nucleus has been identified as the main obstacle for cell migration through confined environments (Calero-Cuenca et al., 2018; Moser et al., 2017). Accordingly, nuclear stiffness might be a key determinant of the migratory potential of a cell. The contribution of nuclear actin to the overall stiffness of the nucleus is not clear to date. The main contribution obviously comes from the nuclear lamina. Recently, a direct crosstalk between the nuclear lamina and intranuclear polymerization of actin via the LINC complex has been demonstrated (Plessner et al., 2015). Surprisingly it seems to be quite common that cells overcome the motility restrictions of the nucleus by rupture of the nuclear envelope, which causes severe DNA damage (Shah et al., 2017). As a consequence nuclear actin levels would of course instantly equilibrate with the cytoplasmic conditions. Upon re-sealing of the ruptured nuclear envelope, the putatively elevated level of nuclear actin could serve the DNA repair mechanisms needed at this stage.

Overall it is very hard to discriminate the functional and morphological contributions of the perinuclear actin cap from that of nuclear actin. It has even been convincingly shown by using electron and super-resolution optical microscopy that the nuclei of cells in 3D culture have deep invaginations encasing actin filaments and mechanically coupling cell adhesion structures to the interior of nuclei (Jorgens et al., 2017). These structures may in part have previously been misinterpreted as nuclear actin fibers.

5. Nuclear actin and cancer: More than meets the eye?

The pathophysiological relevance of actin mutations or other causes of a disturbed equilibrium or polymerization state of nuclear actin for myopathies is relatively well documented (Serebryannyy et al., 2016b). In the context of cancer the connection is much less obvious. Screening through a cancer transcriptome database (https://www.cbioportal.org/) currently elicits 12% amplifications of the ACTB gene and 3% mutations in the Cancer Cell Line Encyclopedia (Ghandi et al., 2019) throughout all tested cell lines with no preference for specific cell provenience. When examining clinical pan cancer studies, actin genes (ACTA1, ACTB, and ACTG1) are altered in approximately 12% of the cases, again without clear association to tumor entities. Amplifications seem to be more numerous than mutations. Thus, dysregulations of actin occur in cancer; however, their functional role is still unclear. A specific involvement of nuclear actin is even more speculative—even if it is tempting to speculate about an involvement in invasion and metastasis. There are several in vitro studies, which hint toward multiple links between nuclear actin and cancer. An interesting case of post-translational modification of actin related to cancer is oxidation at Met44/47 by MICAL2, a flavoprotein monoxygenase, which has already been associated to cancer growth and invasion (Mariotti et al., 2016). MICAL2 is enriched in the nucleus and causes redox modifications of nuclear actin, which induce depolymerization. Surprisingly, this leads to a decrease of nuclear G-actin, which in turn causes nuclear localization of MRTF and transcription of target genes (Lundquist et al., 2014). It still has to be investigated, if oxidation of actin by MICAL2 directly alters actin trafficking through the nuclear envelope. In addition to the MRTF/SRF signaling pathway also the Wnt pathway, which plays an important role in many forms of cancer (Kahn, 2014), is affected by nuclear actin. Upon forced generation of nuclear actin polymerization by overexpressing actin with a NLS, β-catenin accumulates in the nucleus, and the transcription of respective downstream target genes is enhanced (Yamazaki et al., 2016).

Histone acetylation and deacetylation has turned out to be a crucial tool for transcriptional regulation and a major pathway in cancer progression (Johnstone, 2002). Nuclear actin has been shown to directly interact with the class I HDAC complex (most likely in its activated state) and to inhibit activity of HDAC1 and 2 (Serebryannyy et al., 2016a). A mass spectrometry based nuclear interactome analysis enhanced by overexpression of actin with a NLS elicited KAT14 histone acetyl transferase as binding partner of nuclear actin, with a preference of monomeric actin. This interaction causes a reduction of histone acetylation (Viita et al., 2019). Consequently, nuclear actin can cause both, increased or decreased histone acetylation.

A further proteomic approach indicated an interaction between nuclear actin and the kinase AKT/protein kinase B, both in its phosphorylated and unphosphorylated form (Coa et al., 2019). AKT is primarily located in the cytoplasm, but its nuclear translocation seems to be increased in some types of tumor, such as lung, breast, thyroid, prostate and melanoma (Lee et al., 2002; Nicholson et al., 2003; Slipicevic et al., 2005; Van De Sande et al., 2005; Vasko et al., 2004), and its activity has been generally linked to cancer (Bellacosa et al., 2005).

Very recently shuttling of nuclear actin has been linked to tumor biology (Chatzifrangkeskou et al., 2019). RASSF1A (Ras associated domain family 1 isoform A) has been shown to be necessary for binding of expoprtin-6 to RAN GTPase, and, consequently, for the active maintenance of nuclear actin levels. Silencing of RASSF1A caused increased nuclear actin levels, while overexpression of RASSF1A in the highly invasive cancer cell line MDA-MB231, which lacks endogenous expression of RASSF1A, reduced nuclear actin levels. These effects are linked to a modulation of the MRTF/SRF signaling pathway. Since epigenetic silencing of the RASSF1A gene is quite common and associated with poor clinical outcome in all kinds solid tumors (Grawenda and O'neill, 2015), this axis seems to be functionally quite important.

It has become increasingly clear that infections with the human cytomegalovirus (HCMV) play a role in human cancer progression and death (Cobbs, 2019). Astoundingly, nuclear actin is gain involved, as formation of F-actin fibers is induced by virus infection, and nuclear F-actin promotes nuclear egression of the virus capsids (Wilkie et al., 2016). Also in other virus types, like, e.g., baculovirus (Ohkawa and Welch, 2018), similar phenomena, including disruption of the nuclear envelope by nuclear F-actin have been described.

The regulation of cellular quiescence is a key step in stabilizing tissues and in the prevention of malignancies. This process is influenced by the cellular microenvironment, like, e.g., the extracellular matrix or the basal membrane. Laminin 111 (LN1) seems to be the most important constituent of the basal membrane in this context. LN1 causes quiescence of epithelial cells by decreasing the levels of nuclear actin (Spencer et al., 2011). Overexpression of nuclear actin prevented quiescence in the presence of LN1.

This process was specific for non-malignant epithelial cells, while malignant cells did not react to LN1 exposure. The changes in nuclear actin were due to regulation of the nuclear export via exportin-6, and silencing of exportin-6 overcame epithelial quiescence (Fiore et al., 2017). Interestingly, the PI3K/AKT pathway seems to be primary regulator of this process.

In summary, the levels of nuclear actin, its conformation and the functional consequences of its presence are closely linked to many different features of tumors. This might be due to the fact that tumors recapitulate developmental programs, which often are mechano-regulated. Though actin itself is still a poor drug target due to its ubiquitous nature, a better understanding of the nature of nuclear actin, its conformation and its interaction partners, might not only further our knowledge on tumor progression and metastasis, but also open new roads for developing strategies to specifically address nuclear actin—this very peculiar manifestation of an old acquaintance.

References

Baarlink, C., Plessner, M., Sherrard, A., Morita, K., Misu, S., Virant, D., Kleinschnitz, E.M., Harniman, R., Alibhai, D., Baumeister, S., Miyamoto, K., Endesfelder, U., Kaidi, A., Grosse, R., 2017. A transient pool of nuclear F-actin at mitotic exit controls chromatin organization. Nat. Cell Biol. 19, 1389–1399.

Belin, B.J., Cimini, B.A., Blackburn, E.H., Mullins, R.D., 2013. Visualization of actin filaments and monomers in somatic cell nuclei. Mol. Biol. Cell 24, 982–994.

Belin, B.J., Goins, L.M., Mullins, R.D., 2014. Comparative analysis of tools for live cell imaging of actin network architecture. Bioarchitecture 4, 189–202.

Bellacosa, A., Kumar, C.C., Di Cristofano, A., Testa, J.R., 2005. Activation of AKT kinases in cancer: implications for therapeutic targeting. Adv. Cancer Res. 94, 29–86.

Bohnsack, M.T., Stuven, T., Kuhn, C., Cordes, V.C., Gorlich, D., 2006. A selective block of nuclear actin export stabilizes the giant nuclei of Xenopus oocytes. Nat. Cell Biol. 8, 257–263.

Burkel, B.M., Von Dassow, G., Bement, W.M., 2007. Versatile fluorescent probes for actin filaments based on the actin-binding domain of utrophin. Cell Motil. Cytoskeleton 64, 822–832.

Calero-Cuenca, F.J., Janota, C.S., Gomes, E.R., 2018. Dealing with the nucleus during cell migration. Curr. Opin. Cell Biol. 50, 35–41.

Cao, T., Sun, L., Jiang, Y., Huang, S., Wang, J., Chen, Z., 2016. Crystal structure of a nuclear actin ternary complex. Proc. Natl. Acad. Sci. U. S. A. 113, 8985–8990.

Chang, L., Azzolin, L., Di Biagio, D., Zanconato, F., Battilana, G., Lucon Xiccato, R., Aragona, M., Giulitti, S., Panciera, T., Gandin, A., Sigismondo, G., Krijgsveld, J., Fassan, M., Brusatin, G., Cordenonsi, M., Piccolo, S., 2018. The SWI/SNF complex is a mechanoregulated inhibitor of YAP and TAZ. Nature 563, 265–269.

Chatzifrangkeskou, M., Pefani, D.E., Eyres, M., Vendrell, I., Fischer, R., Pankova, D., O'neill, E., 2019. RASSF1A is required for the maintenance of nuclear actin levels. EMBO J. 38, e101168.

Clark, T.G., Merriam, R.W., 1977. Diffusible and bound actin nuclei of Xenopus laevis oocytes. Cell 12, 883–891.

Coa, L.L., Abreu, T.F., Tashima, A.K., Green, J., Pascon, R.C., Vallim, M.A., Machado-JR, J., 2019. AKT/protein kinase B associates with beta-actin in the nucleus of melanoma cells. Biosci. Rep. 39, BSR20181312.

Cobbs, C., 2019. Cytomegalovirus is a tumor-associated virus: armed and dangerous. Curr. Opin. Virol. 39, 49–59.

Dopie, J., Skarp, K.P., Rajakyla, E.K., Tanhuanpaa, K., Vartiainen, M.K., 2012. Active maintenance of nuclear actin by importin 9 supports transcription. Proc. Natl. Acad. Sci. U. S. A. 109, E544–E552.

Douvas, A.S., Harrington, C.A., Bonner, J., 1975. Major nonhistone proteins of rat liver chromatin: preliminary identification of myosin, actin, tubulin, and tropomyosin. Proc. Natl. Acad. Sci. U. S. A. 72, 3902–3906.

Doyle, T., Botstein, D., 1996. Movement of yeast cortical actin cytoskeleton visualized in vivo. Proc. Natl. Acad. Sci. U. S. A. 93, 3886–3891.

Feric, M., Brangwynne, C.P., 2013. A nuclear F-actin scaffold stabilizes ribonucleoprotein droplets against gravity in large cells. Nat. Cell Biol. 15, 1253–1259.

Fiore, A., Spencer, V.A., Mori, H., Carvalho, H.F., Bissell, M.J., Bruni-Cardoso, A., 2017. Laminin-111 and the level of nuclear actin regulate epithelial quiescence via Exportin-6. Cell Rep. 19, 2102–2115.

Fukui, Y., 1978. Intranuclear actin bundles induced by dimethyl sulfoxide in interphase nucleus of Dictyostelium. J. Cell Biol. 76, 146–157.

Gegenfurtner, F.A., Zisis, T., Al Danaf, N., Schrimpf, W., Kliesmete, Z., Ziegenhain, C., Enard, W., Kazmaier, U., Lamb, D.C., Vollmar, A.M., Zahler, S., 2018. Transcriptional effects of actin-binding compounds: the cytoplasm sets the tone. Cell. Mol. Life Sci. 75, 4539–4555.

Ghandi, M., Huang, F.W., Jane-Valbuena, J., Kryukov, G.V., Lo, C.C., McDonald 3rd, E.R., Barretina, J., Gelfand, E.T., Bielski, C.M., Li, H., Hu, K., Andreev-Drakhlin, A.Y., Kim, J., Hess, J.M., Haas, B.J., Aguet, F., Weir, B.A., Rothberg, M.V., Paolella, B.R., Lawrence, M.S., Akbani, R., Lu, Y., Tiv, H.L., Gokhale, P.C., De Weck, A., Mansour, A.A., Oh, C., Shih, J., Hadi, K., Rosen, Y., Bistline, J., Venkatesan, K., Reddy, A., Sonkin, D., Liu, M., Lehar, J., Korn, J.M., Porter, D.A., Jones, M.D., Golji, J., Caponigro, G., Taylor, J.E., Dunning, C.M., Creech, A.L., Warren, A.C., McFarland, J.M., Zamanighomi, M., Kauffmann, A., Stransky, N., Imielinski, M., Maruvka, Y.E., Cherniack, A.D., Tsherniak, A., Vazquez, F., Jaffe, J.D., Lane, A.A., Weinstock, D.M., Johannessen, C.M., Morrissey, M.P., Stegmeier, F., Schlegel, R., Hahn, W.C., Getz, G., Mills, G.B., Boehm, J.S., Golub, T.R., Garraway, L.A., Sellers, W.R., 2019. Next-generation characterization of the cancer cell line encyclopedia. Nature 569, 503–508.

Goldstein, L., Rubin, R., Ko, C., 1977. The presence of actin in nuclei: a critical appraisal. Cell 12, 601–608.

Gonsior, S.M., Platz, S., Buchmeier, S., Scheer, U., Jockusch, B.M., Hinssen, H., 1999. Conformational difference between nuclear and cytoplasmic actin as detected by a monoclonal antibody. J. Cell Sci. 112 (Pt. 6), 797–809.

Grawenda, A.M., O'neill, E., 2015. Clinical utility of RASSF1A methylation in human malignancies. Br. J. Cancer 113, 372–381.

Iida, K., Iida, H., Yahara, I., 1986. Heat shock induction of intranuclear actin rods in cultured mammalian cells. Exp. Cell Res. 165, 207–215.

Jockusch, B.M., Brown, D.F., Rusch, H.P., 1970. Synthesis of a nuclear protein in G2-phase. Biochem. Biophys. Res. Commun. 38, 279–283.

Jockusch, B.M., Schoenenberger, C.A., Stetefeld, J., Aebi, U., 2006. Tracking down the different forms of nuclear actin. Trends Cell Biol. 16, 391–396.

Johnson, M.A., Sharma, M., Mok, M.T., Henderson, B.R., 2013. Stimulation of in vivo nuclear transport dynamics of actin and its co-factors IQGAP1 and Rac1 in response to DNA replication stress. Biochim. Biophys. Acta 1833, 2334–2347.

Johnstone, R.W., 2002. Histone-deacetylase inhibitors: novel drugs for the treatment of cancer. Nat. Rev. Drug Discov. 1, 287–299.

Jorgens, D.M., Inman, J.L., Wojcik, M., Robertson, C., Palsdottir, H., Tsai, W.T., Huang, H., Bruni-Cardoso, A., Lopez, C.S., Bissell, M.J., Xu, K., Auer, M., 2017. Deep nuclear invaginations are linked to cytoskeletal filaments—integrated bioimaging of epithelial cells in 3D culture. J. Cell Sci. 130, 177–189.

Kahn, M., 2014. Can we safely target the WNT pathway? Nat. Rev. Drug Discov. 13, 513–532.

Knoll, K.R., Eustermann, S., Niebauer, V., Oberbeckmann, E., Stoehr, G., Schall, K., Tosi, A., Schwarz, M., Buchfellner, A., Korber, P., Hopfner, K.P., 2018. The nuclear actin-containing Arp8 module is a linker DNA sensor driving INO80 chromatin remodeling. Nat. Struct. Mol. Biol. 25, 823–832.

Lee, S.H., Kim, H.S., Park, W.S., Kim, S.Y., Lee, K.Y., Kim, S.H., Lee, J.Y., Yoo, N.J., 2002. Non-small cell lung cancers frequently express phosphorylated Akt; an immunohistochemical study. APMIS 110, 587–592.

Lestourgeon, W.M., Forer, A., Yang, Y.Z., Bertram, J.S., Pusch, H.P., 1975. Contractile proteins. Major components of nuclear and chromosome non-histone proteins. Biochim. Biophys. Acta 379, 529–552.

Lukinavicius, G., Reymond, L., D'este, E., Masharina, A., Gottfert, F., Ta, H., Guther, A., Fournier, M., Rizzo, S., Waldmann, H., Blaukopf, C., Sommer, C., Gerlich, D.W., Arndt, H.D., Hell, S.W., Johnsson, K., 2014. Fluorogenic probes for live-cell imaging of the cytoskeleton. Nat. Methods 11, 731–733.

Lundquist, M.R., Storaska, A.J., Liu, T.C., Larsen, S.D., Evans, T., Neubig, R.R., Jaffrey, S.R., 2014. Redox modification of nuclear actin by MICAL-2 regulates SRF signaling. Cell 156, 563–576.

Mana-Capelli, S., Paramasivam, M., Dutta, S., McCollum, D., 2014. Angiomotins link F-actin architecture to hippo pathway signaling. Mol. Biol. Cell 25, 1676–1685.

Mariotti, S., Barravecchia, I., Vindigni, C., Pucci, A., Balsamo, M., Libro, R., Senchenko, V., Dmitriev, A., Jacchetti, E., Cecchini, M., Roviello, F., Lai, M., Broccoli, V., Andreazzoli, M., Mazzanti, C.M., Angeloni, D., 2016. MICAL2 is a novel human cancer gene controlling mesenchymal to epithelial transition involved in cancer growth and invasion. Oncotarget 7, 1808–1825.

McDonald, D., Carrero, G., Andrin, C., De Vries, G., Hendzel, M.J., 2006. Nucleoplasmic beta-actin exists in a dynamic equilibrium between low-mobility polymeric species and rapidly diffusing populations. J. Cell Biol. 172, 541–552.

Mcgough, A., Pope, B., Chiu, W., Weeds, A., 1997. Cofilin changes the twist of F-actin: implications for actin filament dynamics and cellular function. J. Cell Biol. 138, 771–781.

Melak, M., Plessner, M., Grosse, R., 2017. Actin visualization at a glance. J. Cell Sci. 130, 525–530.

Moser, C., Rudiger, D., Forster, F., Von Blume, J., Yu, P., Kuster, B., Kazmaier, U., Vollmar, A.M., Zahler, S., 2017. Persistent inhibition of pore-based cell migration by sub-toxic doses of miuraenamide, an actin filament stabilizer. Sci. Rep. 7, 16407.

Mouilleron, S., Langer, C.A., Guettler, S., McDonald, N.Q., Treisman, R., 2011. Structure of a pentavalent G-actin*MRTF-A complex reveals how G-actin controls nucleocytoplasmic shuttling of a transcriptional coactivator. Sci. Signal. 4, ra40.

Nicholson, K.M., Streuli, C.H., Anderson, N.G., 2003. Autocrine signalling through erbB receptors promotes constitutive activation of protein kinase B/Akt in breast cancer cell lines. Breast Cancer Res. Treat. 81, 117–128.

Nishida, E., Iida, K., Yonezawa, N., Koyasu, S., Yahara, I., Sakai, H., 1987. Cofilin is a component of intranuclear and cytoplasmic actin rods induced in cultured cells. Proc. Natl. Acad. Sci. U. S. A. 84, 5262–5266.

Ohkawa, T., Welch, M.D., 2018. Baculovirus actin-based motility drives nuclear envelope disruption and nuclear egress. Curr. Biol. 28, 2153–2159. e4.
Ohnishi, T., Kawamura, H., Yamamoto, T., 1963. Extraction of a protein resembling actin from the cell nucleus of the calf thymus. J. Biochem. 54, 298–300.
Pendleton, A., Pope, B., Weeds, A., Koffer, A., 2003. Latrunculin B or ATP depletion induces cofilin-dependent translocation of actin into nuclei of mast cells. J. Biol. Chem. 278, 14394–14400.
Plessner, M., Melak, M., Chinchilla, P., Baarlink, C., Grosse, R., 2015. Nuclear F-actin formation and reorganization upon cell spreading. J. Biol. Chem. 290, 11209–11216.
Posern, G., Treisman, R., 2006. Actin' together: serum response factor, its cofactors and the link to signal transduction. Trends Cell Biol. 16, 588–596.
Riedl, J., Crevenna, A.H., Kessenbrock, K., Yu, J.H., Neukirchen, D., Bista, M., Bradke, F., Jenne, D., Holak, T.A., Werb, Z., Sixt, M., Wedlich-Soldner, R., 2008. Lifeact: a versatile marker to visualize F-actin. Nat. Methods 5, 605–607.
Schoenenberger, C.A., Buchmeier, S., Boerries, M., Sutterlin, R., Aebi, U., Jockusch, B.M., 2005. Conformation-specific antibodies reveal distinct actin structures in the nucleus and the cytoplasm. J. Struct. Biol. 152, 157–168.
Serebryannyy, L.A., Cruz, C.M., De Lanerolle, P., 2016a. A role for nuclear actin in HDAC 1 and 2 regulation. Sci. Rep. 6, 28460.
Serebryannyy, L.A., Yuen, M., Parilla, M., Cooper, S.T., De Lanerolle, P., 2016b. The effects of disease models of nuclear actin polymerization on the nucleus. Front. Physiol. 7, 454.
Shah, P., Wolf, K., Lammerding, J., 2017. Bursting the bubble—nuclear envelope rupture as a path to genomic instability? Trends Cell Biol. 27, 546–555.
Skarp, K.P., Huet, G., Vartiainen, M.K., 2013. Steady-state nuclear actin levels are determined by export competent actin pool. Cytoskeleton (Hoboken) 70, 623–634.
Slipicevic, A., Holm, R., Nguyen, M.T., Bohler, P.J., Davidson, B., Florenes, V.A., 2005. Expression of activated Akt and PTEN in malignant melanomas: relationship with clinical outcome. Am. J. Clin. Pathol. 124, 528–536.
Spencer, V.A., Costes, S., Inman, J.L., Xu, R., Chen, J., Hendzel, M.J., Bissell, M.J., 2011. Depletion of nuclear actin is a key mediator of quiescence in epithelial cells. J. Cell Sci. 124, 123–132.
Stuven, T., Hartmann, E., Gorlich, D., 2003. Exportin 6: a novel nuclear export receptor that is specific for profilin-actin complexes. EMBO J. 22, 5928–5940.
Van De Sande, T., Roskams, T., Lerut, E., Joniau, S., Van Poppel, H., Verhoeven, G., Swinnen, J.V., 2005. High-level expression of fatty acid synthase in human prostate cancer tissues is linked to activation and nuclear localization of Akt/PKB. J. Pathol. 206, 214–219.
Varland, S., Vandekerckhove, J., Drazic, A., 2019. Actin post-translational modifications: the Cinderella of cytoskeletal control. Trends Biochem. Sci. 44, 502–516.
Vartiainen, M.K., Guettler, S., Larijani, B., Treisman, R., 2007. Nuclear actin regulates dynamic subcellular localization and activity of the SRF cofactor MAL. Science 316, 1749–1752.
Vasko, V., Saji, M., Hardy, E., Kruhlak, M., Larin, A., Savchenko, V., Miyakawa, M., Isozaki, O., Murakami, H., Tsushima, T., Burman, K.D., De Micco, C., Ringel, M.D., 2004. Akt activation and localisation correlate with tumour invasion and oncogene expression in thyroid cancer. J. Med. Genet. 41, 161–170.
Viita, T., Kyheroinen, S., Prajapati, B., Virtanen, J., Frilander, M.J., Varjosalo, M., Vartiainen, M.K., 2019. Nuclear actin interactome analysis links actin to KAT14 histone acetyl transferase and mRNA splicing. J. Cell Sci. 132. pii: jcs226852. https://doi.org/10.1242/jcs.226852.

Wada, A., Fukuda, M., Mishima, M., Nishida, E., 1998. Nuclear export of actin: a novel mechanism regulating the subcellular localization of a major cytoskeletal protein. EMBO J. 17, 1635–1641.

Wilkie, A.R., Lawler, J.L., Coen, D.M., 2016. A role for nuclear F-actin induction in human cytomegalovirus nuclear egress. MBio 7. pii: e01254-16. https://doi.org/10.1128/mBio.01254-16.

Yamazaki, S., Yamamoto, K., De Lanerolle, P., Harata, M., 2016. Nuclear F-actin enhances the transcriptional activity of beta-catenin by increasing its nuclear localization and binding to chromatin. Histochem. Cell Biol. 145, 389–399.

CHAPTER THREE

Nuclear actin and myosin in chromatin regulation and maintenance of genome integrity

Tomas Venit[a], S. Raza Mahmood[a], Martin Endara-Coll[a,b], Piergiorgio Percipalle[a,b,*]
[a]Science Division, Biology Program, New York University Abu Dhabi (NYUAD), Abu Dhabi, United Arab Emirates
[b]Department of Molecular Biosciences, The Wenner-Gren Institute, Stockholm University, Stockholm, Sweden
[*]Corresponding author: e-mail address: pp69@nyu.edu

Contents

Abstract

Cytoskeletal proteins are beginning to be considered as key regulators of nuclear function. Among them, actin and myosin have been implicated in numerous tasks, including chromatin regulation, transcription and assembly of nascent ribonucleoprotein complexes. We also know from work performed by several labs that influx of actin and myosin into the nucleus and out of the nucleus is tightly regulated. In particular, in the case of actin, its nucleocytoplasmic import/export cycle is controlled by the importin/exportin system and it correlates with the transcriptional state of the cell.

International Review of Cell and Molecular Biology, Volume 355
ISSN 1937-6448
https://doi.org/10.1016/bs.ircmb.2020.05.001

These basic molecular functions of both actin and myosin seem to impact key cellular functions, including development and differentiation as well as the cellular response to DNA damage by directly affecting transcriptional reprograming. These observations are beginning to suggest that actin and myosin could play an important role in consolidating the organization of the mammalian genome and that loss of actin and myosin likely leads to a general instability of the genome. In this chapter, we provide a general background on evidence that actin and myosin are important in key nuclear functions. Following this, we will focus on evidence supporting of a role in genome organization and finally we will discuss increasingly striking results on the role of actin and myosin in the maintenance of genome integrity.

1. Introduction

Actin is one of the most highly conserved and abundant proteins in eukaryotic cells and its role in regulating cell shape, motility and cytoskeletal organization has been extensively studied (Dominguez and Holmes, 2011; Krendel and Mooseker, 2005; McConnell and Tyska, 2010; Miyamoto and Gurdon, 2013). Although initially discovered in skeletal muscle cells, actin is now also known to be an essential part of nonmuscle cells where it forms the basis of the cytoskeleton and regulates cell motility and adhesion via controlled polymerization of actin monomers (G-actin) into actin filaments (F-actin) (Kelpsch and Tootle, 2018). The three main actin isoforms include skeletal or α-actin and nonmuscle β and γ-actins. While the role of actin in skeletal muscles is well-established, an increasing number of studies in the last few decades have shown that actin also plays crucial roles inside the nucleus (Bajusz et al., 2018; Miyamoto and Gurdon, 2013; Percipalle and Vartiainen, 2019; Visa and Percipalle, 2010). It has been shown, for example, that the levels of actin inside the nucleus are actively regulated (Stüven et al., 2003) and disruption of proteins involved in nuclear actin transport can lead to embryonic lethality (Dopie et al., 2012; Kelpsch and Tootle, 2018; Stüven et al., 2003). Similarly, more than 30 actin-binding proteins that affect actin polymerization and function are now known to exist inside nuclei (Kristo et al., 2016). One of the best described in terms of its nuclear function is Nuclear Myosin 1 (NM1), an isoform of the unconventional Myosin 1C (Hofmann et al., 2006b; Ihnatovych et al., 2012; Obrdlik et al., 2010)

While it has long been suggested that actin (Clark and Merriam, 1977; Clark and Rosenbaum, 1979; Pederson and Aebi, 2002) and myosin (Hagen et al., 1986) are present in the nucleus of various cell types, discriminating nuclear and cytoplasmic actin and myosin was challenging for many years. However, several studies have provided convincing evidence that both actin and myosin are in the nucleus where they play crucial roles in nuclear processes such as transcriptional regulation (Percipalle et al., 2001, 2003; Pestic-Dragovich et al., 2000), chromatin structure (Zhao et al., 1998) and long-range chromatin movements (Chuang et al., 2006; Dundr et al., 2007; Mehta et al., 2010).

The involvement of actin and myosin in chromatin regulation seems to be a consequence of their physical interactions with chromatin remodelers. Actin directly affects chromatin by associating with or being a component of ATP-dependent chromatin remodelers such as the INO80 and TIP60 complexes in yeast and the BAF (Brg/Brm-associated factors) complex in mammals where actin regulates Brg1 dependent ATPase function (Miyamoto and Gurdon, 2013). In line with the BAF complex having important roles in lineage specification and cell fate reprogramming, actin depletion has also been shown to affect the induction of differentiation gene programs in cells undergoing transcriptional reprogramming during neurogenesis (Xie et al., 2018b). Similarly, NM1 as a part of the chromatin remodeling complex B-WICH with the subunits SNF2h and WSTF is required to recruit histone acetyl-transferases to rRNA genes (Sarshad et al., 2013; Vintermist et al., 2011). Altogether these chromatin remodelers do affect gene expression and, in particular, transcription. Given their association with both remodelers and polymerase machinery, actin and myosin may play an important role in bridging these molecular machines, possibly as dynamic molecular switches (see discussion below) that directly regulate transcription (see Sarshad and Percipalle, 2014), impacting downstream events in the gene expression pathway (Almuzzaini et al., 2015; Fomproix and Percipalle, 2004; Kelpsch and Tootle, 2018; Kukalev et al., 2005; Percipalle et al., 2003; Philimonenko et al., 2004; Sarshad et al., 2013, 2014; Ye et al., 2008). Indeed, actin preferentially associates with euchromatic regions (Percipalle, 2013), it binds the polymerase machinery, remains associated with nascent ribonucleoprotein complexes (RNPs) bound to heterogeneous nuclear ribonucleoproteins (hnRNPs) and accompanies mature RNPs to the cytoplasm (Percipalle, 2013).

An exciting new development in the field is that actin polymerization also appears to be implicated in the regulation of nuclear functions. While early studies have shown that perturbation of actin polymerization inhibits movement of fluorescently labeled transgenes, endogenous chromatin territories and chromosome segregation (Dundr et al., 2007;

Kelpsch and Tootle, 2018), more recent work has convincingly visualized regulated nuclear actin polymerization using advanced microscopy methods (Baarlink et al., 2017). In addition to this, nuclear actin and myosin have now been linked to important cellular processes including cellular differentiation (Miyamoto and Gurdon, 2013; Percipalle and Vartiainen, 2019; Xie et al., 2018b) and cellular response to DNA damage (Caridi et al., 2018; Schrank et al., 2018). These new findings seem compatible with a global role of actin and myosin in regulating the functional architecture of the cell nucleus.

Based on the above considerations, here we will summarize evidence for the presence of actin and myosin inside the nucleus and we will focus on the suggested mechanisms through which they regulate transcription and chromatin structure and how these molecular mechanisms now appear to have a major impact on the three dimensional organization of the genome during differentiation as well as genome stability and integrity.

2. Evidence that actin, myosin and actin-related proteins are *bona fide* nuclear proteins

While actin was first suggested to be present inside the nucleus five decades ago (Lane, 1969), this claim remained controversial due to the absence of experimental methods that could detect actin inside the nucleus and differentiate it from cytoplasmic actin. Initial experiments, for example, showed that the actin-binding phalloidin, a short bicyclic peptide, could not stain nuclear actin and electron microscopy similarly did not detect actin filaments inside the nucleus (Sanger et al., 1980). However, advances in the actin visualization methods such as the 17-amino-acid peptide Lifeact staining polymeric actin (Riedl et al., 2008) or the chromobody nAC-GFP-NLS (Plessner et al., 2015) have enabled the community to study actin's nuclear functions in many different contexts revolutionizing the nuclear actin field.

Specific evidence for the presence of actin inside the nucleus came first through the detection of actin in isolated Xenopus nuclei using SDS-PAGE experiments and then *via* the observation that general inhibition of CRM1-dependent nuclear export resulted in actin accumulation in the nucleus (Collier et al., 2000; Wada et al., 1998). The discovery that proteins such as profilin and cofilin, which regulate actin dynamics, are also present in the cell nucleus, further fueled the idea that actin exists in the nucleus in a highly regulated state (Castano et al., 2010). Indeed, we now know that actin in the nucleus is present in both monomeric and polymeric states and that the levels of nuclear actin are tightly regulated by transporters such as importin 9 and exportin 6 (Dopie et al., 2012; Stüven et al., 2003). These transporters mediate import and export of actin in a monomeric form in complex with cofilin and profilin, respectively. It has also been shown that while the levels of nuclear actin are usually maintained at a constant level, changes in transcription rates and processes such as differentiation, DNA damage and stress can result in an increase in nuclear actin levels (Dopie et al., 2012; Hofmann et al., 2009; Xu et al., 2010), highlighting its importance in the transcriptional response in the context of multiple cellular functions. Once in the nucleus, actin undergoes regulated polymerization and depolymerization depending on external or internal stimuli (Plessner and Grosse, 2019; Plessner et al., 2015). For example, serum addition triggers nuclear actin polymerization by activating G-protein coupled receptors and subsequent release of calcium from endoplasmic reticulum. Calcium activates the actin-regulator INF2 at the inner nuclear membrane leading to burst of short lived actin filaments (Wang et al., 2019). Moreover, after serum stimulation, polymeric actin serves as a dynamic scaffold for Pol II to aggregate around serum-response genes in N-WASP and Arp2/3 dependent manner (Wei et al., 2020). Although not proven at this stage, it is likely that the polymeric actin present inside the nucleus exists in a conformation that prevents phalloidin binding which suggests that nuclear polymeric actin is potentially organized into non-canonical filaments different from those observed in the cytoplasm. Like its tightly controlled levels in the nucleus, the polymerization state of nuclear actin also seems to be highly regulated. While cytoplasmic actin primarily exists as filaments, almost 80% of the nuclear actin pool has been reported to be monomeric or short oligomeric (McDonald et al., 2006), a concentration that under normal conditions should lead to spontaneous polymerization. There is evidence that monomeric or short oligomeric actin is functional in nuclear function in complex with the G-actin binding proteins cofilin and profilin (Obrdlik and Percipalle, 2011; Söderberg et al., 2012). However, the mechanisms that lead to the maintenance of this nuclear actin pool in a largely monomeric state are yet to be fully uncovered.

While canonical actin can perform some nuclear functions independently, many actin-linked protein complexes also contain other actin related proteins (ARPs). ARPs are proteins found in all eukaryotic cells and are thought to have arisen from actin through gene duplication events (Meagher et al., 2009). They share significant structural and functional similarities with actin including a core actin fold domain and an ADP/ATP binding motif (Fenn et al., 2011; Kabsch and Holmes, 1995). Interestingly, unlike actin, Arps cannot polymerize into filaments and are present inside the nucleus as monomers (Fenn et al., 2011). Arps can be divided into 10 sub-families (Arp1 to Arp10) depending on their similarity with actin of which Arp4–9 are known to localize and perform various metabolic functions inside the nucleus (Harata et al., 2002; Poch and Winsor, 1997). The most important nuclear functions of Arps include nucleosome recognition, transcriptional regulation and chromatin organization. Arp4 is the most commonly found Arp in nuclear protein complexes and is associated with a range of DNA interacting complexes including histone deacetylase 2 (Harata et al., 2002; Joshi et al., 2013; Meagher et al., 2009). Arp4 also exists in the nucleus independently and binds histones suggesting a role in regulation of chromatin structure (Sunada et al., 2005). Consistent with its important nuclear roles Arp4 is an essential gene in yeast. Arp5–9 while more structurally different from actin than Arp4, also play important roles in the nucleus. Arp5 is known to be present in smooth muscle cells where it binds and regulates myocardin (Morita and Hayashi, 2014) while Arp6 is known to interact with centromeres, promoters and telomeres with a potential role in telomere silencing (Kato et al., 2001; Ueno et al., 2004). Arp6 also plays a role in the activation of critical developmental genes in *Arabidopsis* in association with the chromatin remodeling complex SWR1 (Maruyama et al., 2012). Similarly, Arp7 is also known to regulate Arabidopsis embryogenesis and development (Kandasamy et al., 2005) while Arp8 localizes to the chromosomes during mitosis with possible roles in chromosome segregations (Aoyama et al., 2008) and together with Arp4 and actin is a component of the INO80 complex (Knoll et al., 2018). Arp9 is the most divergent nuclear Arp and has been shown to be involved in fruiting initiation in some plants (Nakazawa et al., 2016). Recently the full structure of yeast SWI/SNF remodeling complex RSC have been described with and Arp7 and Arp9 have been shown to be a core component of it (Han et al., 2020; Wagner et al., 2020). Together both canonical actin and Arps are therefore critical for the regulation of various metabolic nuclear processes.

Similarly to actin, for many years myosins, a superfamily of motor proteins comprising at least 40 different genes divided in 18 classes (Foth et al., 2006), were believed to be excluded from the cell nucleus. The first myosin was discovered in 1864 (Kühne, 1864), although it was not properly characterized until decades later (Engelhardt and Ljubimowa, 1939; Straub, 1943). Since then, the functions of myosin have extended from the contraction in muscle cells to cell motility, cytokinesis, growth cone extension and vesicle trafficking (Berg et al., 2001). Unlike nuclear actin, studies regarding nuclear myosin are sparse. In 1986, a paper on monoclonal antibodies against myosin I reported a strong reaction with isolated nuclei (Hagen et al., 1986), and in 1997 the first report of myosin in the nucleus confirmed its presence with immunoprecipitation, confocal microscopy and electron microscopy. Being the first myosin discovered in the nucleus, it was termed nuclear myosin I, now commonly referred to as NM1 (Nowak et al., 1997).

The unconventional Myosin 1C (Myo1C) isoform NM1 consists of an actin- and ATP-binding head domain, a linker neck domain and a tail domain, which allows NM1 to bind DNA, chromatin and phospholipids through positively-charged regions (de Lanerolle and Serebryannyy, 2011; Hofmann et al., 2006a; Hokanson and Ostap, 2006). In mammals, 3 isoforms of Myo1C have been described, resulting from the alternative splicing of N-terminal exons carrying different translation start sites (Fig. 1A). The longest isoform A, detected so far only in some cancer cell lines has extra 35 amino acids on its N-terminus and was shown to have distinct functions from the other two isoforms (Ihnatovych et al., 2012; Schwab et al., 2013). Isoform B, or NM1, has an extra 16 amino acid stretch in the N-terminus (Philimonenko et al., 2004). The function of this 16 amino acid stretch is not understood although it is possibly implicated in nuclear retention. The shortest isoform C, often misnamed in older literature as cytoplasmic Myo1C, is sequentially identical to other two isoforms with exception of its N-terminus that lacks the extensions described above for isoforms A and B (Fig. 1B). Deletion studies revealed a functional nuclear localization signal in the neck domain of Myo1C, a sequence that is actually present in all three isoforms (Dzijak et al., 2012). In addition, the N-terminal extensions found in isoforms A and B seem to function in fine-tuning myosin dynamics through its working cycle (Zattelman et al., 2017). Therefore, although it has not been formally demonstrated if and how Myo1C isoforms can dynamically shuttle between nucleus and cytoplasm, all three Myo1C isoforms are likely to be present in both cellular compartments. In line with this, a major fraction of NM1 is located at the plasma membrane, where it participates in the regulation of cell membrane tension (Venit et al., 2016). It is not known how the N-terminal extensions affect the functions of myosin isoforms in specific processes within the cell nucleus and how or if these isoforms cooperate together. Several attempts to prepare double knock-outs of Myo1C and NM1 failed, while specific deletion of NM1 in mouse leads only to mild phenotypes, supporting the idea that two isoforms are at least partially interchangeable in nuclear processes (Venit et al., 2013, 2016).

Other myosins have been found in the cell nucleus as well, involved in different functions including transcription, differentiation and proliferation

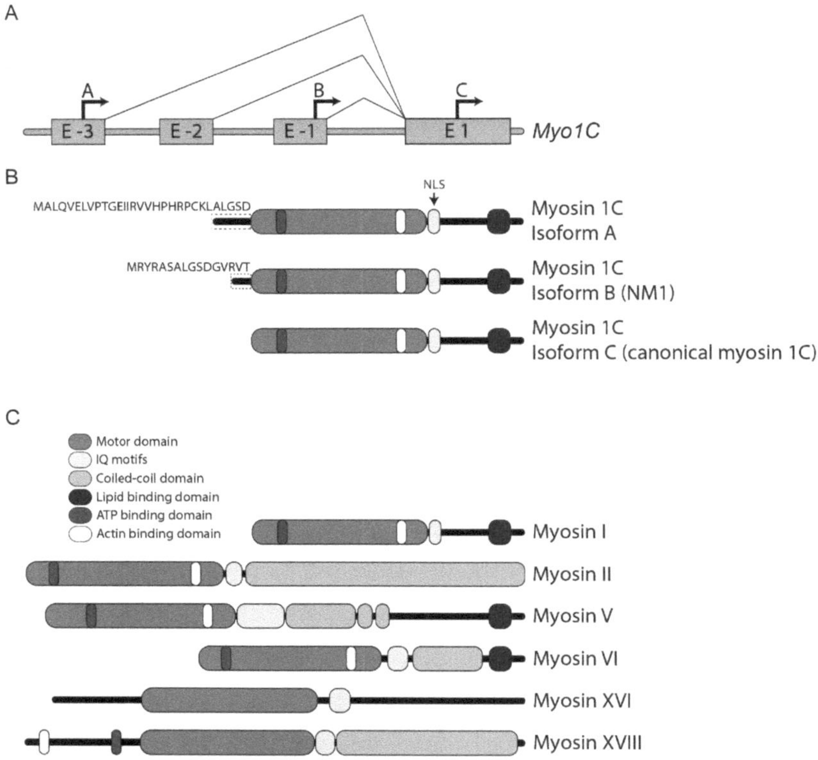

Fig. 1 Structure of nuclear myosins. (A) Alternative splicing of human *Myo1C* gene leads to expression of 3 different isoforms of Myo1C protein. (B) Protein structure of each of the Myo1C isoforms. Canonical myosin 1C (isoform C) shares the same amino acid sequence with other two isoforms with exceptions of N-terminal extensions in Isoforms A and B (NM1). (C) Primary structure and domain organization in the myosin isoforms found in the nucleus. *Adapted from Maravillas-Montero, J.L., Santos-Argumedo, L., 2012. The myosin family: unconventional roles of actin-dependent molecular motors in immune cells. J. Leukoc. Biol. 91, 35–46; Sarshad, A.A., Percipalle, P., 2014. New insight into role of myosin motors for activation of RNA polymerases. Int. Rev. Cell Mol. Biol. 311, 183–230.*

(Fig. 1C). Embryonic myosin 2 heavy chain is present in the nucleus of proliferating myoblasts, although it is downregulated upon differentiation (Rodgers, 2005). Myosin 5a has been detected in nuclear speckles suggesting its function in mRNA splicing (Pranchevicius et al., 2008). Myosin 5b is present in the dense fibrillar component of nucleoli, where it associates with β-actin and Pol I in a transcription-dependent manner (Lindsay and McCaffrey, 2009). Myosin 6 is recruited to promoters and intergenic regions, and its depletion leads to gene repression (Vreugde et al., 2006).

Similarly to NM1, Myosin 6 is involved in regulation of Pol II transcription (Fili et al., 2017), but the fact that Myosin 6 is a backward motor with a large and variable step size suggests that the two proteins do not have synergistic roles (Altman et al., 2004; Buss et al., 2004). Myosin 16 is involved in cell cycle and proliferation. It is localized into the nucleolus by nucleolar localization signal present in its N-terminal region (Altman et al., 2004; Buss et al., 2004). Finally, myosin 18b is a muscle protein, which partially migrates to the nucleus at the beginning of myoblasts differentiation and participates in muscle development (Ajima et al., 2008).

3. Nuclear actin and myosin in transcriptional regulation

The roles of nuclear actin and myosin in transcription regulation have been extensively investigated. Both are associated with all three eukaryotic nuclear RNA polymerases, Pol I, II, and III (Fomproix and Percipalle, 2004; Hofmann et al., 2004; Hu et al., 2004; Kukalev et al., 2005; Philimonenko et al., 2004), with actin being bound to the subunits Rbp6 and Rbp8 common to all three polymerases (Hu et al., 2004). Actin association with the polymerases is functional. It has been shown that anti-actin antibodies inhibit Pol I and Pol II transcription both *in vivo* and *in vitro* while addition of actin to partially purified inactive Pol III *in vitro* can restore its activity (Hofmann et al., 2004; Hu et al., 2004; Philimonenko et al., 2004). Similarly, microscopy experiments and co-immunoprecipitations showed NM1 in association with both Pol I and Pol II during active transcription (Pestic-Dragovich et al., 2000). *In vitro* transcription experiments revealed that Pol II is stimulated by adding purified NM1 which is necessary for the formation of the first phosphodiester bond during initiation of transcription (Hofmann et al., 2006b) and antibodies to NM1 inhibit Pol I transcription activation but do not affect the initiation phase (Percipalle et al., 2006). The synergy between actin and myosin in transcription was first suggested when an actin-myosin complex was co-precipitated with the largest Pol I subunit, RPA194, and found to co-localizes with active rDNA transcription sites in nucleoli (Fomproix and Percipalle, 2004). This actin-myosin complex is likely to be functional in transcription since expression of NM1 mutants with impaired ATPase or actin-binding functions or expression of actin mutants with polymerization defects impair Pol I transcription *in vivo* and *in vitro* (Ye et al., 2008). Altogether, these observations suggest that an actin-based myosin motor potentially accompanies the sliding of the polymerase machinery across the DNA during transcription elongation (Fig. 2).

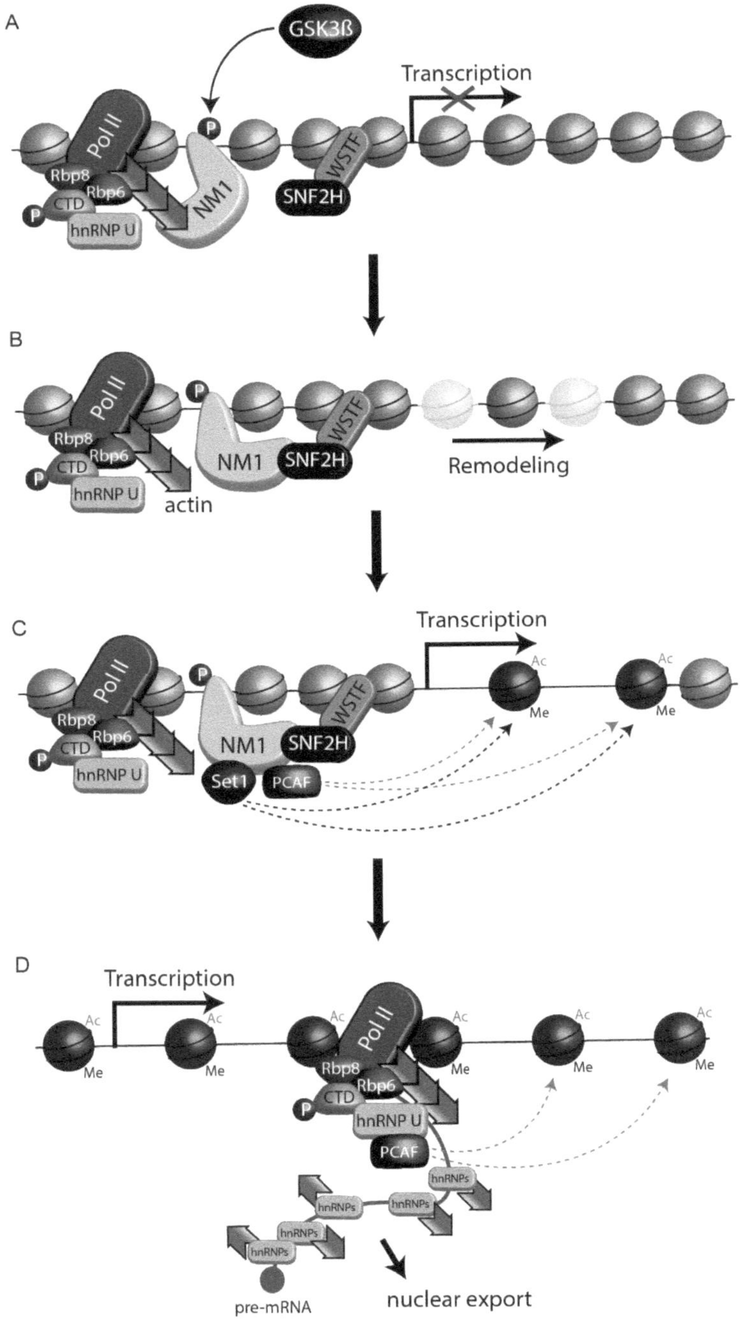

Fig. 2 See legend on opposite page.

However, although, actin and myosin are important for the transcription process, both initiation and elongation (Hofmann et al., 2004; Philimonenko et al., 2004) (Fig. 2), how precisely they cooperate with the polymerase remains unclear. Since actin and myosin affect RNA polymerase activity not only on chromatinized DNA templates (Percipalle et al., 2006) but also on naked DNA (not associated with proteins such as histones) (Hofmann et al., 2004; Hu et al., 2004; Philimonenko et al., 2004), it is likely that they are vital for the polymerase catalytic function. Understanding the precise mechanisms by which actin and NM1 affect RNA polymerase activity requires detailed structural information at atomic resolution.

During the transcriptional process, actin also remains associated with nascent RNA transcripts and this has been observed in both insect and mammalian cells (Percipalle et al., 2001, 2002). Experiments performed in the dipteran insect *C. tentans* show that anti-actin antibodies label isolated polytene chromosomes in an RNA-dependent manner and the entire Balbiani ring (BR) transcription unit (Percipalle et al., 2001), indicating an association of actin with nascent ribonucleoprotein complexes during the elongation process. In line with this milestone observation, chromatin immunoprecipitation (ChIP) experiments including ChIP-seq (ChIP combined with deep sequencing) have shown that actin is bound to coding regions of Pol I and Pol II supporting its presence along actively transcribed genes (Hofmann et al., 2004; Philimonenko et al., 2004). Cotranscriptional association of actin with nascent RNA is functional and it is mediated by specific heterogeneous nuclear ribonucleoproteins (hnRNPs) in both *C. tentans* and mammalian cells. In *C. tentans*, actin directly interacts with the hnRNP A1-like hrp36 protein and with hrp65, a homolog of the transcriptional co-activator PSF/NonA (Percipalle et al., 2001, 2003). In mammals, actin was found to be in complex with a subset of A/B-type hnRNPs, including hnRNP A2, hnRNP A3, the transport mediator CArG-box binding factor

Fig. 2 Nuclear actin and myosin in transcription. (A) During initiation of transcription NM1 is phosphorylated by GSK3β and interacts with the chromatin where it stabilizes actin-Pol II complex (actin is symbolized by red arrows). (B) NM1 switches its conformation, binds and stabilize SNF2h, which together with WSTF remodel the surrounding chromatin. (C) HAT PCAF and HMT Set1 are recruited to DNA by NM1 and acetylate and methylate surrounding histones allowing Polymerase to go through the gene. (D) During the elongation phase hnRNP U, previously bound to phosphorylated CTD domain of Pol II interacts with actin and facilitate histone acetylation by PCAF. Actin is also associated with newly formed pre-mRNA, forming a complex with other hnRNPs and directing mRNA to nuclear export (Almuzzaini et al., 2015; Kukalev et al., 2005; Percipalle et al., 2003, 2006; Sarshad et al., 2014).

(CBF-A) also referred to as hnRNP A/B and hnRNP U (Percipalle et al., 2002). Direct interactions were reported between actin and CBF-A/hnRNP A/B and with hnRNP U (Kukalev et al., 2005; Percipalle et al., 2002). This actin-hnRNP associations are thought to facilitate transcriptional elongation and assembly of the nascent ribonucleoprotein (RNPs) complexes (Fig. 2D). In addition, actin binds hnRNP U through a specific motif that is conserved also in the *C. tentans* hrp65 (Percipalle et al., 2003). The actin-hnRNP U complex binds to the phosphorylated form of the Pol II C-terminal domain (CTD) and this interaction, similarly to the actin-hrp65 interaction, is required for transcription activation and elongation (Fig. 2A) (Kukalev et al., 2005; Obrdlik et al., 2010; Percipalle et al., 2003). The actin-hrp65 and actin-hnRNP U complexes seem to regulate transcription activation by mediating the recruitment of the histone acetyl transferases (HATs) p2D10 and PCAF, respectively, to transcribed genes (Kukalev et al., 2005; Percipalle et al., 2003). In support of this model, disruption of actin-hnRNP interactions *in vivo* reduces histone acetylation, blocks Pol II transcription and affects HAT occupancy along transcribed genes, further illustrating the vital role actin plays in regulating transcription (Kukalev et al., 2005; Percipalle et al., 2003).

4. Chromatin regulation by actin and myosin and the global transcriptional response

As already anticipated above, several studies have shown that actin is a *bona fide* component of several chromatin remodeling complexes including the INO80 (Shen et al., 2000), SWR1 (Mizuguchi et al., 2004), and NuA4 (Ikura et al., 2000) complexes in yeast and the BAF (Zhao et al., 1998) or BAP (Papoulas et al., 1998) complex in mammals and *Drosophila* respectively.

INO80 is a yeast chromatin remodeling complex with roles in DNA repair, replication, cell cycle control and transcription and was initially identified as a co-activator of genes involved in inositol metabolism (Kapoor et al., 2013; Shen et al., 2000). Nuclear actin in combination with other actin-related proteins Arp4 and Arp8 forms the functional module that interacts with chromatin as part of the INO80 complex (Kapoor et al., 2013; Shen et al., 2000). Interestingly, the chromatin remodeling activity of INO80 can be inhibited by introducing a point mutation in subdomain 2 of actin. Since subdomain 2 normally helps keep actin in a monomeric state in the INO80 complex by making its barbed end inaccessible for polymerization (Kapoor et al., 2013), it is thought that the presence of monomeric actin in INO80 is critical for its nuclear remodeling functions. In contrast, mutations that inhibit actin polymerization have no impact on INO80 activity. This was further supported by crystalizing part of the INO80 complex containing monomeric actin, Arp4 and Arp8 showing that this Arp4-Actin-Arp8 complex is responsible for recognition of extranucleosomal linker DNA, coupling ATP-dependent DNA pumping with processive nucleosome translocation by INO80 (Knoll et al., 2018).

SWR1 is another actin containing chromatin remodeling complex that plays an important role in regulating transcription and chromatin structure (Mizuguchi et al., 2004). It is critical for the expression of key developmental genes and is known to regulate plant development and flowering in *Arabidopsis* (Choi et al., 2007). SWR1 performs its regulatory functions by depositing the histone H2A.Z at specific locations between nucleosomes (Mizuguchi et al., 2004). H2A.Z is a non-canonical histone variant enriched at promoters, enhancers, transcription start sites and nucleosome free regions and facilitates transcriptional activation. Like its role in INO80, actin along with Arp4 and Arp6 forms a key component of the SWR1 complex (Wu et al., 2005). Arp4 can recognize and bind histones while Arp6 can mediate binding between chromatin and nuclear pores. Together both these functions are crucial for the ability of SWR1 to interact with chromatin, deposit H2A.Z and regulate gene expression and nucleosome structure (Choi et al., 2007; Mizuguchi et al., 2004; Wu et al., 2005).

In addition to INO80 and SWR1, arguably the most important chromatin remodeling complex containing actin there is also the mammalian Brm/Brg associated factor complex (BAF complex) which is a homolog of the yeast Swi/Snf and Drosophila BAP (Brm associated protein) complex (Papoulas et al., 1998; Zhao et al., 1998). BAF is a multi-subunit complex containing at least 15 subunits, including the ATPase Brg or Brm, of which β-actin is a critical part. The BAF complex can regulate activity of a wide variety of genes and is also considered as a neuronal progenitor, known as neuronal BAF (nBAF), or poly-bromo associated BAF complex (Alfert et al., 2019). The BAF complex regulates chromatin accessibility by sliding or evicting nucleosomes in an ATP-dependent manner and plays critical roles in development and differentiation (Alfert et al., 2019; Zhao et al., 1998). The BAF complex is required to oppose the repressive activity of polycomb repressive complexes and these antagonistic activities between BAF and polycomb complexes control expression of critical developmental genes (Kadoch et al., 2017). Interestingly, unlike the yeast Swi/Snf complex which only contains Arp subunits, both the mammalian and Drosophila BAF/BAP complexes contain β-actin (Zhao et al., 1998) making them similar to INO80 and SWR1. The β-actin present in the BAF/BAP complexes can both polymerize, exhibits ATPase activity and has been shown to be critical for the ATPase activity of Brg1 (Kapoor and Shen, 2014). Recently, loss of β-actin has been shown to negatively regulate Brg1 deposition and association with chromatin (Xie et al., 2018a). This, in turn, leads to increased repressive epigenetic modifications like trimethylation of histone H3 in residues K9 (H3K9me3) and K27 (H3K27me3), defects in localization of heterochromatin and dysregulation of genes involved in neuronal differentiation and angiogenesis (Xie et al., 2018a,b). It is, therefore, likely that the levels of β-actin inside the nucleus may have an impact on the chromatin remodeling activities of the BAF complex.

NM1 also plays a role in chromatin regulation during transcription. NM1 forms a complex with WICH (Cavellan et al., 2006), a chromatin remodeling complex belonging to the ISWI family of chromatin remodelers which slide nucleosomes along DNA *in vitro* (Langst and Becker, 2001). WICH contains the subunits WSTF and SNF2h, and interacts with several RNAs and nuclear proteins. Bound to NM1 this macromolecular complex has been referred to as B-WICH (Cavellan et al., 2006; Percipalle et al., 2006). B-WICH is involved in Pol I, Pol II and Pol III transcription, and its inhibition leads to reduced levels of 5s rRNA, mRNA and 7SL RNA (Cavellan et al., 2006). Furthermore, NM1, as part of B-WICH, interacts with chromatin to activate both Pol I and Pol II transcription and it may even participate in the maintenance of structure across both rDNA and class II genes (Percipalle and Farrants, 2006). Mechanistically, the WSTF-SNF2h complex is stabilized by the interaction between SNF2h and NM1, and the HAT PCAF is next recruited. Interestingly, binding of NM1 to SNF2h is antagonistic to actin binding suggesting an active switching of NM1 binding during transcription (Cavellan et al., 2006; Percipalle et al., 2006; Sarshad et al., 2013). This suggests that NM1 has an actin-related motor function that leads to the induction of a permissive chromatin state compatible with Pol I and Pol II transcription (Almuzzaini et al., 2015; Sarshad et al., 2013, 2014). Since chromatin bound-NM1 can interact with either actin or SNF2h depending on the state of its motor function, we have recently proposed that NM1 functions as a molecular switch that is required to activate transcription (Fig. 2). In the first step the polymerase associated actin binds to the chromatin bound NM1 to recruit and maintain the polymerase machinery at the correct location (Fig. 2A). Once NM1 interacts with SNF2h, the

WICH chromatin remodeling activity is activated (Fig. 2B) (Almuzzaini et al., 2015) and the HAT PCAF is recruited for H3K9 acetylation, promoting transcription activation. This mechanism appears to be general as it is required for both Pol I and Pol II transcription activation. In fact, in the latter case, we showed that NM1 activation of the WICH remodeler is also followed by recruitment of both PCAF and the histone methyl transferase HMT Set1/Ash1 (Fig. 2C) (Almuzzaini et al., 2015). The generality of this mechanism is also supported by the fact that NM1 associates with a significant proportion of the mammalian genome (Almuzzaini et al., 2015) and its loss affects both chromatin accessibility and global transcriptional levels (Venit et al., 2020). Interestingly, association of NM1 with the chromatin is regulated by phosphorylation (Sarshad et al., 2014). During the G1 phase of the cell cycle, NM1 is phosphorylated in Ser-1020 by the glycogen synthase kinase 3β (GSK3β) (Fig. 2A). Phosphorylation by GSK3β activates NM1, promoting its association with the chromatin, which, in turn, leads to epigenetic modifications compatible with Pol I transcription, as shown by the Pol I transcriptional repression phenotype observed in GSK3β knockout cells (Sarshad et al., 2014).

Altogether these observations suggest that the function of NM1 as actin-based molecular switch to activate transcription through a chromatin based mechanism is mediated by essential signaling pathways.

5. Nuclear actin in transcriptional reprogramming during differentiation

Spatially and temporally regulated control of gene expression is critical for development and differentiation. During development, gene expression patterns are restricted to ensure cell specific expression and in fact, abnormalities in expression of developmental genes lead to impairment of developmental processes leading to differentiation phenotypes such as loss of cell identity that are often compatible with a disease. Transcriptional activators and chromatin remodelers, therefore, play crucial roles in regulating lineage-specific gene expression patterns. It is, therefore, not surprising that nuclear actin, a key component of several chromatin remodelers and major regulator of the transcriptional process, is emerging as an essential regulator of development and differentiation.

Transcriptional reprogramming is a process by which transcriptional patterns of specialized cells can be reverted back to those of pluripotent stem cells (iPS) or to different specialized cell types (transdifferentiation) (Miyamoto and Gurdon, 2013). First demonstrated using nuclear transfer experiments, reprogramming can now be performed *via* overexpression of transcription factors that promote pluripotency or tissue specific gene expression and allows researchers to mimic the process of cellular differentiation *in vitro* (Gurdon et al., 1958; Takahashi and Yamanaka, 2006; Vierbuchen et al., 2010). Reprogramming involves the activation of genes specific to a desired cell type which were previously silenced either due to lack of transcriptional activators (activatable genes) or due to chromatin-based repression mechanisms (occluded genes) (Miyamoto and Gurdon, 2013). Several lines of evidence have now suggested that nuclear actin can have a potential role in the activation of both occluded and activatable genes during reprogramming.

An example of the role of nuclear actin in gene activation is the tightly controlled expression of serum response factor (SRF), a transcription factor which induces muscle specific and serum inducible genes (Sotiropoulos et al., 1999). SRF expression itself requires a coactivator called MAL which can bind globular actin (G-actin) inside the nucleus. Nuclear actin binding inhibits MAL's function as a coactivator and promotes its export to the cytoplasm. An increase in polymerized F-actin in the cytoplasm, however, can free MAL from G-actin binding, restore its activity and hence trigger SRF expression (Sotiropoulos et al., 1999). This is an example of transcriptional activation of a key gene that depends on both the balance of actin levels between the nucleus and cytoplasm and actin polymerization state. Similarly, nuclear actin is also involved in triggering the expression of the *HoxB* gene upon retinoic acid treatment and it has been shown that overexpression of nonpolymerizable actin inhibits *HoxB* transcription (Ferrai et al., 2009). Since *HoxB* expression requires the transcriptional activator Prep-Pbx1 which is known to bind both Poll II and nuclear b-actin, it is thought that actin polymerization is involved in the recruitment of transcriptional machinery required for *HoxB* activation.

Another example of the role of nuclear actin in transcriptional activation is the expression of Toll-like receptor responsive (TLR) genes (Huang et al., 2011). Nuclear receptor co-repressor complexes (NCoR) which occupy the promoters of TLR genes, inhibit their transcriptional activation and must be cleared from the chromatin for gene activation. The transcriptional activation of these genes is achieved by the binding of nuclear actin to an NCoR complex component Coronin2A which triggers the removal of these complexes from gene promoters.

While SRF and *HoxB* represent examples of transcription from activatable genes, nuclear actin may also have a role in transcriptional activation from occluded genes *i.e.* genes repressed *via* chromatin-based mechanisms (Fig. 3). An example of one such occluded gene is the critical regulator of differentiation *Oct4*. Nuclear transplant experiments in *Xenopus* oocytes have shown while *Oct4* is occluded in somatic cells (Foshay et al., 2012), its expression can be induced soon after nuclear transfer suggesting that cells undergoing reprogramming have the ability to reverse its occlusion. Importantly, the transplanted nuclei contain a large amount of actin and inhibition of actin polymerization in these nuclei inhibits *Oct4* transcription suggesting that nuclear actin may have a role in reversing gene occlusion (Miyamoto et al., 2011).

More recently, we reported a more global role for actin in transcriptional reprograming during neurogenesis. In this study we performed transcriptional profiling of β-actin null fibroblasts obtained from an embryonic lethal knockout mouse using RNAseq. Results from this study demonstrated that loss of β-actin dysregulate global transcription with plenty of differentially expressed genes involved in neuronal development being either up- or down-regulated (Xie et al., 2018a). Since fibroblasts share a neuroectodermal lineage with neuronal cells, they are known to express neuronal mRNAs and they can be reprogrammed to neurons. Therefore, to find out if the transcriptional role of β-actin is important for expression of subsets of neuronal genes and contributes to development, we applied a recently published reprogramming protocol that induces neurons from fibroblasts using a cocktail of small molecules that drive direct lineage reprogramming (Li et al., 2015; Xie et al., 2018b). We discovered that although β-actin null fibroblasts can be converted into *bona fide* neurons, these neurons develop into different neuronal subtypes compared to the wild-type condition. Neurons induced from wild-type embryonic fibroblasts and β-actin null embryonic fibroblasts tend to exhibit increased number of inhibitory and excitatory synapses, respectively. We show evidence that this is a consequence of local alterations in chromatin compaction that affects expression of neuronal and proneuronal genes during the reprograming of fibroblasts to neurons. In fact, at transcription start sites of subsets of neuronal and pro-neuronal genes, we show that chromatin exhibits increased histone methylation and decreased Brg1 deposition which impacts expression.

In summary, the above results indicate that one of the primary functions of nuclear actin is to maintain a chromatin landscape compatible with transcription and this is important to consolidate gene expression during

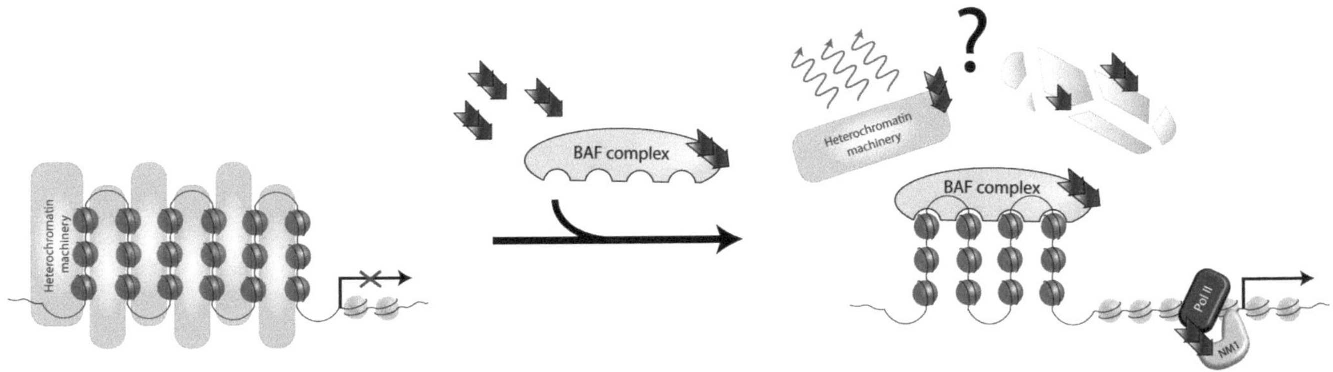

Fig. 3 A model for actin-mediated transcriptional reprogramming. Transcription of some genes is occluded because of inaccessibility of chromatin to Pol II machinery. Actin-based chromatin remodelers as BAF can lead to reorganization of nucleosomes and deposition of heterochromatin machinery allowing occluded genes for transcription. The role of actin in direct deposition or degradation of heterochromatin remains to be elucidated. Actin is symbolized by red arrows (Miyamoto and Gurdon, 2013).

transcriptional reprograming. This idea opens up a novel paradigm that may explain how activatable and occluded genes are regulated during development and differentiation by nuclear actin as hypothesized earlier by the Gurdon laboratory (see Miyamoto and Gurdon, 2013). Below, we discuss how nuclear actin, possibly together with nuclear myosin, is likely to play a fundamental role in the functional architecture of the mammalian genome and lay the foundation for an entirely new area of research.

6. Potential role of nuclear actin and myosin in regulating the 3D genome: a future direction

The chromosomes of eukaryotes are folded into a multilayered hierarchical structure with each layer playing a role in regulating genome activity and function. Recent studies have utilized chromosome conformation capture techniques like 3-C and Hi-C to infer 3D genome organization by analyzing interaction frequencies between different regions of the genome (Bonev and Cavalli, 2016). Chromosome conformation capture studies together with high throughput microscopy have revealed a hierarchical genome organization in which chromosomes occupy distinct territories in nuclear space. They then fold into gene-rich and transcriptionally active or gene-poor and transcriptionally repressed compartments which are referred to as A and B compartment respectively (Buchwalter et al., 2018). The A and B compartments can be further divided into topologically associated domains (TADs) defined as regions enriched for local genomic interactions and conserved across cell types and species. TADs further give rise to sub-TADs and chromatin loops that facilitate long range enhancer-promoter interactions and often vary in a tissue specific manner (Bonev and Cavalli, 2016).

While there is considerable evidence for the role of nuclear actin in regulating chromatin structure at a local level, recent work hints at the possibility that actin might also regulate higher order chromatin organization. It has been shown, for instance, that inhibition of actin polymerization can increase the volume and surface area of chromosomal territories (Ondrej et al., 2008). Similarly, recent work shows that β-actin deficient cells exhibit changes in spatial organization of H3K9Me3/HP1α—positive heterochromatin (Xie et al., 2018a) suggesting a role for actin in heterochromatin segregation and in the organization of specific heterochromatin compartments.

A potential role for nuclear actin in regulating the 3D genome is also in line with mechanistic models proposed to underlie higher order chromatin organization. Many studies have suggested that the 3D arrangement of chromatin fibers and separation of genomic regions into compartments arises from a phase separation based mechanism which allows self-organization of chromatin domains into functionally distinct compartments (Stam et al., 2019). In this model, molecules containing multiple interaction domains induce the formation of supramolecular assemblies of proteins and nucleic acids that underlie the formation of higher order chromatin structures. Liquid-like droplets of chromatin which are thought to underlie higher order structure can arise from the tendency of such molecules to interact with each other and with DNA. Interestingly, two such molecules known to be involved in regulating genome organization are potentially affected by nuclear actin levels. These include HP1α and Polycomb Repressive Complex 1 (PRC1).

HP1α and PRC1 are characteristic of constitutive and facultative heterochromatin respectively and both have been shown to form liquid-like droplets *in vivo* and *in vitro* (Stam et al., 2019). HP1α protein can homodimerize and bridge H3K9 methylated nucleosomes while PRC1 proteins are thought to induce clustering of polycomb protein-bound chromatin and nucleosomes (Fig. 4) (Stam et al., 2019). Such phase separated "polycomb bodies" are implicated in the regulation of higher order genome architecture (Illingworth, 2019) and maintenance of long-range chromatin interactions (Bantignies et al., 2011). Recent studies have shown that β-actin depletion can potentially affect the regulation of both HP1α and PRC1 proteins (Xie et al., 2018a,b). β-actin knockout cells, for instance, exhibit redistribution of HP1a-positive heterochromatin and loss of chromatin association of Brg1; the essential ATPase subunit of the BAF complex (Xie et al., 2018a). Since Brg1 directly opposes the deposition of polycomb proteins (Kadoch et al., 2017) these observations support a model where loss of β-actin may trigger increased recruitment of PRC complexes to specific genomic loci and potentially induce compartment level changes in heterochromatin organization (Fig. 4).

In addition to potential changes in compartment level organization, loss of Brg1 chromatin binding in β-actin KO cells may also affect genome architecture at the level of TADs and loops (Alexander et al., 2015; Barutcu et al., 2016; Wang et al., 2017). It has been shown that more than 20% of Brg1 binding sites overlap with TAD boundaries and Brg1 loss at these boundaries can lead to a decrease in TAD insulation (Barutcu et al., 2016).

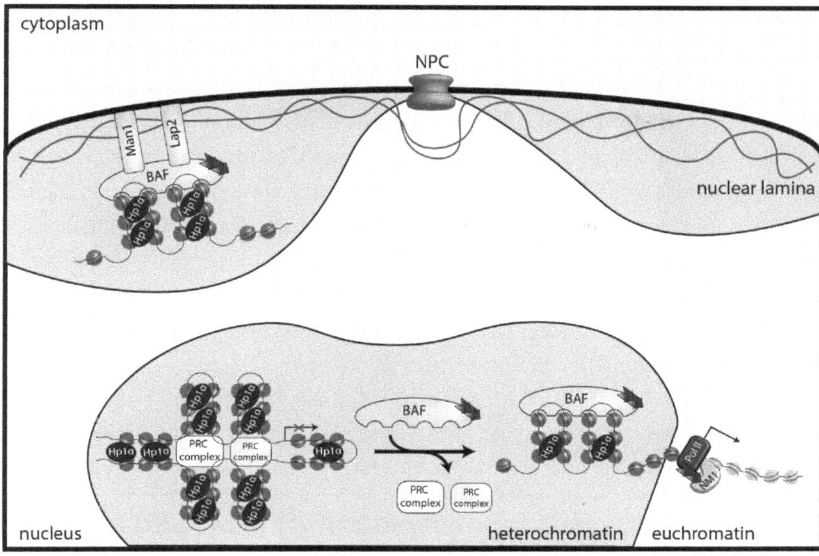

Fig. 4 A speculative model of β-actin controlling spatial distribution of heterochromatin and global euchromatin levels. Recent work suggests that β-actin regulates Brg1-dependent ATPase activity of the BAF complex and its interplay with both nuclear envelope and Polycomb repressive complexes in the nuclear interior. Actin is symbolized by red arrows (Xie and Percipalle, 2018).

This decrease in TAD strength is thought to result from defects in Brg1 mediated chromatin accessibility changes which may affect recruitment of transcription factors and topoisomerase proteins to TAD boundaries (Barutcu et al., 2016). Similarly, it has also been shown that Brg1 binding sites can overlap with distal enhancers and loss of Brg1 binding can impair enhancer activation and cell differentiation (Alexander et al., 2015). β-actin mediated changes in Brg1 chromatin binding may therefore constitute another potential pathway through which actin may affect 3D genome organization. Interestingly, recent study by Barisic et al. compares the effect of SWI/SNF remodeler Brg1 and ISWI remodeler SNF2H on transcription factor binding and TADs organization in mouse embryonic stem cells. They showed that each chromatin remodeler mediates binding of distinct transcription factors and point out that SNF2H is critical for proper localization of CTCF protein and insulation of TADs (Barisic et al., 2019). Therefore, NM1 as a direct binding partner of SNF2H which stabilize it on Pol I and Pol II promoters (Almuzzaini et al., 2015; Sarshad et al., 2013;

Vintermist et al., 2011), could be functionally linked to genome-wide nuclear topology as well. Another study links directly cytoskeletal and nuclear functions of actin and myosin through binding to Lamin A/C and Emerin. Co-depletion of both proteins leads to destabilization of cytoskeletal organization and changes in spatial positioning of chromosome territories, dependent on NM1 and actin polymerization (Ranade et al., 2019).

While recent evidence hints at a role for nuclear actin and myosin in 3D genome organization, this role is yet to be directly investigated. Experiments utilizing conformation capture techniques like Hi-C in actin and NM1 knockout models therefore offer an exciting avenue for future research. The development of better experimental systems for precise manipulation of nuclear actin levels and replication of these studies in animal models are critical for elucidating actin's role in development and differentiation.

7. Evidence on the involvement of actin and myosin in genome integrity

A potential role for actin and myosin in consolidating the architecture of the mammalian genome would be compatible with a general function in the stability of the genome. We would, therefore, expect that loss of actin and myosin is compatible with loss of genome integrity.

Genome integrity is the safeguarding of genome structure and the preservation of the information encoded by it. There are several factors that might threaten genome integrity by causing DNA lesions. DNA damaging factors can be classified as exogenous, endogenous and spontaneous (Hoeijmakers, 2009). Exogenous factors include all kinds of ionizing radiation, drugs and chemicals. On the other hand, examples of endogenous DNA damaging factors include reactive oxygen species (ROS) and replication stress (Ciccia and Elledge, 2010). Furthermore, the DNA molecule itself is chemically unstable and highly reactive, which leads to spontaneous decay, even in ideal temperature and pH conditions (Lindahl, 1993).

The most common DNA lesions include intra-strand cross-links, inter-strand cross-links, nucleotide modifications and breaks of the phosphodiester bonds, like single-strand breaks (SSBs) and double-strand breaks (DSBs) (Manova and Gruszka, 2015). DSBs are considered the most dangerous form of DNA damage, since they can result in genomic aberrations, apoptosis, disease and cancer (Jackson and Bartek, 2009). Cells respond to DNA damage by activating the DNA Damage Response (DDR), which aims to mitigate, evaluate and repair the possible damage. The DDR involves transcription, cell cycle regulation, apoptosis and several DNA repair pathways that are activated depending of the kind of lesion (Ciccia and Elledge, 2010).

There are two major pathways to repair DSBs: nonhomologous end-joining (NHEJ) and homologous recombination (HR). NHEJ is active throughout the cell cycle and is a fast repair mechanism, although it is considered to be error prone. Since NHEJ relies on the quick relegation of the DNA ends with minimum end processing, it can give rise to deletions, translocations and telomere fusions (Espejel et al., 2002; Shibata et al., 2018). On the other hand, HR starts with resection of the DNA ends, and the repair is performed using a homologous sequence as a template to prevent the loss of information. Typically, the homologous sequence used for repair is the sister chromatid, which limits the activity of HR to the S and G2 phases of the cell cycle (Ceccaldi et al., 2016; Gospodinov and Herceg, 2013; Shibata and Jeggo, 2014). However, the choice between HR and NHEJ is not entirely dependent on the moment of the cell cycle. Other factors like the chromatin context, the sequence of the damage site and the presence of DNA:RNA hybrids are known to influence the decision (Jimeno et al., 2019; Marcomini et al., 2018). While both pathways have been extensively studied, new factors and variations are still being revealed (Cohen et al., 2018; Domingo-Prim et al., 2019; Shibata et al., 2018).

The damage caused by DSBs goes beyond the loss of genetic information encoded in the DNA sequence. Chromatin topology is affected when the DNA strand is broken, and the epigenetic changes that happen around DSBs could lead to the loss of epigenetic information surrounding the break. A recent study by Ochs et al. suggests that the cell has a mechanism to avoid this loss of epigenetic information *via* preservation of topologically associated domains (TADs). By using super-resolution microscopy to study the formation of DSBs, it was proposed that 53BP1 and RIF1 work together to form a circular arrangement that stabilizes TAD-like structures. This organization, in turn, would form a barrier that limits the function of repair enzymes and prevents further damage derived from repair. Moreover, this mechanism is independent of the repair pathway choice (Ochs et al., 2019). These findings suggest that chromatin remodeling and the safeguarding of DNA topology plays a role in DNA damage and repair.

Preservation of DNA topology, in turn, suggests that maintaining and consolidating a stable architecture of the genome is mandatory to efficiently protect cells from DNA damage. Genome integrity can be achieved by regulating the spatial organization of chromatin and by coordinating the transcriptional response to DNA damage with DNA replication and cell division. In this context, recent work has highlighted the importance of nuclear actin and myosin which are now beginning to be considered as gatekeepers of genome integrity. *In vitro* studies have shown that actin is associated with several DNA damage and repair factors (Samwer et al., 2013). Treatment with drugs that block actin polymerization as Cytochalasin D or Latrunculin A or overexpression of mutant actin unable to polymerize leads to an increase of double strand breaks suggesting that filamentous actin has a role in DNA damage sensing and/or repair (Andrin et al., 2012). Similarly, decreasing the nuclear actin pool by knocking down importin-9 which mediates nuclear import of actin, leads to impairment of DNA damage clearance (Dopie et al., 2012). In contrast, treating cells with DNA damage-inducing drug methyl methansulfonate (MMS) increases the amount of F-actin inside the nucleus, even though these filaments do not co-localize with DNA damage marker γH2AX (Belin et al., 2015). Similarly, deletion of nuclear myosin I (NM1) leads to increased DNA damage, deregulation of cell cycle and increased cellular apoptosis (Venit et al., 2020). The myosin 1C gene locus has been also found to be mutated in several types of cancers and the Myo1C protein has been suggested to be a potential tumor suppressor (Hedberg Oldfors et al., 2015).

One can speculate that from a mechanistic point of view actin and myosin could regulate genome integrity as part of different chromatin remodelers, they can actively contribute to chromosomal movements and organization of chromatin domains such as TADs working as transporters or by regulating expression of genes important for maintenance of genome integrity. These functions are not mutually exclusive and are discussed below.

8. Actin- and myosin-dependent chromatin remodelers playing a role in genome integrity

Activation of specific DNA repair mechanisms is dependent on cell cycle phase and also on the spatial organization of DNA within the nucleus. In fact, nuclear position of double strand breaks and accessibility of chromatin dictates which DNA repair pathway is selected by the cell (Lemaitre et al., 2014) and initiation of DSB repair is associated with unfolding of higher-order, condensed chromatin structures (Kruhlak et al., 2006). Therefore, increased heterochromatization in both β-actin (Xie et al., 2018a,b) and NM1 (Venit et al., 2020) knockout cells could affect global accessibility of DNA to DNA damaging sources but also to DNA damage repair proteins. This may be consequence of the impairment of actin- and myosin-dependent chromatin remodelers involved in DNA damage-dependent DNA repair. For instance, the mammalian INO80 complex is recruited to DNA damage sites in an ARP8 dependent manner (Kashiwaba et al., 2010) and promotes large-scale movements of DNA loci during homologous recombination (Neumann et al., 2012). During DNA replication, depletion of human INO80 impairs DNA synthesis (Hur et al., 2010) and in *S. cerevisiae* INO80 moves with the replication fork to evict improperly deposited H2A.Z (Au et al., 2011). INO80 and SWI/SNF have also been shown to be recruited to centromeres and have been implicated in chromosome segregation (Ogiwara et al., 2007; Xue et al., 2000).

Likewise, subunits of the B-WICH complex—SNF2H and WSTF—are recruited to DSBs where WSTF regulates DNA damage response by phosphorylation of histone H2A.X (Cook et al., 2009; Ogiwara et al., 2011; Xiao et al., 2009). During DNA replication, there are natural replication fork barriers such as actively transcribing genes or highly repetitive sequences found in the heterochromatin which need to be resolved before a cell enters mitosis (Branzei and Foiani, 2010). It is known that SNF2h and WSTF interact with PCNA and localize to heterochromatin to protect newly replicated DNA from heterochromatinization during S phase (Collins et al., 2002; Poot et al., 2004). However, the direct role of actin and myosin in these processes remains to be elucidated.

9. Actin and myosin as nuclear transporters to maintain genome integrity

Eukaryotic nuclei are dynamic organelles which undergo severe rearrangements that depend on both extracellular and intracellular cues. On the bases of their cytoplasmic functions as motors, actin and myosin have also been suggested to play a role in intranuclear movements. Both have been described to have a function in the relocation of chromosome territories upon DNA damage (Kulashreshtha et al., 2016), serum starvation (Mehta et al., 2010), or upon gene activation (Chuang et al., 2006; Dundr et al., 2007; Gieni and Hendzel, 2008). Recent study in yeast proposed that myosin anchored to activated DNA gene loci, with help of ARP-dependent chromatin remodelers, slides along the meshwork of short actin fibers and move gene toward the environment more favorable for transcription (Wang et al., 2020). Similarly, nuclear actin and NM1 are found to be responsible for estrogen-induced changes of interchromosome interactions and reorganization of nuclear territories (Hu et al., 2008).

Interactions of actin with several actin-binding proteins have been described to have a direct role at sites of DNA damage. Two modes of action have been suggested, depending on the DNA environment in which damage occurs. In actively transcribed genes, DNA damage clusters into large foci (Aymard et al., 2017) and this clustering is actin-dependent (Schrank et al., 2018), while in the context of repressive heterochromatin actin facilitates the movement of damaged foci toward an environment that enables repair (Caridi et al., 2018; Chiolo et al., 2011; Tsouroula et al., 2016).

The movement of DNA breaks seems to be dependent on the type of mechanism used for the DNA repair. In budding yeast, homology directed repair (HDR) which is the preferred way to repair DNA breaks, induces rapid DNA damage foci movement and clustering (Dion et al., 2012; Lisby et al., 2003). In contrast, in mammalian cells where nonhomologous end joining (NHEJ) is the primary repair mechanism, movement of DNA damage foci is limited (Soutoglou et al., 2007). However, during G2 and S phase, HDR is used to fix DSBs clustered together (Aymard et al., 2017). In this context, a recent study by Schrank et al. has shown a direct involvement of actin and actin-binding proteins in HDR repair mechanisms. Mass spectrometry analysis of cell free S-phase extracts derived from *Xenopus* eggs as well as immunofluorescent microscopy of human U2OS cells revealed that actin and the actin nucleators WASP and ARP2/3 co-localize with double strand breaks (Fig. 5). While WASP binds to DSBs independently of the type of DNA repair pathway, ARP2/3 seems to be specific for HDR (Schrank et al., 2018). Mobility and clustering of S/G2-specific Rad52 damage foci were compromised in cells with inactivated ARP2/3 proteins, whereas G1-specific 53BP1 foci were unaffected. Notably, U2OS cells expressing the nuclear actin-chromobody TagGFP revealed actin containing foci and short rods after treatment with the DNA damage-causing drug neocarzinostatin. These actin foci co-localize with RAD51 foci and in contrast to actin rods are dependent on ARP2/3. The disruption of WASP, ARP2/3 or actin leads to decreased ssDNA generation at DSBs, indicating that clustering facilitates DSB end-resection in HDR (Schrank et al., 2018).

While in the gene-rich euchromatin environment, Brownian movement of DNA breaks could serve to find and unwind homologous DNA, in heterochromatin we have an opposite scenario. Highly repetitive sequences, even from different chromosomes can serve as homologues to broken DNA and can lead to irreversible chromosomal rearrangements. Therefore, relocalization and separation of DSB foci from heterochromatin toward euchromatin by long directional movement could protect cells from unwanted ectopic recombination (Fig. 5) (Caridi et al., 2017). In *Drosophila*, ionizing radiation-induced repair sites leave heterochromatin and accumulate at the nuclear periphery. Inhibition of actin polymerization

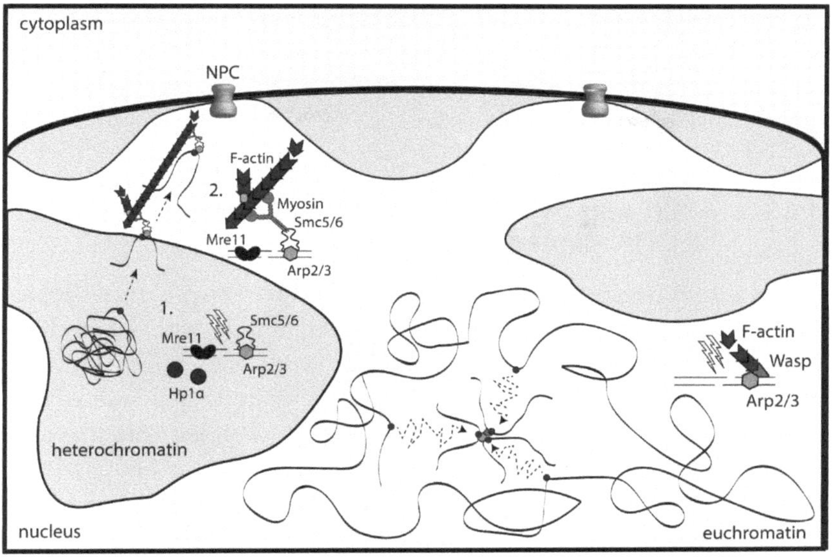

Fig. 5 A model for actin-mediated DSB movement. In euchromatin, Arp2/3 localizes to sites of DNA damage together with Wasp protein which serves as a nucleating factor for actin polymerization. DNA breaks then move by Brownian motion and cluster together to be fixed by homology directed repair (Schrank et al., 2018). In heterochromatin, binding of Arp2/3 to the site of DNA damage is dependent on Hp1α and Mre11. Arp2/3 forms a complex with Smc5/6 and myosin which slides along the actin filament toward nuclear periphery (Caridi et al., 2018). Actin is symbolized by red arrows.

with Latrunculin B or inhibition of Arp2/3 complex leads to accumulation of γH2AX foci in the heterochromatin, without affecting the total number of foci (Caridi et al., 2018). Arp2/3 co-localization with the γH2AX foci in heterochromatin is dependent on the DSB signaling protein Mre11 and the heterochromatin component HP1α. Arp2/3 then serves as nucleator for actin polymerization from the repair site within heterochromatin. These newly generated actin polymers have been shown to be highly dynamic and progressively disappear during relocalization of DSB to the nuclear periphery. DSBs are believed to travel along those filaments and this relocalization is driven by nuclear myosins 1A, 1B and V (Fig. 5) (Caridi et al., 2018). However, results from another study in human cells show that DNA damage induced at heterochromatic regions close to the nuclear membrane (within lamina-associated domains or LADs) fail to rapidly activate the DNA damage response, do not move toward more permissive environment and are repaired by alternative end-joining (Lemaitre et al., 2014). This discrepancy could be a consequence of using different model systems, or more interestingly it can suggest a high variability of actin forms and functions dependent on the actual context in which actin is present. Cell cycle phase, chromatin condensation and position, vicinity of specific actin-binding proteins could be just one of the few factors affecting actin structure and function in these processes.

10. Actin and myosin in the transcriptional response to DNA damage

In addition to the repair process, survival after DNA damage depends on activation of the DNA damage checkpoint pathway, which halts the cell cycle and coordinates repair of damage with cell cycle transitions (Papamichos-Chronakis and Peterson, 2013). Unprocessed ends of DSBs are recognized by Mre11-Rad50-Nbs1 (MRN) complex which activates ataxia telangiectasia mutated (ATM)—first kinase in the cascade of phosphorylation events leading to activation of downstream targets and arresting the cell cycle (Abraham, 2001). Among others, this leads to activation of the tumor suppressor p53. After phosphorylation, p53 is released from its inhibitory complex with MDM2 and serves as a transcription factor activating several p53-target genes including *Cdkn1A*. p21, the product of the *Cdkn1A* gene, in turn represses cell cycle progression by inhibiting cyclin E and CyclinA/CDK2 required for G1/S transition (Brugarolas et al., 1999).

In NM1 KO cells, we recently reported occurrence of constitutive DNA damage foci (Venit et al., 2020). Under these conditions, we found that naturally occurring DNA damage leads to initiation of DNA damage checkpoint pathways and upregulation of genes which are part of the phosphorylation cascade upstream of p53. However, although expression of p53 is not attenuated, p21 expression is abolished and cells fail to stop and properly repair the damage (Venit et al., 2020). Cells accumulate DNA breaks over several generations, which leads to a general instability of the genome and apoptosis. p53 itself is not sufficient for activation of p21 expression and an interplay with additional factors is needed. Indeed, under DNA damage conditions, NM1 was found as part of the same complex with p53, occupying promoter and the transcription start site of the Cdkn1A gene. We reported that NM1 recruits the HAT PCAF and the HMT Set1 to acetylate and methylate histone H3, relaxing the chromatin and allowing Pol II to transcribe the *Cdkn1A* gene. As p53 binds only to promoter region but not to transcription start site of *Cdkn1A*, we speculate that p53 maintains a poised polymerase at the gene promoter ready for a rapid response to NM1-dependent transcription activation upon DNA damage (Fig. 6) (Venit et al., 2020). Whether this mechanism

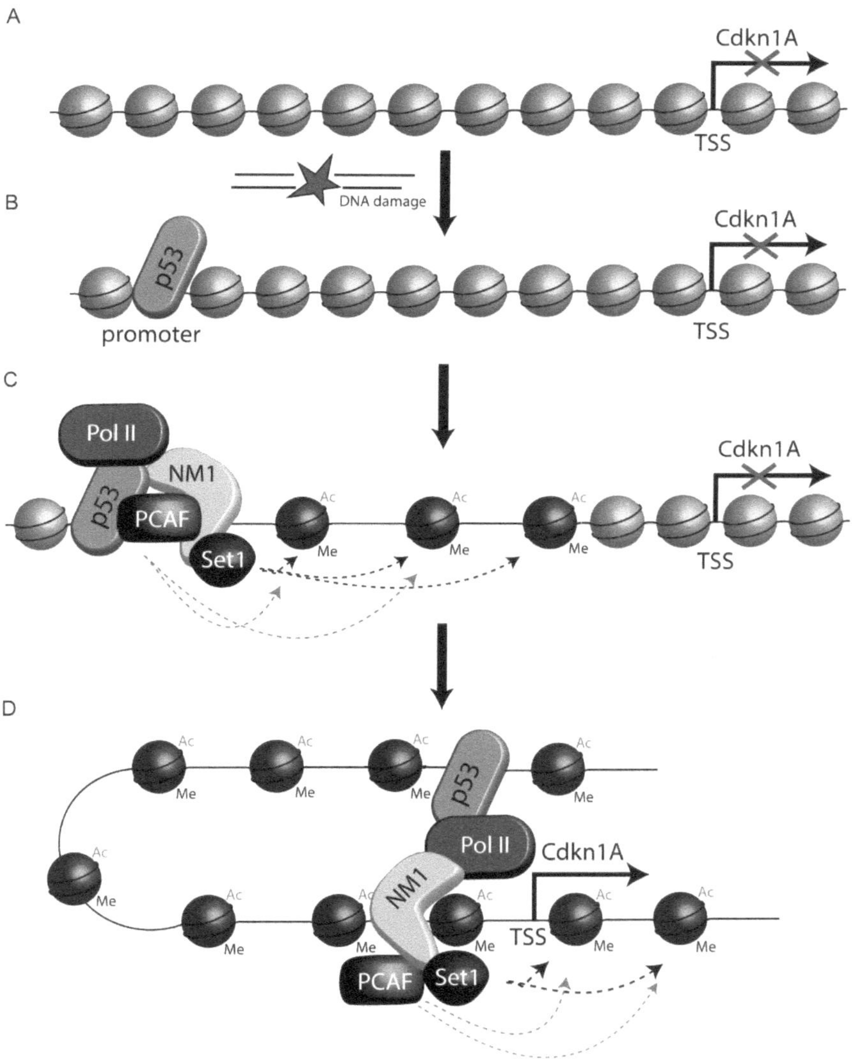

Fig. 6 A model for transcriptional activation of p21 expression in response to DNA damage. (A) In undamaged cells, *Cdkn1A* gene is inaccessible for transcription. (B) After DNA damage, p53 binds to the promoter region of *Cdkn1a* independently from NM1. (C) NM1 binds to the promoter of *Cdkn1a*, forms a complex with p53 and recruits PCAF and Set1 for histone acetylation and methylation and opening of chromatin. (D) In contrast to p53 which binds only to promoter region of *Cdkn1a*, NM1 binds along the whole gene suggesting a model in which Pol II is initially recruited to gene promoter by p53 and then slides along the DNA with NM1 to transcribe *Cdkn1a* (Venit et al., 2020).

also entails changes in the topological organization of chromatin remains to be established. In more general terms, given the importance of NM1 in transcriptional regulation, this study suggests that NM1 is required for the transcriptional response to DNA damage and may be considered a potential tumor suppressor.

Actin has also been shown to affect p53 function. Another layer of p53 regulation, apart from its directed phosphorylation and release from MDM2 complex, is reversible nucleocytoplasmic shuttling which depends on actin polymerization. p53 binds to polymeric actin in the cytoplasm (Metcalfe et al., 1999) and also in the nucleus (Okorokov et al., 2002). Both increased actin polymerization (Wang et al., 2013) and actin depletion (Qi et al., 2020) negatively regulate p53 nuclear import after DNA damage suggesting a joint translocation of actin monomers and p53 through the nuclear membrane. Similarly to NM1 (Venit et al., 2020), depletion of β-actin leads to downregulation of p21 expression (Qi et al., 2020). This could be caused primarily by decreased levels of p53 in the nucleus, but we cannot exclude the possibility that actin forms a complex with p53 and NM1 on the p21 promoter, directly regulating its expression as a part of the RNA Polymerase II transcription machinery.

11. Concluding remarks

The localization of cytoskeletal proteins in the cell nucleus is emerging as an important cellular feature. Actin, actin-binding proteins, including myosin isoforms as well as actin-related proteins (ARPs), are probably the best characterized in terms of their nuclear functions. However, evolving evidence that also components of the microtubules system are found in the cell nucleus (Akoumianaki et al., 2009) reflects an ever increasing complexity in the way cells work. In fact, seeing that components of microfilaments, microtubules and intermediate filaments are present in the cell nucleus will have to make us think differently about limiting the notion of a cytoskeleton only to the cytoplasm. There is considerable cross-talk between microfilaments and microtubules in regulating critical processes such as cell migration, cell division, cell polarity and neuronal shape and function (Dogterom and Koenderink, 2019). Not much is known about the role of microtubules and how a potential crosstalk with polymeric actin is established in the cell nucleus. In contrast, intermediate filaments such as lamins are key components of the nuclear envelope and we are beginning to understand that there is a crosstalk that engages the nuclear lamina with nuclear actin and myosin through interactions with lamina-associated proteins. In addition, as we have discussed in this chapter, the involvement of actin and myosin in nuclear functions such as chromatin regulation, transcription and cotranscriptional ribonucleoprotein assembly has been extensively studied. Recent exciting observations indicate that these actin- and myosin-based molecular mechanisms regulate cellular functions, including development and differentiation as well as the cellular response to DNA damage.

In terms of development and differentiation, it is likely that actin and myosin, will have a big impact on the way the genome is organized, affecting its architecture and the way gene programs are expressed. This will ultimately affect the establishment of cellular identities and therefore, physiology at the organismal level. Clearly, by maintaining the organization of the genome, actin and myosin will also play an important role in genome stability. A prediction is that in a loss-of-function model where localization of actin and myosin in the cell nucleus is compromised, there will be considerable loss of genome integrity. Several recent studies are in line with this prediction since loss of nuclear actin and myosin lead to specific impairment of the cellular response to DNA damage. It is indeed remarkable that actin and myosin seem to perform this function in different ways. On one hand they seem to maintain a conserved function (compared to their classical cytoplasmic mode of action) as transporters shuffling around DNA damage foci. On the other hand, a remarkable and unexpected observation is that embryonic fibroblasts from an NM1 KO mouse model exhibit constitutive γH2AX-positive DNA damage foci, loss of p53 function and changes in growth and proliferation rates. In the absence of NM1 these observations seem to be a direct consequence of a globally altered transcriptional response to DNA damage. This might be at the site of DNA damage itself or at other nuclear locations. In any case, these observations lay down in front of us a completely new area of research that might end up toward the identification of novel factors and pathways involved in the transcriptional response to DNA damage. To achieve this, it will be necessary to clarify how actin and NM1 control the architecture of the genome, whether this is by stabilizing chromatin compartments, topologically associating domains (TADs) or facilitating DNA looping all geared toward the establishment of hubs for transcriptional activation or repression. An important avenue for future studies will be to understand the connection between actin, myosin and factors that are known to stabilize the genome including the Cohesin complex and CTCF. Importantly, understanding how loss of nuclear actin and myosin compromises genome architecture and underlies loss of genome integrity will give us insights into their potential as markers for disease. While it is unlikely that loss of actin and myosin is compatible with a disease phenotype rather than lethality, it is possible that mild mutations in the respective genes or regulatory regions may indeed provide susceptibility to disease. Integrating metabolomics studies with transcriptional profiling and with conformation capture technologies and deep sequencing will be a way forward to start identifying the implications of dysregulated nuclear actin and myosin functions in physiology and their impact on disease.

Acknowledgments

This work is primarily supported by grants from New York University Abu Dhabi, the Sheikh Hamdan Bin Rashid Al Maktoum Award for Medical Sciences, the Swedish Research Council (Vetenskapsrådet) and the Swedish Cancer Society (Cancerfonden) to PP.

References

Abraham, R.T., 2001. Cell cycle checkpoint signaling through the ATM and ATR kinases. Genes Dev. 15, 2177–2196.

Ajima, R., Akazawa, H., Kodama, M., Takeshita, F., Otsuka, A., Kohno, T., Komuro, I., Ochiya, T., Yokota, J., 2008. Deficiency of Myo18B in mice results in embryonic lethality with cardiac myofibrillar aberrations. Genes Cells 13, 987–999.

Akoumianaki, T., Kardassis, D., Polioudaki, H., Georgatos, S.D., Theodoropoulos, P.A., 2009. Nucleocytoplasmic shuttling of soluble tubulin in mammalian cells. J. Cell. Sci. 122, 1111–1118.

Alexander, J.M., Hota, S.K., He, D., Thomas, S., Ho, L., Pennacchio, L.A., Bruneau, B.G., 2015. Brg1 modulates enhancer activation in mesoderm lineage commitment. Development 142, 1418–1430.

Alfert, A., Moreno, N., Kerl, K., 2019. The BAF complex in development and disease. Epigenetics Chromatin 12, 19, 1–15.

Almuzzaini, B., Sarshad, A.A., Farrants, A.K., Percipalle, P., 2015. Nuclear myosin 1 contributes to a chromatin landscape compatible with RNA polymerase II transcription activation. BMC Biol. 13, 35.

Altman, D., Sweeney, H.L., Spudich, J.A., 2004. The mechanism of myosin VI translocation and its load-induced anchoring. Cell 116, 737–749.

Andrin, C., McDonald, D., Attwood, K.M., Rodrigue, A., Ghosh, S., Mirzayans, R., Masson, J.Y., Dellaire, G., Hendzel, M.J., 2012. A requirement for polymerized actin in DNA double-strand break repair. Nucleus 3, 384–395.

Aoyama, N., Oka, A., Kitayama, K., Kurumizaka, H., Harata, M., 2008. The actin-related protein hArp8 accumulates on the mitotic chromosomes and functions in chromosome alignment. Exp. Cell Res. 314, 859–868.

Au, T.J., Rodriguez, J., Vincent, J.A., Tsukiyama, T., 2011. ATP-dependent chromatin remodeling factors tune S phase checkpoint activity. Mol. Cell. Biol. 31, 4454–4463.

Aymard, F., Aguirrebengoa, M., Guillou, E., Javierre, B.M., Bugler, B., Arnould, C., Rocher, V., Iacovoni, J.S., Biernacka, A., Skrzypczak, M., Ginalski, K., Rowicka, M., Fraser, P., Legube, G., 2017. Genome-wide mapping of long-range contacts unveils clustering of DNA double-strand breaks at damaged active genes. Nat. Struct. Mol. Biol. 24, 353–361.

Baarlink, C., Plessner, M., Sherrard, A., Morita, K., Misu, S., Virant, D., Kleinschnitz, E.M., Harniman, R., Alibhai, D., Baumeister, S., Miyamoto, K., Endesfelder, U., Kaidi, A., Grosse, R., 2017. A transient pool of nuclear F-actin at mitotic exit controls chromatin organization. Nat. Cell Biol. 19, 1389–1399.
Bajusz, C., Borkuti, P., Kristo, I., Kovacs, Z., Abonyi, C., Vilmos, P., 2018. Nuclear actin: ancient clue to evolution in eukaryotes? Histochem. Cell Biol. 150, 235–244.
Bantignies, F., Roure, V., Comet, I., Leblanc, B., Schuettengruber, B., Bonnet, J., Tixier, V., Mas, A., Cavalli, G., 2011. Polycomb-dependent regulatory contacts between distant Hox loci in Drosophila. Cell 144, 214–226.
Barisic, D., Stadler, M.B., Iurlaro, M., Schubeler, D., 2019. Mammalian ISWI and SWI/SNF selectively mediate binding of distinct transcription factors. Nature 569, 136–140.
Barutcu, A.R., Lajoie, B.R., Fritz, A.J., McCord, R.P., Nickerson, J.A., Van Wijnen, A.J., Lian, J.B., Stein, J.L., Dekker, J., Stein, G.S., Imbalzano, A.N., 2016. SMARCA4 regulates gene expression and higher-order chromatin structure in proliferating mammary epithelial cells. Genome Res. 26, 1188–1201.
Belin, B.J., Lee, T., Mullins, R.D., 2015. DNA damage induces nuclear actin filament assembly by Formin-2 and Spire-(1/2) that promotes efficient DNA repair. Elife, 4, e07735.
Berg, J.S., Powell, B.C., Cheney, R.E., 2001. A millennial myosin census. Mol. Biol. Cell 12, 780–794.
Bonev, B., Cavalli, G., 2016. Organization and function of the 3D genome. Nat. Rev. Genet. 17, 661.
Branzei, D., Foiani, M., 2010. Maintaining genome stability at the replication fork. Nat. Rev. Mol. Cell Biol. 11, 208–219.
Brugarolas, J., Moberg, K., Boyd, S.D., Taya, Y., Jacks, T., Lees, J.A., 1999. Inhibition of cyclin-dependent kinase 2 by p21 is necessary for retinoblastoma protein-mediated G1 arrest after gamma-irradiation. Proc. Natl. Acad. Sci. U. S. A. 96, 1002–1007.
Buchwalter, A., Kaneshiro, J.M., Hetzer, M.W., 2018. Coaching from the sidelines: the nuclear periphery in genome regulation. Nat. Rev. Genet. 20, 39.
Buss, F., Spudich, G., Kendrick-Jones, J., 2004. Myosin VI: cellular functions and motor properties. Annu. Rev. Cell Dev. Biol. 20, 649–676.
Caridi, P.C., Delabaere, L., Zapotoczny, G., Chiolo, I., 2017. And yet, it moves: nuclear and chromatin dynamics of a heterochromatic double-strand break. Philos. Trans. R. Soc. Lond. B Biol. Sci. 372, 1–14.
Caridi, C.P., D'Agostino, C., Ryu, T., Zapotoczny, G., Delabaere, L., Li, X., Khodaverdian, V.Y., Amaral, N., Lin, E., Rau, A.R., Chiolo, I., 2018. Nuclear F-actin and myosins drive relocalization of heterochromatic breaks. Nature 559, 54–60.
Castano, E., Philimonenko, V.V., Kahle, M., Fukalova, J., Kalendova, A., Yildirim, S., Dzijak, R., Dingova-Krasna, H., Hozak, P., 2010. Actin complexes in the cell nucleus: new stones in an old field. Histochem. Cell Biol. 133, 607–626.
Cavellan, E., Asp, P., Percipalle, P., Farrants, A.K., 2006. The WSTF-SNF2h chromatin remodeling complex interacts with several nuclear proteins in transcription. J. Biol. Chem. 281, 16264–16271.
Ceccaldi, R., Rondinelli, B., D'Andrea, A.D., 2016. Repair pathway choices and consequences at the double-strand break. Trends Cell Biol. 26, 52–64.
Chiolo, I., Minoda, A., Colmenares, S.U., Polyzos, A., Costes, S.V., Karpen, G.H., 2011. Double-strand breaks in heterochromatin move outside of a dynamic HP1a domain to complete recombinational repair. Cell 144, 732–744.
Choi, K., Park, C., Lee, J., Oh, M., Noh, B., Lee, I., 2007. Arabidopsis homologs of components of the SWR1 complex regulate flowering and plant development. Development 134, 1931–1941.
Chuang, C.H., Carpenter, A.E., Fuchsova, B., Johnson, T., De Lanerolle, P., Belmont, A.S., 2006. Long-range directional movement of an interphase chromosome site. Curr. Biol. 16, 825–831.

Ciccia, A., Elledge, S.J., 2010. The DNA damage response: making it safe to play with knives. Mol. Cell 40, 179–204.
Clark, T.G., Merriam, R.W., 1977. Diffusible and bound actin nuclei of Xenopus laevis oocytes. Cell 12, 883–891.
Clark, T.G., Rosenbaum, J.L., 1979. An actin filament matrix in hand-isolated nuclei of X. laevis oocytes. Cell 18, 1101–1108.
Cohen, S., Puget, N., Lin, Y.L., Clouaire, T., Aguirrebengoa, M., Rocher, V., Pasero, P., Canitrot, Y., Legube, G., 2018. Senataxin resolves RNA:DNA hybrids forming at DNA double-strand breaks to prevent translocations. Nat. Commun. 9, 533.
Collier, S., Chan, H.Y., Toda, T., McKimmie, C., Johnson, G., Adler, P.N., O'Kane, C., Ashburner, M., 2000. The Drosophila embargoed gene is required for larval progression and encodes the functional homolog of schizosaccharomyces Crm1. Genetics 155, 1799–1807.
Collins, N., Poot, R.A., Kukimoto, I., Garcia-Jimenez, C., Dellaire, G., Varga-Weisz, P.D., 2002. An ACF1-ISWI chromatin-remodeling complex is required for DNA replication through heterochromatin. Nat. Genet. 32, 627–632.
Cook, P.J., Ju, B.G., Telese, F., Wang, X., Glass, C.K., Rosenfeld, M.G., 2009. Tyrosine dephosphorylation of H2AX modulates apoptosis and survival decisions. Nature 458, 591–596.
De Lanerolle, P., Serebryannyy, L., 2011. Nuclear actin and myosins: life without filaments. Nat. Cell Biol. 13, 1282–1288.
Dion, V., Kalck, V., Horigome, C., Towbin, B.D., Gasser, S.M., 2012. Increased mobility of double-strand breaks requires Mec1, Rad9 and the homologous recombination machinery. Nat. Cell Biol. 14, 502–509.
Dogterom, M., Koenderink, G.H., 2019. Actin-microtubule crosstalk in cell biology. Nat. Rev. Mol. Cell Biol. 20, 38–54.
Domingo-Prim, J., Endara-Coll, M., Bonath, F., Jimeno, S., Prados-Carvajal, R., Friedlander, M.R., Huertas, P., Visa, N., 2019. EXOSC10 is required for RPA assembly and controlled DNA end resection at DNA double-strand breaks. Nat. Commun. 10, 2135.
Dominguez, R., Holmes, K.C., 2011. Actin structure and function. Annu. Rev. Biophys. 40, 169–186.
Dopie, J., Skarp, K.P., Rajakyla, E.K., Tanhuanpaa, K., Vartiainen, M.K., 2012. Active maintenance of nuclear actin by importin 9 supports transcription. Proc. Natl. Acad. Sci. U. S. A. 109, E544–E552.
Dundr, M., Ospina, J.K., Sung, M.H., John, S., Upender, M., Ried, T., Hager, G.L., Matera, A.G., 2007. Actin-dependent intranuclear repositioning of an active gene locus in vivo. J. Cell Biol. 179, 1095–1103.
Dzijak, R., Yildirim, S., Kahle, M., Novak, P., Hnilicova, J., Venit, T., Hozak, P., 2012. Specific nuclear localizing sequence directs two myosin isoforms to the cell nucleus in calmodulin-sensitive manner. PLoS One 7, e30529.
Engelhardt, W.A., Ljubimowa, M.N., 1939. Myosine and adenosinetriphosphatase. Nature 144, 668–669.
Espejel, S., Franco, S., Rodriguez-Perales, S., Bouffler, S.D., Cigudosa, J.C., Blasco, M.A., 2002. Mammalian Ku86 mediates chromosomal fusions and apoptosis caused by critically short telomeres. EMBO J. 21, 2207–2219.
Fenn, S., Breitsprecher, D., Gerhold, C.B., Witte, G., Faix, J., Hopfner, K.P., 2011. Structural biochemistry of nuclear actin-related proteins 4 and 8 reveals their interaction with actin. EMBO J. 30, 2153–2166.
Ferrai, C., Naum-Onganía, G., Longobardi, E., Palazzolo, M., Disanza, A., Diaz, V.M., Crippa, M.P., Scita, G., Blasi, F., 2009. Induction of HoxB transcription by retinoic acid requires actin polymerization. Mol. Biol. Cell 20, 3543–3551.

Fili, N., Hari-Gupta, Y., Dos Santos, A., Cook, A., Poland, S., Ameer-Beg, S.M., Parsons, M., Toseland, C.P., 2017. NDP52 activates nuclear myosin VI to enhance RNA polymerase II transcription. Nat. Commun. 8, 1871.

Fomproix, N., Percipalle, P., 2004. An actin-myosin complex on actively transcribing genes. Exp. Cell Res. 294, 140–148.

Foshay, K.M., Looney, T.J., Chari, S., Mao, F.F., Lee, J.H., Zhang, L., Fernandes, C.J., Baker, S.W., Clift, K.L., Gaetz, J., Di, C.G., Xiang, A.P., Lahn, B.T., 2012. Embryonic stem cells induce pluripotency in somatic cell fusion through biphasic reprogramming. Mol. Cell 46, 159–170.

Foth, B.J., Goedecke, M.C., Soldati, D., 2006. New insights into myosin evolution and classification. Proc. Natl. Acad. Sci. U. S. A. 103, 3681–3686.

Gieni, R.S., Hendzel, M.J., 2008. Mechanotransduction from the ECM to the genome: are the pieces now in place? J. Cell. Biochem. 104, 1964–1987.

Gospodinov, A., Herceg, Z., 2013. Chromatin structure in double strand break repair. DNA Repair (Amst.) 12, 800–810.

Gurdon, J.B., Elsdale, T.R., Fischberg, M., 1958. Sexually mature individuals of Xenopus laevis from the transplantation of single somatic nuclei. Nature 182, 64–65.

Hagen, S.J., Kiehart, D.P., Kaiser, D.A., Pollard, T.D., 1986. Characterization of monoclonal antibodies to Acanthamoeba myosin-I that cross-react with both myosin-II and low molecular mass nuclear proteins. J. Cell Biol. 103, 2121–2128.

Han, Y., Reyes, A.A., Malik, S., He, Y., 2020. Cryo-EM structure of SWI/SNF complex bound to a nucleosome. Nature 579, 452–455.

Harata, M., Zhang, Y., Stillman, D.J., Matsui, D., Oma, Y., Nishimori, K., Mochizuki, R., 2002. Correlation between chromatin association and transcriptional regulation for the Act3p/Arp4 nuclear actin-related protein of Saccharomyces cerevisiae. Nucleic Acids Res. 30, 1743–1750.

Hedberg Oldfors, C., Dios, D.G., Linder, A., Visuttijai, K., Samuelson, E., Karlsson, S., Nilsson, S., Behboudi, A., 2015. Analysis of an independent tumor suppressor locus telomeric to Tp53 suggested Inpp5k and Myo1c as novel tumor suppressor gene candidates in this region. BMC Genet. 16, 80.

Hoeijmakers, J.H., 2009. DNA damage, aging, and cancer. N. Engl. J. Med. 361, 1475–1485.

Hofmann, W.A., Stojiljkovic, L., Fuchsova, B., Vargas, G.M., Mavrommatis, E., Philimonenko, V., Kysela, K., Goodrich, J.A., Lessard, J.L., Hope, T.J., Hozak, P., de Lanerolle, P., 2004. Actin is part of pre-initiation complexes and is necessary for transcription by RNA polymerase II. Nat. Cell Biol. 6, 1094–1101.

Hofmann, W.A., Johnson, T., Klapczynski, M., Fan, J.L., De Lanerolle, P., 2006a. From transcription to transport: emerging roles for nuclear myosin I. Biochem. Cell Biol. 84, 418–426.

Hofmann, W.A., Vargas, G.M., Ramchandran, R., Stojiljkovic, L., Goodrich, J.A., de Lanerolle, P, 2006b. Nuclear myosin I is necessary for the formation of the first phosphodiester bond during transcription initiation by RNA polymerase II. J. Cell. Biochem. 99, 1001–1009.

Hofmann, W.A., Arduini, A., Nicol, S.M., Camacho, C.J., Lessard, J.L., Fuller-Pace, F.V., De Lanerolle, P., 2009. SUMOylation of nuclear actin. J. Cell Biol. 186, 193–200.

Hokanson, D.E., Ostap, E.M., 2006. Myo1c binds tightly and specifically to phosphatidylinositol-4,5-bisphosphate and inositol-1,4,5-trisphosphate. Proc. Natl. Acad. Sci. U. S. A. 103, 3118–3123.

Hu, P., Wu, S., Hernandez, N., 2004. A role for β-actin in RNA polymerase III transcription. Genes Dev. 18, 3010–3015.

Hu, Q., Kwon, Y.S., Nunez, E., Cardamone, M.D., Hutt, K.R., Ohgi, K.A., Garcia-Bassets, I., Rose, D.W., Glass, C.K., Rosenfeld, M.G., Fu, X.D., 2008. Enhancing nuclear receptor-induced transcription requires nuclear motor and LSD1-dependent gene networking in interchromatin granules. Proc. Natl. Acad. Sci. U. S. A. 105, 19199–19204.

Huang, W., Ghisletti, S., Saijo, K., Gandhi, M., Aouadi, M., Tesz, G.J., Zhang, D.X., Yao, J., Czech, M.P., Goode, B.L., Rosenfeld, M.G., Glass, C.K., 2011. Coronin 2A mediates actin-dependent de-repression of inflammatory response genes. Nature 470, 414–418.

Hur, S.K., Park, E.J., Han, J.E., Kim, Y.A., Kim, J.D., Kang, D., Kwon, J., 2010. Roles of human INO80 chromatin remodeling enzyme in DNA replication and chromosome segregation suppress genome instability. Cell. Mol. Life Sci. 67, 2283–2296.

Ihnatovych, I., Migocka-Patrzalek, M., Dukh, M., Hofmann, W.A., 2012. Identification and characterization of a novel myosin Ic isoform that localizes to the nucleus. Cytoskeleton (Hoboken) 69, 555–565.

Ikura, T., Ogryzko, V.V., Grigoriev, M., Groisman, R., Wang, J., Horikoshi, M., Scully, R., Qin, J., Nakatani, Y., 2000. Involvement of the TIP60 histone acetylase complex in DNA repair and apoptosis. Cell 102, 463–473.

Illingworth, R.S., 2019. Chromatin folding and nuclear architecture: PRC1 function in 3D. Curr. Opin. Genet. Dev. 55, 82–90.

Jackson, S.P., Bartek, J., 2009. The DNA-damage response in human biology and disease. Nature 461, 1071–1078.

Jimeno, S., Prados-Carvajal, R., Huertas, P., 2019. The role of RNA and RNA-related proteins in the regulation of DNA double strand break repair pathway choice. DNA Repair (Amst.) 81, 102662.

Joshi, P., Greco, T.M., Guise, A.J., Luo, Y., Yu, F., Nesvizhskii, A.I., Cristea, I.M., 2013. The functional interactome landscape of the human histone deacetylase family. Mol. Syst. Biol. 9, 672.

Kabsch, W., Holmes, K.C., 1995. The actin fold. FASEB J. 9, 167–174.

Kadoch, C., Williams, R.T., Calarco, J.P., Miller, E.L., Weber, C.M., Braun, S.M., Pulice, J.L., Chory, E.J., Crabtree, G.R., 2017. Dynamics of BAF-Polycomb complex opposition on heterochromatin in normal and oncogenic states. Nat. Genet. 49, 213–222.

Kandasamy, M.K., McKinney, E.C., Deal, R.B., Meagher, R.B., 2005. Arabidopsis ARP7 is an essential actin-related protein required for normal embryogenesis, plant architecture, and floral organ abscission. Plant Physiol. 138, 2019–2032.

Kapoor, P., Shen, X., 2014. Mechanisms of nuclear actin in chromatin-remodeling complexes. Trends Cell Biol. 24, 238–246.

Kapoor, P., Chen, M., Winkler, D.D., Luger, K., Shen, X., 2013. Evidence for monomeric actin function in INO80 chromatin remodeling. Nat. Struct. Mol. Biol. 20, 426–432.

Kashiwaba, S., Kitahashi, K., Watanabe, T., Onoda, F., Ohtsu, M., Murakami, Y., 2010. The mammalian INO80 complex is recruited to DNA damage sites in an ARP8 dependent manner. Biochem. Biophys. Res. Commun. 402, 619–625.

Kato, M., Sasaki, M., Mizuno, S., Harata, M., 2001. Novel actin-related proteins in vertebrates: similarities of structure and expression pattern to Arp6 localized on Drosophila heterochromatin. Gene 268, 133–140.

Kelpsch, D.J., Tootle, T.L., 2018. Nuclear actin: from discovery to function. Anat. Rec. (Hoboken) 301, 1999–2013.

Knoll, K.R., Eustermann, S., Niebauer, V., Oberbeckmann, E., Stoehr, G., Schall, K., Tosi, A., Schwarz, M., Buchfellner, A., Korber, P., Hopfner, K.P., 2018. The nuclear actin-containing Arp8 module is a linker DNA sensor driving INO80 chromatin remodeling. Nat. Struct. Mol. Biol. 25, 823–832.

Krendel, M., Mooseker, M.S., 2005. Myosins: tails (and heads) of functional diversity. Physiology (Bethesda) 20, 239–251.

Kristo, I., Bajusz, I., Bajusz, C., Borkuti, P., Vilmos, P., 2016. Actin, actin-binding proteins, and actin-related proteins in the nucleus. Histochem. Cell Biol. 145, 373–388.

Kruhlak, M.J., Celeste, A., Dellaire, G., Fernandez-Capetillo, O., Muller, W.G., McNally, J.G., Bazett-Jones, D.P., Nussenzweig, A., 2006. Changes in chromatin structure and mobility in living cells at sites of DNA double-strand breaks. J. Cell Biol. 172, 823–834.

Kühne, W., 1864. Untersuchungen über das Protoplasma und die Contractilität. W. Engelmann.

Kukalev, A., Nord, Y., Palmberg, C., Bergman, T., Percipalle, P., 2005. Actin and hnRNP U cooperate for productive transcription by RNA polymerase II. Nat. Struct. Mol. Biol. 12, 238–244.

Kulashreshtha, M., Mehta, I.S., Kumar, P., Rao, B.J., 2016. Chromosome territory relocation during DNA repair requires nuclear myosin 1 recruitment to chromatin mediated by upsilon-H2AX signaling. Nucleic Acids Res. 44, 8272–8291.

Lane, N.J., 1969. Intranuclear fibrillar bodies in actinomycin D-treated oocytes. J. Cell Biol. 40, 286–291.

Langst, G., Becker, P.B., 2001. ISWI induces nucleosome sliding on nicked DNA. Mol. Cell 8, 1085–1092.

Lemaitre, C., Grabarz, A., Tsouroula, K., Andronov, L., Furst, A., Pankotai, T., Heyer, V., Rogier, M., Attwood, K.M., Kessler, P., Dellaire, G., Klaholz, B., Reina-San-Martin, B., Soutoglou, E., 2014. Nuclear position dictates DNA repair pathway choice. Genes Dev. 28, 2450–2463.

Li, X., Zuo, X., Jing, J., Ma, Y., Wang, J., Liu, D., Zhu, J., Du, X., Xiong, L., Du, Y., Xu, J., Xiao, X., Wang, J., Chai, Z., Zhao, Y., Deng, H., 2015. Small-molecule-driven direct reprogramming of mouse fibroblasts into functional neurons. Cell Stem Cell 17, 195–203.

Lindahl, T., 1993. Instability and decay of the primary structure of DNA. Nature 362, 709–715.

Lindsay, A.J., McCaffrey, M.W., 2009. Myosin Vb localises to nucleoli and associates with the RNA polymerase I transcription complex. Cell Motil. Cytoskeleton 66, 1057–1072.

Lisby, M., Mortensen, U.H., Rothstein, R., 2003. Colocalization of multiple DNA double-strand breaks at a single Rad52 repair centre. Nat. Cell Biol. 5, 572–577.

Manova, V., Gruszka, D., 2015. DNA damage and repair in plants—from models to crops. Front. Plant Sci. 6, 885.

Marcomini, I., Shimada, K., Delgoshaie, N., Yamamoto, I., Seeber, A., Cheblal, A., Horigome, C., Naumann, U., Gasser, S.M., 2018. Asymmetric processing of DNA ends at a double-strand break leads to unconstrained dynamics and ectopic translocation. Cell Rep. 24, 2614–2628. e4.

Maruyama, E.O., Hori, T., Tanabe, H., Kitamura, H., Matsuda, R., Tone, S., Hozak, P., Habermann, F.A., Von Hase, J., Cremer, C., Fukagawa, T., Harata, M., 2012. The actin family member Arp6 and the histone variant H2A.Z are required for spatial positioning of chromatin in chicken cell nuclei. J. Cell Sci. 125, 3739–3743.

McConnell, R.E., Tyska, M.J., 2010. Leveraging the membrane—cytoskeleton interface with myosin-1. Trends Cell Biol. 20, 418–426.

McDonald, D., Carrero, G., Andrin, C., De Vries, G., Hendzel, M.J., 2006. Nucleoplasmic beta-actin exists in a dynamic equilibrium between low-mobility polymeric species and rapidly diffusing populations. J. Cell Biol. 172, 541–552.

Meagher, R.B., Kandasamy, M.K., McKinney, E.C., Roy, E., 2009. Chapter 5. Nuclear actin-related proteins in epigenetic control. Int. Rev. Cell Mol. Biol. 277, 157–215.

Mehta, I.S., Amira, M., Harvey, A.J., Bridger, J.M., 2010. Rapid chromosome territory relocation by nuclear motor activity in response to serum removal in primary human fibroblasts. Genome Biol. 11, R5.

Metcalfe, S., Weeds, A., Okorokov, A.L., Milner, J., Cockman, M., Pope, B., 1999. Wild-type p53 protein shows calcium-dependent binding to F-actin. Oncogene 18, 2351–2355.

Miyamoto, K., Gurdon, J.B., 2013. Transcriptional regulation and nuclear reprogramming: roles of nuclear actin and actin-binding proteins. Cell. Mol. Life Sci. 70, 3289–3302.

Miyamoto, K., Pasque, V., Jullien, J., Gurdon, J.B., 2011. Nuclear actin polymerization is required for transcriptional reprogramming of Oct4 by oocytes. Genes Dev. 25, 946–958.

Mizuguchi, G., Shen, X., Landry, J., Wu, W.H., Sen, S., Wu, C., 2004. ATP-driven exchange of histone H2AZ variant catalyzed by SWR1 chromatin remodeling complex. Science 303, 343–348.

Morita, T., Hayashi, K., 2014. Arp5 is a key regulator of myocardin in smooth muscle cells. J. Cell Biol. 204, 683–696.

Nakazawa, T., Ando, Y., Hata, T., Nakahori, K., 2016. A mutation in the Cc.arp9 gene encoding a putative actin-related protein causes defects in fruiting initiation and asexual development in the agaricomycete Coprinopsis cinerea. Curr. Genet. 62, 565–574.

Neumann, F.R., Dion, V., Gehlen, L.R., Tsai-Pflugfelder, M., Schmid, R., Taddei, A., Gasser, S.M., 2012. Targeted INO80 enhances subnuclear chromatin movement and ectopic homologous recombination. Genes Dev. 26, 369–383.

Nowak, G., Pestic-Dragovich, L., Hozak, P., Philimonenko, A., Simerly, C., Schatten, G., De Lanerolle, P., 1997. Evidence for the presence of myosin I in the nucleus. J. Biol. Chem. 272, 17176–17181.

Obrdlik, A., Percipalle, P., 2011. The F-actin severing protein cofilin-1 is required for RNA polymerase II transcription elongation. Nucleus 2, 72–79.

Obrdlik, A., Louvet, E., Kukalev, A., Naschekin, D., Kiseleva, E., Fahrenkrog, B., Percipalle, P., 2010. Nuclear myosin 1 is in complex with mature rRNA transcripts and associates with the nuclear pore basket. FASEB J. 24, 146–157.

Ochs, F., Karemore, G., Miron, E., Brown, J., Sedlackova, H., Rask, M.B., Lampe, M., Buckle, V., Schermelleh, L., Lukas, J., Lukas, C., 2019. Stabilization of chromatin topology safeguards genome integrity. Nature 574, 571–574.

Ogiwara, H., Enomoto, T., Seki, M., 2007. The INO80 chromatin remodeling complex functions in sister chromatid cohesion. Cell Cycle 6, 1090–1095.

Ogiwara, H., Ui, A., Otsuka, A., Satoh, H., Yokomi, I., Nakajima, S., Yasui, A., Yokota, J., Kohno, T., 2011. Histone acetylation by CBP and p300 at double-strand break sites facilitates SWI/SNF chromatin remodeling and the recruitment of non-homologous end joining factors. Oncogene 30, 2135–2146.

Okorokov, A.L., Rubbi, C.P., Metcalfe, S., Milner, J., 2002. The interaction of p53 with the nuclear matrix is mediated by F-actin and modulated by DNA damage. Oncogene 21, 356–367.

Ondrej, V., Lukasova, E., Krejci, J., Matula, P., Kozubek, S., 2008. Lamin A/C and polymeric actin in genome organization. Mol. Cells 26, 356–361.

Papamichos-Chronakis, M., Peterson, C.L., 2013. Chromatin and the genome integrity network. Nat. Rev. Genet. 14, 62–75.

Papoulas, O., Beek, S.J., Moseley, S.L., McCallum, C.M., Sarte, M., Shearn, A., Tamkun, J.W., 1998. The Drosophila trithorax group proteins BRM, ASH1 and ASH2 are subunits of distinct protein complexes. Development 125, 3955–3966.

Pederson, T., Aebi, U., 2002. Actin in the nucleus: what form and what for? J. Struct. Biol. 140, 3–9.

Percipalle, P., 2013. Co-transcriptional nuclear actin dynamics. Nucleus 4, 43–52.

Percipalle, P., Farrants, A.K., 2006. Chromatin remodelling and transcription: be-WICHed by nuclear myosin 1. Curr. Opin. Cell Biol. 18, 267–274.

Percipalle, P., Vartiainen, M., 2019. Cytoskeletal proteins in the cell nucleus: a special nuclear actin perspective. Mol. Biol. Cell 30, 1781–1785.
Percipalle, P., Zhao, J., Pope, B., Weeds, A., Lindberg, U., Daneholt, B., 2001. Actin bound to the heterogeneous nuclear ribonucleoprotein Hrp36 is associated with Balbiani ring mRNA from the gene to polysomes. J. Cell Biol. 153, 229–236.
Percipalle, P., Jonsson, A., Nashchekin, D., Karlsson, C., Bergman, T., Guialis, A., Daneholt, B., 2002. Nuclear actin is associated with a specific subset of hnRNP A/B-type proteins. Nucleic Acids Res. 30, 1725–1734.
Percipalle, P., Fomproix, N., Kylberg, K., Miralles, F., Bjorkroth, B., Daneholt, B., Visa, N., 2003. An actin-ribonucleoprotein interaction is involved in transcription by RNA polymerase II. Proc. Natl. Acad. Sci. U. S. A. 100, 6475–6480.
Percipalle, P., Fomproix, N., Cavellan, E., Voit, R., Reimer, G., Kruger, T., Thyberg, J., Scheer, U., Grummt, I., Farrants, A.K., 2006. The chromatin remodelling complex WSTF-SNF2h interacts with nuclear myosin 1 and has a role in RNA polymerase I transcription. EMBO Rep. 7, 525–530.
Pestic-Dragovich, L., Stojiljkovic, L., Philimonenko, A.A., Nowak, G., Ke, Y., Settlage, R.E., Shabanowitz, J., Hunt, D.F., Hozak, P., De Lanerolle, P., 2000. A myosin I isoform in the nucleus. Science 290, 337–341.
Philimonenko, V.V., Zhao, J., Iben, S., Dingova, H., Kysela, K., Kahle, M., Zentgraf, H., Hofmann, W.A., De Lanerolle, P., Hozak, P., Grummt, I., 2004. Nuclear actin and myosin I are required for RNA polymerase I transcription. Nat. Cell Biol. 6, 1165–1172.
Plessner, M., Grosse, R., 2019. Dynamizing nuclear actin filaments. Curr. Opin. Cell Biol. 56, 1–6.
Plessner, M., Melak, M., Chinchilla, P., Baarlink, C., Grosse, R., 2015. Nuclear F-actin formation and reorganization upon cell spreading. J. Biol. Chem. 290, 11209–11216.
Poch, O., Winsor, B., 1997. Who's who among the Saccharomyces cerevisiae actin-related proteins? A classification and nomenclature proposal for a large family. Yeast 13, 1053–1058.
Poot, R.A., Bozhenok, L., Van Den Berg, D.L., Steffensen, S., Ferreira, F., Grimaldi, M., Gilbert, N., Ferreira, J., Varga-Weisz, P.D., 2004. The Williams syndrome transcription factor interacts with PCNA to target chromatin remodelling by ISWI to replication foci. Nat. Cell Biol. 6, 1236–1244.
Pranchevicius, M.C.S., Baqui, M.M., Ishikawa-Ankerhold, H.C., Lourenço, E.V., Leão, R.M., Banzi, S.R., Dos Santos, C.T., Barreira, M.C.R., Espreafico, E.M., Larson, R.E., 2008. Myosin Va phosphorylated on Ser1650 is found in nuclear speckles and redistributes to nucleoli upon inhibition of transcription. Cell Motil. Cytoskeleton 65, 441–456.
Qi, W., Li, J., Pei, X., Ke, Y., Bu, Q., Ni, X., 2020. beta-Actin facilitates etoposide-induced p53 nuclear import. J. Biosci. 45, 34.
Ranade, D., Pradhan, R., Jayakrishnan, M., Hegde, S., Sengupta, K., 2019. Lamin A/C and Emerin depletion impacts chromatin organization and dynamics in the interphase nucleus. BMC Mol. Cell. Biol. 20, 11.
Riedl, J., Crevenna, A.H., Kessenbrock, K., Yu, J.H., Neukirchen, D., Bista, M., Bradke, F., Jenne, D., Holak, T.A., Werb, Z., Sixt, M., Wedlich-Soldner, R., 2008. Lifeact: a versatile marker to visualize F-actin. Nat. Methods 5, 605–607.
Rodgers, B.D., 2005. Insulin-like growth factor-I downregulates embryonic myosin heavy chain (eMyHC) in myoblast nuclei. Growth Horm. IGF Res. 15, 377–383.
Samwer, M., Dehne, H.J., Spira, F., Kollmar, M., Gerlich, D.W., Urlaub, H., Gorlich, D., 2013. The nuclear F-actin interactome of Xenopus oocytes reveals an actin-bundling kinesin that is essential for meiotic cytokinesis. EMBO J. 32, 1886–1902.

Sanger, J.W., Sanger, J.M., Kreis, T.E., Jockusch, B.M., 1980. Reversible translocation of cytoplasmic actin into the nucleus caused by dimethyl sulfoxide. Proc. Natl. Acad. Sci. U. S. A. 77, 5268–5272.

Sarshad, A.A., Percipalle, P., 2014. New insight into role of myosin motors for activation of RNA polymerases. Int. Rev. Cell Mol. Biol. 311, 183–230.

Sarshad, A., Sadeghifar, F., Louvet, E., Mori, R., Bohm, S., Al-Muzzaini, B., Vintermist, A., Fomproix, N., Ostlund, A.K., Percipalle, P., 2013. Nuclear myosin 1c facilitates the chromatin modifications required to activate rRNA gene transcription and cell cycle progression. PLoS Genet. 9, e1003397.

Sarshad, A.A., Corcoran, M., Al-Muzzaini, B., Borgonovo-Brandter, L., Von Euler, A., Lamont, D., Visa, N., Percipalle, P., 2014. Glycogen synthase kinase (GSK) 3beta phosphorylates and protects nuclear myosin 1c from proteasome-mediated degradation to activate rDNA transcription in early G1 cells. PLoS Genet. 10, e1004390.

Schrank, B.R., Aparicio, T., Li, Y., Chang, W., Chait, B.T., Gundersen, G.G., Gottesman, M.E., Gautier, J., 2018. Nuclear ARP2/3 drives DNA break clustering for homology-directed repair. Nature 559, 61–66.

Schwab, R.S., Ihnatovych, I., Yunus, S.Z., Domaradzki, T., Hofmann, W.A., 2013. Identification of signals that facilitate isoform specific nucleolar localization of myosin IC. Exp. Cell Res. 319, 1111–1123.

Shen, X., Mizuguchi, G., Hamiche, A., Wu, C., 2000. A chromatin remodelling complex involved in transcription and DNA processing. Nature 406, 541–544.

Shibata, A., Jeggo, P.A., 2014. DNA double-strand break repair in a cellular context. Clin. Oncol. (R. Coll. Radiol.) 26, 243–249.

Shibata, A., Jeggo, P., Lobrich, M., 2018. The pendulum of the Ku-Ku clock. DNA Repair (Amst.) 71, 164–171.

Söderberg, E., Hessle, V., von Euler, A., Visa, N., 2012. Profilin is associated with transcriptionally active genes. Nucleus 3, 290–299.

Sotiropoulos, A., Gineitis, D., Copeland, J., Treisman, R., 1999. Signal-regulated activation of serum response factor is mediated by changes in actin dynamics. Cell 98, 159–169.

Soutoglou, E., Dorn, J.F., Sengupta, K., Jasin, M., Nussenzweig, A., Ried, T., Danuser, G., Misteli, T., 2007. Positional stability of single double-strand breaks in mammalian cells. Nat. Cell Biol. 9, 675–682.

Stam, M., Tark-Dame, M., Fransz, P., 2019. 3D genome organization: a role for phase separation and loop extrusion? Curr. Opin. Plant Biol. 48, 36–46.

Straub, F.B., 1943. Actin, ii. J. Stud. Inst. Med. Chem. Univ. Szeged 3, 23–37.

Stüven, T., Hartmann, E., Görlich, D., 2003. Exportin 6: a novel nuclear export receptor that is specific for profilin·actin complexes. EMBO J. 22, 5928–5940.

Sunada, R., Gorzer, I., Oma, Y., Yoshida, T., Suka, N., Wintersberger, U., Harata, M., 2005. The nuclear actin-related protein Act3p/Arp4p is involved in the dynamics of chromatin-modulating complexes. Yeast 22, 753–768.

Takahashi, K., Yamanaka, S., 2006. Induction of pluripotent stem cells from mouse embryonic and adult fibroblast cultures by defined factors. Cell 126, 663–676.

Tsouroula, K., Furst, A., Rogier, M., Heyer, V., Maglott-Roth, A., Ferrand, A., Reina-San-Martin, B., Soutoglou, E., 2016. Temporal and spatial uncoupling of DNA double strand break repair pathways within mammalian heterochromatin. Mol. Cell 63, 293–305.

Ueno, M., Murase, T., Kibe, T., Ohashi, N., Tomita, K., Murakami, Y., Uritani, M., Ushimaru, T., Harata, M., 2004. Fission yeast Arp6 is required for telomere silencing, but functions independently of Swi6. Nucleic Acids Res. 32, 736–741.

Venit, T., Dzijak, R., Kalendova, A., Kahle, M., ROhozkova, J., Schmidt, V., Rulicke, T., Rathkolb, B., Hans, W., Bohla, A., Eickelberg, O., Stoeger, T., Wolf, E., Yildirim, A.O., Gailus-Durner, V., Fuchs, H., De Angelis, M.H., Hozak, P., 2013. Mouse nuclear myosin I knock-out shows interchangeability and redundancy of myosin isoforms in the cell nucleus. PLoS One 8, e61406.

Venit, T., Kalendova, A., Petr, M., Dzijak, R., Pastorek, L., Rohozkova, J., Malohlava, J., Hozak, P., 2016. Nuclear myosin I regulates cell membrane tension. Sci. Rep. 6, 30864.

Venit, T., Semesta, K., Farrukh, S., Endara-Coll, M., Havalda, R., Hozak, P., Percipalle, P., 2020. Nuclear myosin 1 activates p21 gene transcription in response to DNA damage through a chromatin-based mechanism. Commun. Biol. 3, 115.

Vierbuchen, T., Ostermeier, A., Pang, Z.P., Kokubu, Y., Sudhof, T.C., Wernig, M., 2010. Direct conversion of fibroblasts to functional neurons by defined factors. Nature 463, 1035–1041.

Vintermist, A., Bohm, S., Sadeghifar, F., Louvet, E., Mansen, A., Percipalle, P., Ostlund Farrants, A.K., 2011. The chromatin remodelling complex B-WICH changes the chromatin structure and recruits histone acetyl-transferases to active rRNA genes. PLoS One 6, e19184.

Visa, N., Percipalle, P., 2010. Nuclear functions of actin. Cold Spring Harb. Perspect. Biol. 2, a000620.

Vreugde, S., Ferrai, C., Miluzio, A., Hauben, E., Marchisio, P.C., Crippa, M.P., Bussi, M., Biffo, S., 2006. Nuclear myosin VI enhances RNA polymerase II-dependent transcription. Mol. Cell 23, 749–755.

Wada, A., Fukuda, M., Mishima, M., Nishida, E., 1998. Nuclear export of actin: a novel mechanism regulating the subcellular localization of a major cytoskeletal protein. EMBO J. 17, 1635–1641.

Wagner, F.R., Dienemann, C., Wang, H., Stutzer, A., Tegunov, D., Urlaub, H., Cramer, P., 2020. Structure of SWI/SNF chromatin remodeller RSC bound to a nucleosome. Nature 579, 448–451.

Wang, L., Wang, M., Wang, S., Qi, T., Guo, L., Li, J., Qi, W., Ampah, K.K., Ba, X., Zeng, X., 2013. Actin polymerization negatively regulates p53 function by impairing its nuclear import in response to DNA damage. PLoS One 8, e60179.

Wang, X., Lee, R.S., Alver, B.H., Haswell, J.R., Wang, S., Mieczkowski, J., Drier, Y., Gillespie, S.M., Archer, T.C., Wu, J.N., Tzvetkov, E.P., Troisi, E.C., Pomeroy, S.L., Biegel, J.A., Tolstorukov, M.Y., Bernstein, B.E., Park, P.J., Roberts, C.W., 2017. SMARCB1-mediated SWI/SNF complex function is essential for enhancer regulation. Nat. Genet. 49, 289–295.

Wang, Y., Sherrard, A., Zhao, B., Melak, M., Trautwein, J., Kleinschnitz, E.M., Tsopoulidis, N., Fackler, O.T., Schwan, C., Grosse, R., 2019. GPCR-induced calcium transients trigger nuclear actin assembly for chromatin dynamics. Nat. Commun. 10, 5271.

Wang, A., Kolhe, J.A., Gioacchini, N., Baade, I., Brieher, W.M., Peterson, C.L., Freeman, B.C., 2020. Mechanism of long-range chromosome motion triggered by gene activation. Dev. Cell 52, 309–320 e5.

Wei, M., Fan, X., Ding, M., Li, R., Shao, S., Hou, Y., Meng, S., Tang, F., Li, C., Sun, Y., 2020. Nuclear actin regulates inducible transcription by enhancing RNA polymerase II clustering. Sci. Adv. 6, 1–12.

Wu, W.H., Alami, S., Luk, E., Wu, C.H., Sen, S., Mizuguchi, G., Wei, D., Wu, C., 2005. Swc2 is a widely conserved H2AZ-binding module essential for ATP-dependent histone exchange. Nat. Struct. Mol. Biol. 12, 1064–1071.

Xiao, A., Li, H., Shechter, D., Ahn, S.H., Fabrizio, L.A., Erdjument-Bromage, H., Ishibe-Murakami, S., Wang, B., Tempst, P., Hofmann, K., Patel, D.J., Elledge, S.J., Allis, C.D., 2009. WSTF regulates the H2A.X DNA damage response via a novel tyrosine kinase activity. Nature 457, 57–62.

Xie, X., Percipalle, P., 2018. An actin-based nucleoskeleton involved in gene regulation and genome organization. Biochem. Biophys. Res. Commun. 506, 378–386.

Xie, X., Almuzzaini, B., Drou, N., Kremb, S., Yousif, A., Farrants, A.O., Gunsalus, K., Percipalle, P., 2018a. Beta-actin-dependent global chromatin organization and gene expression programs control cellular identity. FASEB J. 32, 1296–1314.

Xie, X., Jankauskas, R., Mazari, A.M.A., Drou, N., Percipalle, P., 2018b. Beta-actin regulates a heterochromatin landscape essential for optimal induction of neuronal programs during direct reprograming. PLoS Genet. 14, e1007846.

Xu, Y.Z., Thuraisingam, T., Morais, D.A., Rola-Pleszczynski, M., Radzioch, D., 2010. Nuclear translocation of beta-actin is involved in transcriptional regulation during macrophage differentiation of HL-60 cells. Mol. Biol. Cell 21, 811–820.

Xue, Y., Canman, J.C., Lee, C.S., Nie, Z., Yang, D., Moreno, G.T., Young, M.K., Salmon, E.D., Wang, W., 2000. The human SWI/SNF-B chromatin-remodeling complex is related to yeast rsc and localizes at kinetochores of mitotic chromosomes. Proc. Natl. Acad. Sci. U. S. A. 97, 13015–13020.

Ye, J., Zhao, J., Hoffmann-Rohrer, U., Grummt, I., 2008. Nuclear myosin I acts in concert with polymeric actin to drive RNA polymerase I transcription. Genes Dev. 22, 322–330.

Zattelman, L., Regev, R., Usaj, M., Reinke, P.Y.A., Giese, S., Samson, A.O., Taft, M.H., Manstein, D.J., Henn, A., 2017. N-terminal splicing extensions of the human MYO1C gene fine-tune the kinetics of the three full-length myosin IC isoforms. J. Biol. Chem. 292, 17804–17818.

Zhao, K., Wang, W., Rando, O.J., Xue, Y., Swiderek, K., Kuo, A., Crabtree, G.R., 1998. Rapid and phosphoinositol-dependent binding of the SWI/SNF-like BAF complex to chromatin after T lymphocyte receptor signaling. Cell 95, 625–636.

CHAPTER FOUR

The actin-bundling protein L-plastin—A double-edged sword: Beneficial for the immune response, maleficent in cancer

Elisabeth Schaffner-Reckinger*, Raquel A.C. Machado
Cancer Cell Biology and Drug Discovery Group, Faculty of Science, Technology and Medicine, University of Luxembourg, Esch-sur-Alzette, Luxembourg
*Corresponding author: e-mail address: elisabeth.schaffner@uni.lu

Contents

Abstract

The dynamic organization of the actin cytoskeleton into bundles and networks is orchestrated by a large variety of actin-binding proteins. Among them, the actin-bundling protein L-plastin is normally expressed in hematopoietic cells, where it is involved in the immune response. However, L-plastin is also often ectopically expressed in malignant cancer cells of non-hematopoietic origin and is even considered as a marker for cancer progression. Post-translational modification modulates L-plastin activity. In particular, L-plastin Ser5 phosphorylation has been shown to be important for the immune response in leukocytes as well as for invasion and metastasis formation of carcinoma cells.

International Review of Cell and Molecular Biology, Volume 355
ISSN 1937-6448
https://doi.org/10.1016/bs.ircmb.2020.05.004

This chapter discusses the physiological and pathological role of L-plastin with a special focus on the importance of L-plastin Ser5 phosphorylation for the protein functions. The potential use of Ser5 phosphorylated L-plastin as a biomarker and/or therapeutic target will be evoked.

Abbreviations

ABD actin-binding domain
APC antigen-presenting cell
CLL chronic lymphocytic leukemia
fMLP *N*-formylmethionyl-leucyl-phenylalanine
PBMC peripheral blood mononuclear cells
PMA phorbol 12-myristate 13-acetate
PMN polymorphonuclear neutrophil
PTM post-translational modification

1. The actin cytoskeleton and actin-binding proteins

Together with microtubules and intermediate filaments, actin filaments form the skeleton of a cell. Actin filaments (F-actin) are assembled by reversible polymerization of monomeric globular actin (G-actin) into long polymers. Two such polymers, that are wound around each other, form an actin filament (Pollard and Borisy, 2003).

F-actin formation and organization are controlled by a plethora of actin-binding proteins. Among them, the actin-crosslinking proteins are responsible for assembling and stabilizing higher-order structures, such as contractile, loosely packed actin bundles in stress fibers or parallel, tightly packed actin bundles in microvilli and filopodia or actin networks in lamellipodia and the cell cortex. To fulfill their task of physically connecting actin filaments, actin-crosslinking proteins must have two actin-binding domains (ABD), either within one molecule or obtained by dimerization of two polypeptide chains. The actin cytoskeleton is a highly dynamic structure that gives the cell structural plasticity, helps force generation and thereby contributes to many cellular functions including cell morphology, intracellular transport, exo- and endocytosis, cytokinesis, cell motility, and wound healing (Pollard, 2016).

2. Discovery of plastins

Among the actin-binding proteins, plastins constitute a family of actin-bundling proteins that non-covalently crosslink actin filaments to organize them into tight bundles.

The first plastin isoform was identified by Matsudaira and Burgess (1979) as a 68 kDa protein in the microvilli of chicken intestine brush borders. This protein was named fimbrin (from the Latin "fimbria," which means border, fringe) as it was mainly detected in cell surface structures, such as in microvilli of intestinal epithelial cells, in membrane ruffles of tissue culture cells, at cell-substrate attachment sites as well as in other structures containing highly ordered microfilament bundles such as microspikes and inner ear stereocilia (Bretscher and Weber, 1980; Flock et al., 1982; Matsudaira and Burgess, 1979). Purification and biochemical characterization of fimbrin showed that it binds calcium and acts as an F-actin-binding and -bundling protein. Furthermore, electron microscopy revealed that fimbrin leads to tightly packed bundles of parallel actin filaments, *i.e.,* all filaments have the same polarity (Glenney et al., 1981). Compared to other F-actin-bundling proteins such as villin or fascin, fimbrin leads to the formation of more uniform and straight bundles than villin (Bretscher, 1981), whereas fascin-F-actin bundles are characterized by a striking 11-nm banding pattern (DeRosier et al., 1977; Otto et al., 1979). Later, X-ray diffraction patterns have determined the interfilament separation to be in the range of 11–12 nm in fimbrin-mediated actin bundles (Matsudaira et al., 1983).

Since then, plastins/fimbrins have been identified and studied in a variety of model organisms. The strong evolutionary functional conservation of the plastin gene family is underscored by the observation that human plastin could compensate *in vivo* for defects caused by the *Saccharomyces cerevisiae sac6* null mutation, which leads to budding yeast lacking the plastin ortholog Sac6p (Adams et al., 1995). Plastins were also characterized in plants such as *Arabidopsis thaliana*, in fission yeast *Schizosaccharomyces pombe*, in the ciliate *Tetrahymena*, as well as in vertebrates such as zebrafish, chicken, mouse or human.

3. The plastin family in humans

In mammals, three plastin paralogs have been identified and the expression of these paralogs was shown to be tissue-dependent. As early as in 1985, plastin was observed as one of the most abundant lymphocyte proteins and was also associated with neoplastic human fibroblasts (Goldstein et al., 1985). Goldstein et al. have emitted, for the first time, the hypothesis that plastin may either be involved in cytoarchitectural rearrangement or in anchorage independence, two phenotypes that are common to both lymphocytes and transformed fibroblasts. Molecular cloning, sequencing and subsequent characterization led to the identification of L-plastin and T-plastin (Lin et al., 1988). Sequence analysis performed by De Arruda et al. suggested that chicken intestine fimbrin is homologous to L-plastin, the 65 kDa phosphorylated protein expressed in leukocytes and transformed fibroblasts (De Arruda et al., 1990). Regarding the normal tissue-specific expression of the three paralogs, I-plastin or plastin 1, which is the ortholog of the chicken intestinal brush border fimbrin, is specifically expressed in microvilli of the small intestine, the colon and the kidney (Lin et al., 1994). L-plastin, plastin 2 or LCP1 is predominantly found in leukocytes and T-plastin or plastin 3 occurs in all non-hematopoietic cells of solid tissues that have replicative potential (Lin et al., 1988).

Comparative studies have suggested that transfected L-plastin and T-plastin differentially interact with actin filaments resulting in distinct F-actin organization, both in fibroblasts and in polarized epithelial cells (Arpin et al., 1994). The differential localization of L- and T-plastin in mammalian cells, with L-plastin being localized in the cytoplasm and the nucleus and T-plastin predominantly in the cytoplasm, results from the presence of a highly efficient nuclear export signal within the N-terminus of T-plastin and a less well conserved nuclear export signal within L-plastin. Although no nuclear function for L-plastin was provided, the authors suggested that L-plastin might play an indirect role in transcription regulation either through its interaction with actin or with vimentin (Delanote et al., 2005).

4. L-plastin structure

De Arruda et al. first described that chicken fimbrin is composed of two structural domains, a headpiece and a core that are linked by a protease-sensitive region, with the headpiece containing two calcium-binding domains and the core embracing two ABDs (De Arruda et al., 1990). In parallel, a sequence within L-plastin was identified that highly matched calcium-binding sequences of calmodulin and troponin that are two well-known calcium-binding proteins. This finding confirmed the existence of a potential calcium-binding domain in L-plastin and an analogous sequence was also found in T-plastin (Lin et al., 1990). At the same time, a 65 kDa protein that is phosphorylated in human T cells was sequenced and found to be identical with L-plastin. A multidomain structure of this protein was shown, confirming the existence of the two N-terminal calcium-binding sites and the two ABDs at the C-terminus proposed by De Arruda and coworkers. In addition, the presence of a potential calmodulin-binding site between the calcium-binding sites and the core region of the protein was revealed (Zu et al., 1990a,b). Later, the complete sequence of murine L-plastin was provided by Shinomiya and coworkers who also identified two calcium-binding domains, one potential calmodulin-binding domain and a tandem repeat of ABDs similar to the human L-plastin (Shinomiya et al., 1995) (Fig. 1). In 1997, the crystal structure of the N-terminal ABD in human T-plastin was solved, showing for the first time the structure of an ABD (Goldsmith et al., 1997). In 2004, the crystal structure of *Arabidopsis thaliana* and *Schizosaccharomyces pombe* fimbrin cores were obtained and compared, highlighting that, for both orthologs, a direct contact between CH1 of ABD1 and CH4 of ABD2 is made, leading to a compact organization of the fimbrin core with an overall horseshoe appearance. Notably, significant variability in the positioning of CH2 would contribute to conformational plasticity of the ABD1 domain. The derived model suggested that ABD1 and ABD2 have non-identical interactions with F-actin (Klein et al., 2004). X-ray scattering in solution applied as a

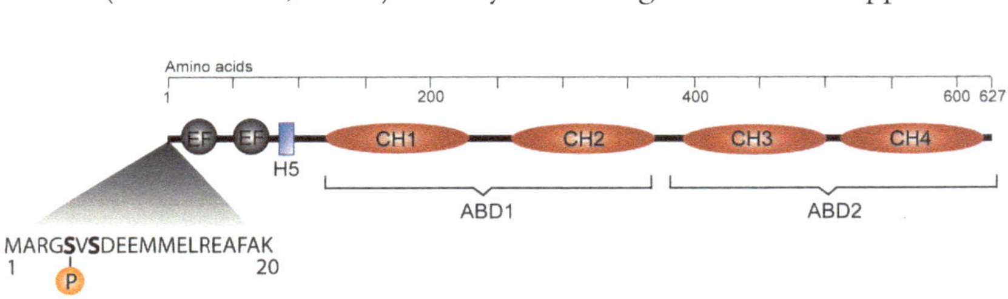

Fig. 1 Functional domains of the human L-plastin protein. L-plastin possesses two actin-binding domains (ABD) within a single polypeptide chain. Each ABD consists of a pair of ~125 residue calponin-homology (CH) domains. This unique composition confers L-plastin the ability to crosslink actin filaments into higher-order assemblies like bundles. The N-terminal headpiece of ~100 amino acids comprises two EF-hand calcium-binding motifs and a switch helix region (H5), overlapping with a potential calmodulin-binding motif and acting as a regulatory switch. In the presence of high Ca^{2+} concentrations, the EF hands have been suggested to interact with the potential calmodulin-binding motif, decreasing the actin bundling efficiency of L-plastin. Additionally, L-plastin has two phosphorylation sites (Ser5 and Ser7) near its N-terminus. In particular, Ser5 phosphorylation has been extensively shown to be critical for L-plastin activity. A scale (in amino acids) is drawn on top of the figure.

technique to elucidate the structure of human L-plastin yielded a similar conclusion: full-length L-plastin adopts a compact horseshoe-like shape with an attached projection that could be the N-terminal headpiece. Notably, Ca^{2+}-dependent conformational changes were noted within this structure (Shinomiya et al., 2007).

5. L-plastin activity regulation by Ca^{2+}

In 1987, Pacaud and Harricane described a 70 kDa protein in macrophages whose actin-bundling activity decreased with increasing intracellular Ca^{2+} concentrations (Pacaud and Harricane, 1987). This protein was later identified as L-plastin (Pacaud and Derancourt, 1993). In their study, Namba and coworkers confirmed the strict dependence of L-plastin bundling activity on free Ca^{2+} concentration, with L-plastin being able to bundle β-actin isolated from human T cells at Ca^{2+} concentration pCa 7, but not at pCa 6 (Namba et al., 1992). Circular dichroism spectra showed that the EF-hand domains of both L- and T-plastin are α-helix rich and that they undergo a conformational change following Ca^{2+} binding. However, isothermal titration calorimetry measurements showed that the EF-hand domains of T-plastin have two Ca^{2+}-binding sites that differ by their binding affinity, one being a low-affinity site and the other one being a high-affinity site, whereas the two Ca^{2+}-binding sites within the EF-hand domains of L-plastin appear to be both high-affinity sites (Miyakawa et al., 2012). As a consequence, the sensitivity to Ca^{2+} is lower for T-plastin as for L-plastin, suggesting that L-plastin activity is more tightly controlled by intracellular Ca^{2+} concentrations and thus, L-plastin would be well-adapted to regulate a dynamic cytoskeleton in highly motile cells. On the other hand, T-plastin, which is expressed in immotile cells would be suitable to maintain static cytoskeletons (Miyakawa et al., 2012). Fourier-transform infrared spectroscopy was used to further characterize the coordination structures of Ca^{2+} and Mg^{2+} bound by T- and L-plastin headpiece. This approach suggested that both headpieces have a higher affinity for Ca^{2+} than for Mg^{2+} and that T-plastin headpiece is more prone to aggregating than L-plastin headpiece, presumably because of the lower Ca^{2+} sensitivity of T-plastin (Nara et al., 2019).

Nuclear magnetic resonance spectroscopy studies allowed Ishida and coworkers to discover the existence of a switch helix region (residues 85–94) between the EF-hand calcium-sensor headpiece region and the four CH-domains in the core region of L-plastin. In the presence of Ca^{2+}, this switch helix bound tightly to the EF hand motifs. The inhibition of this interaction *via* peptide inhibitors resulted in deregulation of the bundling activity of L-plastin (Ishida et al., 2017). Knowing that the bundling activity of plastins is negatively regulated by Ca^{2+} (Namba et al., 1992) and in line with the findings of Miyakawa et al., Schwebach and coworkers showed that the bundling capacities of I-plastin and L-plastin are similarly sensitive to Ca^{2+} whereas T-plastin is less sensitive (Schwebach et al., 2017). They also demonstrated that the ABD1 of plastins can bind to actin independent of Ca^{2+}, but that the interaction of ABD2 with actin, and, thus, actin-bundling activity, is inhibited by increasing Ca^{2+} concentrations. They suggest a model where, at high Ca^{2+} concentrations, binding of ABD2 to actin is inhibited by the interaction of EF hands with the calmodulin-binding motif (residues 83–97), which encompasses the switch helix region described by Ishida and coworkers. In contrast, at low Ca^{2+} concentrations, the EF hand region is associated with ABD2 instead of the calmodulin-binding motif and ABD2 is able to bind to F-actin. Thus, these findings corroborate that actin-bundling activity of L-plastin is enabled at low Ca^{2+} concentrations (Schwebach et al., 2017).

6. L-plastin functions: Impact of post-translational modification

6.1 L-plastin primordial function: F-actin-binding and -bundling

As mentioned before, the primordial function of all plastins consists in F-actin-binding and -bundling. Several studies performed by Hanein and coworkers provided a first atomic model of actin filaments bundled by human T-plastin. They provided evidence that plastin-bundled actin arrays have a more regular organization than plastin-free actin arrays. In addition, plastin does not bind well to preformed actin arrays, but the formation of plastin-bundled actin arrays rather requires coupling of actin polymerization with crosslink assembly (Hanein et al., 1997, 1998; Volkmann et al., 2001).

The plastin ortholog AtFim1, present in *Arabidopsis thaliana*, was shown to bundle actin filaments in plant cells. Interestingly, the AtFim1 ABD1 and ABD2 domains seemed to have different affinities for F-actin (Kovar et al., 2000), as confirmed later by Lebart et al. for human L-plastin (Lebart et al., 2004). Indeed, the biochemical characterization of the interaction between human L-plastin and actin performed by Lebart et al. corroborated the model proposed by Klein et al., which suggested that ABD1 and ABD2 have non-identical interactions with F-actin (Klein et al., 2004). According to Lebart and coworkers, both ABD1 and ABD2 were able to bind immobilized actin, but ABD1 bound actin with a higher affinity than ABD2. Furthermore, it was suggested that ABD1 would be the first domain to interact with actin. Finally, their results provided evidence for a lateral binding of L-plastin to F-actin along the filament axis and showed that L-plastin prevents F-actin depolymerization (Lebart et al., 2004). This functional difference between L-plastin ABD1 and ABD2 is further underlined by the discovery that L-plastin interacts with vimentin, a protein of the intermediate cytoskeleton (Correia et al., 1999). Actually, the vimentin binding site within L-plastin has been mapped to the first CH domain of the ABD1 domain. In adherent macrophages, L-plastin interacts with a vimentin subunit tetramer, but not with vimentin incorporated in intermediate filaments, and both proteins co-localize in filopodia, retraction fibers and podosomes. Since their association was shown to be adhesion-dependent, a mechanism was suggested that this interaction might be involved in mediating the assembly of the intermediate cytoskeleton at cell adhesion sites (Correia et al., 1999).

In a more recent study using cryo-EM and a single particle approach to helical reconstruction, the interaction between ABD2 and F-actin was proposed to be different from any interaction previously described for tandem CH-domain proteins. Indeed, the tandem CH-domains within ABD2 of L-plastin bound F-actin in a rather fixed and stoichiometric manner. In contrast, the binding of ABD1 to actin was more polymorphic and only partial. As opposed to previous models, the authors proposed as a mechanism that ABD2 binds to a first actin filament, leading to activation of ABD1, which then binds to a second actin filament (Galkin et al., 2008).

Importantly, in 2006, Janji and coworkers showed that Ser5 phosphorylation of L-plastin increased its avidity for F-actin in cells and its F-actin-bundling activity *in vitro*, promoting its localization to actin-rich structures, such as microspikes, ruffling membranes, focal adhesions and stress fibers (Janji et al., 2006). In a follow-up study, fluorescence recovery after photobleaching experiments were carried out to investigate the effect of L-plastin expression and/or Ser5 phosphorylation on actin turnover. The obtained results provided evidence that L-plastin presence and, in particular, L-plastin Ser5 phosphorylation leads to enhanced actin stability in focal adhesions. The observed association of L-plastin with cortactin substantiated the role of L-plastin in actin dynamics regulation (Al Tanoury et al., 2010).

6.2 Associated functions of L-plastin

6.2.1 During embryonic development

Together with other actin-binding proteins such as villin and myosin I, plastin was shown to be involved in the assembly of microvilli during the differentiation of the mouse intestinal epithelium (Ezzell et al., 1992). More precisely, the three plastin paralogs were sequentially expressed and differentially localized during intestinal epithelium differentiation, suggesting that they play different roles during this differentiation. Given its localization to the basal surface of the epithelium from day 10.5 to day 14.5, L-plastin would primarily be involved in controlling cell adhesion during the early stages of development. By day 16.5, L-plastin expression was reduced and, in the adult epithelium, it could not be detected anymore (Chafel et al., 1995). More recent work performed using the zebrafish model further analyzed the expression of L-plastin during early embryonic development. This study revealed that the spatial expression of L-plastin was restricted to the enveloping layer, which is an extraembryonic tissue that has epithelial characteristics, and to the periderm (Baumgartner et al., 2019).

6.2.2 Endocytosis

In *Saccharomyces cerevisiae*, the plastin ortholog Sac6p has been shown to be required for receptor-mediated endocytosis of α-factor, presumably because actin-bundling is required for this endocytosis event (Kübler and Riezman, 1993). Similarly, a critical role for endocytosis has been demonstrated for L- and T-plastin in mammalian Cos-1 cells, as both paralogs interacted with the small GTPase Rab5 in its activated state, *i.e.*, when it is bound to GTP. Since Rab5 is important for the endocytic machinery, it could be shown that overexpression of L- or T-plastin promoted fluid-phase endocytosis either by activating or by inducing a sustained activity of Rab5 (Hagiwara et al., 2011).

6.2.3 Stress response

Association of L- or T-plastin with the polyglutamine protein ataxin-2, which is a component of stress granules, has been reported (Ralser et al., 2005). Moreover, a role for the plastin ortholog Sac6 was demonstrated for the regulation of oxidative stress response, morphogenesis and pathogenicity in *Candida albicans* (Zhang et al., 2016).

6.2.4 L-plastin as a player in osteoclastogenesis

Enoxacin, a small molecule inhibitor of the binding of the vacuolar H^+-ATPase to microfilaments, inhibited osteoclastogenesis without affecting cell viability, but rather by affecting protein transport and protein proteolysis, with L-plastin being among the affected proteins (Toro et al., 2012). Similar to the cancer-secreted mediator of osteoclastogenesis peroxiredoxin-4, L-plastin was identified in conditioned media of the breast cancer cells MDA-MB-231. In addition, L-plastin was detected in cancer-derived exosomes. Silencing of L-plastin in MDA-MB-231 breast cancer cells strongly reduced their osteoclastogenic capacity, whereas addition of recombinant L-plastin to osteoclast precursors primed with RANKL (receptor activator of nuclear factor κB ligand) induced osteoclastogenesis, demonstrating that L-plastin functions as an activator of bone-resorbing osteoclasts. Most interestingly, the results of an experimental breast cancer bone metastasis model suggested that L-plastin and peroxiredoxin-4 cooperate to stimulate osteolysis *in vivo* (Tiedemann et al., 2019).

A role for L-plastin has been reported in the early sealing ring formation in osteoclasts, presumably by formation of actin aggregates that would then act as a core in assembling signaling molecules necessary for sealing ring maturation (Ma et al., 2010). In follow-up studies, the same group provided evidence that phosphorylation of L-plastin is necessary for the L-plastin-mediated effects in osteoclasts, such as nascent sealing zone and sealing ring formation and thus bone resorption, but not for osteoclast podosome formation and migration or osteoblast bone forming activity (Chellaiah et al., 2018a,b). In their most recent study, the researchers extended these results by showing that L-plastin −/− osteoclasts did not form nascent sealing zones and functional resorption pits, but were able to form actin-enriched podosomes and to adhere *in vitro* and *in vivo*. Notably, defective functioning of L-plastin −/− osteoclasts could be rescued by transduction of full-length L-plastin, but not of the S5AS7A phosphorylation-defective variant, pointing to L-plastin Ser phosphorylation being critical for L-plastin functioning in osteoclasts, possibly due to abnormal actin dynamics regulation (Chellaiah et al., 2020). In another study, Si et al. showed that Ser5 phosphorylation of L-plastin was reduced in osteoclasts deficient for the kinase Lrrk1, which is a PKC-like serine/threonine protein kinase that is required for bone resorption. The authors suggested that L-plastin Ser5 phosphorylation might account, in part, for Lrrk1-signaled actin cytoskeleton rearrangements in mature osteoclasts (Si et al., 2018).

6.2.5 L-plastin in podosomes

The first report about localization of fimbrin in podosomes of osteoclasts was provided by Marchisio et al. By performing electron microscopy studies, they observed a colocalization of fimbrin with F-actin and α-actinin in short ventral protrusions of cultured avian osteoclasts, that were later called podosomes (Marchisio et al., 1984). A localization of L-plastin in podosomes of macrophages could be shown later by Messier and coworkers (Messier et al., 1993). Adhesion of FcγRII receptor expressing neutrophils to immune complexes also induced L-plastin localization to podosomes (Jones and Brown, 1996). Whereas I-plastin could not be observed in these structures, immunofluorescence microscopy revealed that L- and T-plastin were localized in podosomes of mature rabbit osteoclasts as well as in differentiating avian monocyte-derived osteoclasts (Babb et al., 1997).

De Clercq et al. used nanobodies specific for L-plastin to knock-out functional domains of L-plastin without affecting the expression level. Nanobodies are single immunoglobulin (IgG) domains, which correspond to the variable domain of heavy-chain antibodies that were first discovered in *Camelidae*. They found that L-plastin function is necessary for the formation, stability and the proper functioning of podosomes in macrophage-like THP-1 cells. Indeed, nanobody-dependent blocking of L-plastin actin-bundling function or sequestering of L-plastin in an inactive conformation led to reduced matrix degradation abilities. In addition, they showed that L-plastin in podosomes is, at least partly, Ser5 phosphorylated (De Clercq et al., 2013a). Furthermore, by analyzing L-plastin −/− macrophages isolated from L-plastin null mice, a critical role for L-plastin could be demonstrated for podosome longevity as well as for polarization and migration of macrophages (Zhou et al., 2016).

6.2.6 Immune response

A major function for L-plastin has been described in the immune response, as suggested already by the strong inherent presence of L-plastin in leukocytes. Furthermore, L-plastin was found to be activated by phosphorylation during the immune response.

Indeed, already in the 1980s, many studies reported phosphorylation of proteins with molecular masses of approximately 65 kDa in hematopoietic cells. Evidence for such phosphorylation was provided following mitogen stimulation of human peripheral blood lymphocytes (Chaplin et al., 1980), in phorbol 12-myristate 13-acetate (PMA) treated human promyelocytic HL-60 leukemia cells (Anderson et al., 1985), in adherent monocytes (Goldstein et al., 1985), in interleukin-3 or diacylglycerol-stimulated FDC-P1 mouse myeloid progenitor cells (Evans et al., 1986), in anti-IGM or PMA treated intact B lymphocytes (Hornbeck and Paul, 1986), in interleukin-1 stimulated human peripheral blood mononuclear cells (PBMC) (Matsushima et al., 1988) or during chemotactic activation with interleukin-8 or N-formylmethionyl-leucyl-phenylalanine (fMLP) of human polymorphonuclear leukocytes (Shibata et al., 1993). The 65 kDa protein that was phosphorylated within a few minutes upon tumor necrosis factor or interleukin-1 stimulation of human PBMCs was presumed to be identical to L-plastin already in 1990 (Shiroo and Matsushima, 1990). Bacterial lipopolysaccharide stimulation of murine peritoneal macrophages also led to the phosphorylation on serine residues of a 65 kDa protein, suggesting its crucial role in macrophage activation (Shinomiya et al., 1991). This pp65 protein was later confidently identified as L-plastin (Shinomiya et al., 1994) and the phosphorylation was described to be MAPK-independent (Shinomiya et al., 1995).

In 1993, L-plastin was shown to be constitutively phosphorylated in macrophages and the phosphorylation was mapped to serine residues in the N-terminal headpiece of L-plastin. The authors even hypothesized already at that time that this phosphorylation could regulate the actin-bundling properties of L-plastin or its interaction with partner proteins (Messier et al., 1993). In 1994, phosphorylation of a protein that the authors presumed to be identical with L-plastin, was described as an early signaling event following stimulation of human T lymphocytes (Henning et al., 1994). Moreover, two phosphoproteins with a molecular weight of 65 kDa and nearly identical isoelectric points were detected in lymphokine-activated killer cells following tumor target binding. The apparent molecular weight, isoelectric point, phosphorylation properties as well as the cross-reaction of the proteins on a 2D gel with a polyclonal rabbit antiserum specific for L-plastin suggested altogether again that these proteins might be phosphorylated forms of L-plastin (Frederick et al., 1996). And finally, Jones and Brown showed in 1996 that L-plastin was phosphorylated in neutrophils in response to FcγRII ligation (Jones and Brown, 1996).

Given that various inflammatory stimuli such as chemokines, cytokines, formylated bacterial peptides or PMA induce both integrin β2-mediated adhesion of polymorphonuclear neutrophils (PMN) or monocytes and L-plastin phosphorylation, Jones and coworkers tested whether L-plastin phosphorylation was a prerequisite for integrin activation. By introducing synthetic L-plastin-derived peptides into PMNs, they could show that Ser5 phosphorylation of L-plastin is fundamentally involved in αMβ2-dependent leukocyte adhesion (Jones et al., 1998). Three years later, the same group published a study where they showed that a peptide derived from the L-plastin N-terminus or the entire L-plastin headpiece domain were also able to induce αvβ3-mediated cell adhesion to vitronectin. This adhesion was independent of β3 integrin cytoplasmic domain tyrosine phosphorylation, but required actin depolymerization. In addition, the L-plastin-derived peptide cooperated with RGD peptide ligand to induce a conformational change of integrin αvβ3, as measured by ligand-induced binding site expression. Altogether, these data point to a mechanism where L-plastin would contribute to integrin αvβ3-mediated adhesion by linking integrin diffusion leading to increased integrin avidity with integrin conformational change leading to increased integrin affinity. These L-plastin-dependent events would be downstream of the integrin tyrosine signal in the integrin activation cascade (Wang et al., 2001). To further analyze the effect of L-plastin on integrin-dependent functions, mice lacking L-plastin were generated and PMNs devoid of all plastin isoforms were investigated regarding host defense against *Staphylococcus aureus*. While leukocyte maturation, PMN migration to the site of infection, binding and ingestion of opsonized bacteria and secretion of primary granule contents were normal, integrin-dependent respiratory burst was defective in L-plastin −/− PMNs (Chen et al., 2003). Altogether, these studies underline that the loss of one single actin-binding protein can lead to disruption of integrin-triggered intracellular signal transduction cascades, most probably through changes of the cortical actin cytoskeleton. Additionally, an interaction between the core domain of L-plastin containing the two ABDs and the cytoplasmic domain of β1 integrins as well as β1 and β2 peptides was shown by Le Goff and coworkers. This interaction was demonstrated to be regulated by μ-calpain cleavage of β integrin (Le Goff et al., 2010).

Notably, the interaction of L-plastin with a subset of other binding partners has also been shown to be important for the immune response. As an example, an interaction was identified between L-plastin and grancalcin (Lollike et al., 2001), which was later shown to play a role in the host defense of leukocytes (Liu et al., 2004). Furthermore, Iba1, a microglia/macrophage-specific actin-bundling protein, was demonstrated to associate with L-plastin. Notably, both proteins co-localized in membrane ruffles and phagocytic cups, and binding of Iba1 enhanced the actin-bundling capacity of L-plastin. These observations led to the hypothesis that the two actin-bundlers cooperate to regulate actin cytoskeleton reorganization and promote cell migratory and phagocytic capacities of microglia (Ohsawa et al., 2004). Up-regulation of L-plastin expression could be observed on day three following prothymosin α-stimulation of PBMCs, corroborating again a role of L-plastin in the immune response. More precisely, day three was the time point of enforced adherence of PBMCs to target cells leading to enhanced cytotoxic activity (Skopeliti et al., 2007).

Importantly, Ser5 was the only L-plastin phospho-acceptor site identified in human peripheral blood T lymphocytes following costimulation of TCR/CD3 plus CD2 or CD28. Although both wild type L-plastin and the non-phosphorylatable S5A variant were equally able to relocalize to the immunological synapse between peripheral blood T lymphocytes and antigen-presenting cells (APC), expression of the S5A variant resulted in defective transport of the activation molecules CD25 and CD69 to the cell surface (Wabnitz et al., 2007). The localization of L-plastin was shown to occur predominantly in the outer supramolecular activation cluster during immunological synapse formation between T cells and APCs. Together with calmodulin, L-plastin was shown to be a prerequisite for the sustained accumulation of integrin αLβ2 and talin to the outer supramolecular activation cluster, and therefore for the proper activation of T cells (Wabnitz et al., 2010). In a follow-up study, phosphorylation of L-plastin could be shown to be important for immune synapse maturation, since the unphosphorylatable S5A plastin variant resulted in reduced accumulation of integrin αLβ2 at the immunological synapse and to reduced F-actin content in stimulated primary T cells (Wabnitz et al., 2011). Interestingly, treatment with the glucocorticoid dexamethasone inhibited L-plastin Ser5 phosphorylation concomitant with prevention of immune synapse maturation, revealing a novel mode of function of this immunosuppressive glucocorticoid (Wabnitz et al., 2011). In 2017, Wabnitz and coworkers confirmed that not only the presence of L-plastin, but also L-plastin Ser5 phosphorylation is necessary for the sustained accumulation of integrin $\alpha L\beta 2$ at immunological synapses formed between T cells and APCs, thus regulating immunological synapse stability and proper T cell activation (Wabnitz et al., 2017).

In line, the use of L-plastin-specific nanobodies to knock-out functional domains of L-plastin resulted in delayed L-plastin Ser5 phosphorylation, ineffective immune synapse formation and disturbed T cell activation (De Clercq et al., 2013b).

Using L-plastin null mice, Morley and coworkers carried out a variety of studies, all of them providing strong evidence for the requirement of L-plastin in the immune response. Indeed, a requirement for L-plastin could be noticed for the establishment and maintenance of T cell polarity and directed chemotaxis (Morley et al., 2010). As mentioned above, Wabnitz et al. showed that L-plastin is a prerequisite for the proper activation of T cells (Wabnitz et al., 2010). Further evidence was provided that L-plastin is necessary for T cell receptor-mediated T cell activation as L-plastin −/− T cells were defective for normal lamellipodia formation and stability, two prerequisites for the formation of immunological synapses between T cells and APCs (Wang et al., 2011). In addition, L-plastin was shown to be essential for B cell motility and marginal zone B cell development (Todd et al., 2011), for modulating T and B cell function necessary for functional germinal center formation and for T cell-dependent antibody generation (Todd et al., 2013) as well as for the development of alveolar macrophages and thus, for an efficient pulmonary innate immune response (Deady et al., 2014). In the latter regard, a requirement for L-plastin was also proven for perinatal production of alveolar macrophages through the support of two actin-based processes: transmigration of macrophage precursors into the alveoli followed by engraftment of the monocytes and mature macrophages in the alveoli (Todd et al., 2016).

A link between L-plastin and the immune response was suggested in a study demonstrating an interaction between L-plastin and the IFN-γ-inducible GBP-1 (GTPase guanylate binding protein 1) in T cells.

However, although the authors demonstrated that GBP-1 acts as a negative regulator of early T cell receptor signaling through modulation of IL-2 expression, clear evidence for an involvement of L-plastin in this negative regulation function could not be provided (Forster et al., 2014).

Finally, in the zebrafish model *Danio rerio*, the involvement of L-plastin in myelopoiesis had already been suggested more than 15 years ago (Berman et al., 2005; Crowhurst et al., 2002). More recent studies, using the CRISPR/Cas9 methodology for establishing 3 independent L-plastin knock-out zebrafish lines, have shown that the development and reproduction of the null animals was normal, but that their survival within the first 6–7 weeks was reduced by 30%, presumably due to a compromised innate immune system (Kell et al., 2018).

It is noteworthy that, in addition to Ser5 phosphorylation, further post-translational modifications (PTM) of L-plastin in leukocytes have been described more recently (Table 1). Another phosphorylation site has been identified within L-plastin by Xu et al., who showed that residue Thr89 may be phosphorylated by Ste20–like kinase 1 (Mst1) *in vitro* and that both proteins can interact in cells. Moreover, an L-plastin T89A mutant exhibited impaired localization to lamellipodia in T cells and, when transfected into L-plastin deficient T cells isolated from L-plastin −/− mice, this mutant could not rescue T cell migration. Notably, phosphorylation of residue Thr89 was neither necessary for the actin-bundling activity of L-plastin nor for binding to calmodulin, although Thr89 is located within the calmodulin-binding motif (Xu et al., 2016).

Moreover, oxidative stress was reported to induce S-glutathionylation of L-plastin residue Cys460 in PMNs (Dubey et al., 2015). Further redox-sensitive L-plastin cysteine residues were identified upon treatment of peripheral blood T cells with H_2O_2: this study revealed the oxidation of Cys42 and Cys101 and, at minor levels, of Cys283, Cys336 and Cys460 (Balta et al., 2019).

Finally, by studying the effect of allicin (diallylthiosulfinate), which is produced upon damage of garlic cells, on the post-translational thiol-modification of proteins in Jurkat T cells, it was shown that L-plastin is found among the proteins that are S-thioallylated (Gruhlke et al., 2019). In addition, allicin treatment led to actin cytoskeleton disruption in Jurkat T cells. Given that allicin has anti-cancer effects, it has been proposed that these effects might be due to inactivation of crucial proteins, such as L-plastin (Gruhlke et al., 2019).

Table 1 Post-translational modifications in L-plastin.

Modification	Residue	Established functions	References
Phosphorylation	Ser5	In hematopoietic cells: ↑ immune synapse maturation in T cells ↑ activation of T cells ↑ sealing ring formation and bone resorption in osteoclasts	Wabnitz et al.(2011), Wabnitz et al. (2017), De Clercq et al. (2013b), Chellaiah et al. (2018a,b), and Chellaiah et al. (2020)
		In cancer cells: ↑ translocation to actin-rich structures ↑ actin bundling ↑ cell motility ↑ cell invasion ↑ metastasis formation	Janji et al. (2006), Klemke et al. (2007), Al Tanoury et al. (2010), Riplinger et al. (2014), and Lommel et al. (2016)
		In podocytes: ↑ filopodia formation ↑ filopodia length	Schenk et al. (2017)
	Thr89	In leukocytes: ↑ translocation to the cell periphery in T cells ↑ T cell motility	Xu et al. (2016)
Thiol oxidation	Cys460	In cancer cells: ↓ cell invasion	Dubey et al. (2015)
	Cys101	In cancer cells: ↓ actin bundling ↓ cell spreading ↓ cell motility ↓ cell invasion ↓ extracellular matrix degradation	Balta et al. (2019)

The table summarizes the main PTMs in L-plastin with functional consequences in cells.

6.2.7 L-plastin and disease

Variation in L-plastin expression, secretion or PTMs such as phosphorylation or thiol oxidation has been linked to a variety of diseases. For instance, L-plastin was significantly enriched in the secretome of HIV-1 infected human macrophages as compared to non-infected macrophages (Ciborowski et al., 2007). Furthermore, proteomic analysis of PBMCs from relapsing minimal-change nephrotic syndrome patients identified four proteins including L-plastin as being up-regulated in relapse patients as compared to control groups. Given that three of the four identified proteins were cytoskeleton-associated proteins and taking into consideration that L-plastin is involved in T-cell activation, the authors hypothesized that T-cell cytoskeleton rearrangements might be associated with the pathogenesis of the disease (González et al., 2009).

Another group compared allogeneic kidney grafts leading to interstitial fibrosis and tubular atrophy with syngeneic grafts by 2D gel analysis and mass spectrometry. Their approach allowed to identify 10 differentially expressed proteins, including L-plastin, which was up-regulated (Reuter et al., 2010). In a follow-up study, the same group characterized the phosphoproteome of human podocytes in response to angiotensin II treatment. To this end, they performed SILAC labeling (Stable Isotope Labeling with Amino acids in Cell culture) coupled with phosphopeptide enrichment and liquid chromatography–tandem mass spectrometry followed by Western blot analysis and immunofluorescence experiments. They showed that angiotensin II treatment induced Ser5 phosphorylation of L-plastin and L-plastin redistribution to podocyte margins, concomitant with increased membrane ruffles and lamellipodia formation. Moreover, L-plastin Ser5 phosphorylation was shown to increase both filopodia formation and filopodia length in podocytes, whereas increased membrane ruffling and L-plastin redistribution to these ruffles was independent of Ser5 phosphorylation (Schenk et al., 2017). The authors suggested that high level angiotensin II-triggered actin cytoskeleton reorganization impairs podocyte functioning and viability, partly by promoting L-plastin Ser5 phosphorylation. Thus, drugs targeting cytoskeleton turnover mediated by actin-binding proteins might be a promising approach for podocyte disease treatment.

Performing label-free liquid chromatography/mass spectrometry to compare gingival crevicular fluid from healthy or periodontal disease suffering subjects, Bostanci et al. identified L-plastin as a protein present in gingival crevicular fluid only in periodontal disease affected subjects (Bostanci et al., 2010). Proteomic analysis also identified L-plastin as a significantly up-regulated protein in rheumatoid arthritis synovial tissues compared to osteoarthritis used as a control (Hayashi et al., 2015) and in pancreatic islet tissues of enterovirus-associated fulminant type 1 diabetes mellitus patients (Nishida et al., 2014). Taking a chromatography/mass spectrometry approach, L-plastin could be identified in the saliva of type 1 diabetes mellitus patients with chronic complications such as retinopathy or nephropathy (Caseiro et al., 2013). In PMNs, L-plastin residues Cys-206, Cys-283, and Cys-460 were identified to be susceptible to oxidative stress-induced S-glutathionylation. Among them, residue Cys-460 presented the maximum score and was suggested to play a role in the redox regulation of cell invasion. Since enhanced L-plastin S-glutathionylation was also observed in PMNs of diabetic patients, the authors suggested that this oxidative stress-dependent L-plastin PTM might contribute to neutrophil dysfunction in diabetes mellitus, which is known to cause major diabetic complications (Dubey et al., 2015). Together with metalloproteinase 1 and regenerating islet-derived protein 3-α, L-plastin has been proposed to be a plasma-derived predictive marker for acute graft-*versus*-host-disease and non-relapse mortality in patients that have undergone allogeneic hematopoietic stem cell transplantation (Shin et al., 2019). Finally, L-plastin was suggested to be a biomarker for hypereosinophilia and to mediate eosinophil differentiation. In this system, L-plastin was described to be activated through Ser5 phosphorylation and to be crucial for AKT Ser473 phosphorylation by the mTORC2 complex (Ma et al., 2020).

7. L-plastin and cancer

In addition to the normal expression of L-plastin in cells of the hematopoietic lineage, this protein is also often ectopically expressed in cancer cells derived from solid tissues.

7.1 The L-plastin promoter region and cancer gene therapy based on the activity of this region

Ectopic L-plastin expression in neoplastic cells led to the hypothesis that this expression may be governed by regulatory elements. Indeed, regulatory elements such as four potential Ets-1 binding motifs, one SP-1 binding sequence and one AP1-binding site as well as two potential progesterone and one potential estrogen responsive element have been found upstream of the L-plastin gene (Lin et al., 1993a). In line, L-plastin synthesis was shown to be up-regulated when cultured uterine endometrial stromal cells were incubated with the ovarian steroids estradiol or progesterone (Leavitt et al., 1994). In a follow-up study, Lin et al. identified three androgen receptor (AR)-binding elements (ARE) in the L-plastin promoter region. Although these sequences are imperfect as compared with the consensus

ARE, testosterone treatment of breast or prostate cancer cells increased L-plastin expression, most probably through cooperative binding of the androgen testosterone to the three imperfect AREs in the L-plastin promoter region (Lin et al., 2000).

Based on a study of Zheng et al. showing that, in the human prostate carcinoma cells LNCaP, stimulation with estradiol or the androgen dihydrotestosterone led to the up-regulation of L-plastin expression (Zheng et al., 1997), a more recent study provided further insight into the regulation of L-plastin expression in prostate cancer. This study provided evidence that, at position −1687 in the L-plastin promoter, a single nucleotide polymorphism is responsible for differential affinity for the transcriptional suppressor NKX3.1. This polymorphism, thus, affected L-plastin expression with increased expression being linked to prostate cancer progression (Chen et al., 2016). The same authors showed that the transcription factor AP4 directly binds to the L-plastin promoter region and, as a consequence, increases L-plastin expression. The AP4-dependent L-plastin expression was regulated by the PI3K/AKT pathway and was independent of steroid hormone-dependent regulation. Importantly, L-plastin was shown to be a critical mediator of AP4-dependent prostate cancer progression, as revealed by proliferation, invasion and migration studies (Chen et al., 2017).

Given the particular characteristics of the L-plastin promotor sequence, researchers have raised the idea that it could be used for tissue-specific gene therapy targeting cells that express estrogen- and progesterone specific receptors, such as the cells of steroid hormone-regulated female reproductive tissues (Patterson and Harris, 1999).

Another study has exploited these characteristics and tested the potential use of adenoviral vector-mediated delivery of therapeutic genes driven by the L-plastin promoter for the targeted treatment of ovarian cancer (Chung et al., 1999). This group provided evidence that this approach allows targeting preferentially carcinoma cells as compared to normal cells and that L-plastin promoter-driven therapeutic gene expression reduced tumor growth in mice. Further results confirmed the potential value of adenoviral vectors carrying therapeutic genes controlled by the L-plastin promoter for the intraperitoneal therapy of ovarian cancer (Peng et al., 2001) and for breast cancer treatment (Zhang et al., 2002). In a follow-up study, a replication-competent adenoviral vector that carried the prodrug activation transcription units E1A and cytosine deaminase was constructed under the control of the L-plastin promoter. Whereas cytosine deaminase catalyzes the conversion of the harmless prodrug 5-fluorocytosine into the cytotoxic agent 5-fluorouracil, the early E1a protein stimulates viral genome replication and has host cell cytotoxic effects. The study showed the specific cytotoxicity of this vector on breast, ovary and colon carcinoma cells as compared to normal cells *in vitro* and its reducing effect on the volume of tumor nodules obtained by subcutaneous injection of colon cancer cells in nude mice. Most importantly, these effects were potentiated by the parallel administration of the prodrug 5-fluorocytosine (Akbulut et al., 2003). This so-called "genetic combination therapy" showed significantly increased *in vitro* therapeutic activity as compared to conventional chemotherapy and significant response, survival and toxicity advantages in tumor-bearing mice (Akbulut et al., 2004). The vector was then further modified on a transductional level by inserting the tripeptide RGD as an integrin recognition sequence. This modification helped to increase the transduction efficiency of breast cancer cell lines *in vitro* and to reduce the tumor volume of xenotransplanted human breast cancer cell lines in mice (Liu et al., 2006). Although these results seemed promising, no follow-up studies have been reported and, to our knowledge, the approach appears to have been abandoned.

Finally, L-plastin expression was recently proposed to be regulated by the transcription factor HIF1α. In line, two hypoxia-response element sequences, which are controlled by the transcription factor HIF1α, have been shown at positions −1527/−1531 and +237/+241 from the transcription start site of the L-plastin gene (Bosseler et al., 2018).

7.2 The potential use of L-plastin as cancer biomarker and/or therapeutic target

7.2.1 Studies conducted using cancer cell lines

2D gel Western blot analysis and Northern blot analysis were performed on 59 human tumor-derived cell lines and strains. In addition to abundant L-plastin expression by all the cell lines derived from leukemias, 66% of the carcinoma cell lines expressed L-plastin at varying levels (Lin et al., 1993b). This observation was confirmed by Park et al., who even found that >90% of transformed human cell lines exhibited L-plastin gene activity, although at widely differing degrees (Park et al., 1994). As mentioned before, in the human prostate carcinoma cells LNCaP, L-plastin expression was up-regulated following stimulation with estradiol or the androgen dihydrotestosterone (Zheng et al., 1997). Moreover, the same group showed that L-plastin expression in prostate cancer cells might play a role in cell invasion and metastasis. Indeed, L-plastin silencing in the prostate carcinoma cell line PC-3 and its isogenic metastatic variant PC-3M led to an almost complete suppression of *in vitro* invasion and motility as well as of *in vivo* invasion capacities of both cell lines (Zheng et al., 1999). This study established for the first time a link between L-plastin expression and the motility and invasiveness properties of cancer cells. Foran and coworkers compared the colorectal cancer cell line SW480 and its isogenic metastatic counterpart SW620, which was established from a lymph node metastasis of the same colon adenocarcinoma patient. They found an up-regulation of L-plastin in the metastatic cells as compared to the primary tumor cells. Remarkably, when they overexpressed L-plastin in SW480 cells, they obtained higher proliferation rates, down-regulation of E-cadherin exposure at the plasma membrane and increased invasion, indicating a mechanistic role for L-plastin in colorectal cancer metastasis (Foran et al., 2006).

Knowing that L-plastin Ser5 phosphorylation is important for the functional role of L-plastin in the immune response, research efforts were undertaken to analyze potential effects of L-plastin Ser5 phosphorylation on cancer progression. The first results obtained in this context provided evidence that, in HEK cells, L-plastin-dependent collagen invasion required the presence of residue Ser5 and active PKA (Janji et al., 2006). In a study of Klemke and coworkers, a link between L-plastin phosphorylation and melanoma cell invasive properties was established (Klemke et al., 2007). At first, the authors could not detect a correlation between L-plastin expression and tumor stages or penetration depths of malignant melanomas. This negative result led to the hypothesis that additional factors, *e.g.*, L-plastin phosphorylation, might play a role. Indeed, whereas L-plastin expression was sufficient to enhance haptotactic melanoma cell migration on matrigel, collagen I or fibronectin, L-plastin phosphorylation was necessary in order to increase melanoma cell invasion through a matrigel matrix toward conditioned medium or fetal calf serum. Notably chemotactic migration of melanoma cells was not influenced neither by L-plastin expression nor phosphorylation (Klemke et al., 2007).

In cultured cholangiocarcinoma cells, it was shown that the addition of a matrigel matrix led to up-regulated expression of L-plastin concomitant with increased cell invasion capacities (Chaijan et al., 2014). In line, silencing of L-plastin led to decreased cholangiocarcinoma cell invasion events. Interestingly, L-plastin was localized in the nucleus and cytoplasm of those cholangiocarcinoma cells that had a mesenchymal morphology and invaded the basement membrane, whereas L-plastin was absent from granular cells (Chaijan et al., 2014). Mimicking tumor microenvironment by culturing a cholangiocarcinoma cell line in 3D instead of monolayer culture led to increased L-plastin expression also in the secretome. Furthermore, L-plastin expression was only identified in the secretome of HuCCA-1 cholangiocarcinoma cell line and not in the secretomes of HepG2, SK-HEP-1 and HCC-S102 hepatocellular cell lines, leading to the hypothesis that L-plastin might be a promising biomarker for differentially diagnosing cholangiocarcinoma *versus* hepatocellular carcinoma (Tit-Oon et al., 2014). In an effort to analyze the protein-modulatory ability of curcumin treatment of cholangiocarcinoma induced by *Opisthorchis viverrini* infection, protein expression profiles of liver tissues of cholangiocarcinoma hamsters were compared between conditions with and without curcumin dietary supplementation. Interestingly, L-plastin was found to be highly up-regulated in liver tissues of cholangiocarcinoma hamsters and curcumin treatment led to significantly reduced L-plastin levels in these tissues (Khoontawad et al., 2018). This study also pinpointed L-plastin as a potential therapeutic target for treating cholangiocarcinoma.

As previously shown in leukocytes, a hint for a potential role of L-plastin in regulating integrin activation was also obtained in cancer cells. By performing 2D-fluorescence difference in-gel electrophoresis (2D-DIGE) followed by mass spectrometry to analyze L-plastin effects on integrins in cancer cells, Langereis et al. compared an adherent (L1210-A) and a suspension (L1210-S) variant of the mouse acute lymphocytic leukemia cell line L1210, with the variants functionally differing in integrin activation status. They identified, among others, L-plastin as a protein whose expression was increased 1.6-fold in L1210A cells as compared to L1210S cells, suggesting a potential role of L-plastin in integrin regulation, either by increasing integrin affinity or integrin avidity (Langereis et al., 2009). More recently, L-plastin was identified as a potential target in chronic lymphocytic leukemia (CLL). L-plastin was found to be highly expressed in CLL cells and L-plastin knockdown impaired *in vitro* chemotaxis and *in vivo* bone marrow migration in a xenotransplantation model of leukemia. Notably, inhibition of Bruton's kinase or PI3K blocked B-cell receptor driven Ser5 phosphorylation, and hence, activation of L-plastin (Dubovsky et al., 2013). In a study assessing the intra-tumoral genetic heterogeneity of CLL, mutated L-plastin was uncovered as a novel CLL driver possibly by impacting the DNA damage response (Wang et al., 2017).

An up-regulation of L-plastin expression and Ser5 phosphorylation was observed both in the multiple myeloma cell line MOLP8 grown under hypoxic conditions as compared to normoxia, and in MOLP8/R, which is a drug-resistant cell line that has been derived from doxorubicin-treated

MOLP8 cells. The overexpression was shown to be controlled by HIF1α and the authors proposed that, in addition to HIF1α, L-plastin might be a potential target to overcome drug resistance in multiple myeloma. This hypothesis was supported by the high levels of Ser5 phosphorylated L-plastin that were observed in multidrug-resistant multiple myeloma patients as compared to the control MOLP8 cells (Bosseler et al., 2018).

An interesting approach to analyze L-plastin functions in cancer cells was taken by Delanote et al. as they raised nanobodies against the L-plastin protein (Delanote et al., 2010). In this study, nanobodies were raised against the EF hand domains and against ABD1 + 2 of L-plastin and their effect on L-plastin functional activity was evaluated. Both EF hand- and ABD-directed nanobodies impaired invasion and motility of L-plastin expressing prostate cancer PC-3 cells, whereas only the ABD-specific nanobody was able to strongly reduce actin-bundling activity of L-plastin and concomitant filopodia formation (Delanote et al., 2010). Later, the same approach allowed to also highlight a functional role of L-plastin in the formation of cancer cell invadopodia. Whereas fascin, another actin-bundling protein, was mostly responsible for the rigidity of filopodia and invadopodia by forming tightly packed, rigid and straight actin bundles of about 14 filaments, L-plastin formed bundles that were thinner and more flexible. Therefore, L-plastin was suggested to be involved in the elongation step of these protrusive structures (Van Audenhove et al., 2016) (Fig. 2).

In MCF7 breast cancer cells, the expression and Ser5 phosphorylation of L-plastin conferred resistance to TNF-α treatment (Janji et al., 2010). A function for L-plastin has also been demonstrated in chemokine-dependent migration of breast cancer cells. Given that increased expression of GLI1 transcription factor of Hedgehog signaling is associated with metastasis and low survival rates in breast cancer patients, it was shown that GLI1 enhances metastasis of breast cancer cells concomitant with up-regulation of the expression of the two CXCR4 and CXCR7 chemokine receptors for CXCL12 as well as of L-plastin. Since the up-regulation of the three proteins was necessary for CXCL12-induced ERK/MAPK pathway activation and migration, these findings provide a mechanism linking GLI1 expression and the CXCR4/CXCR7 axis to metastasis formation and poor prognosis in breast cancer patients (Inaguma et al., 2015). Moreover, EGF-triggered activation of the ERK/MAPK pathway led to enhanced L-plastin Ser5 phosphorylation, formation of migratory structures and L-plastin enrichment in these structures in breast cancer cells (Fig. 3). In contrast, inhibition of RSK1 and RSK2, which are two kinases that are located downstream of

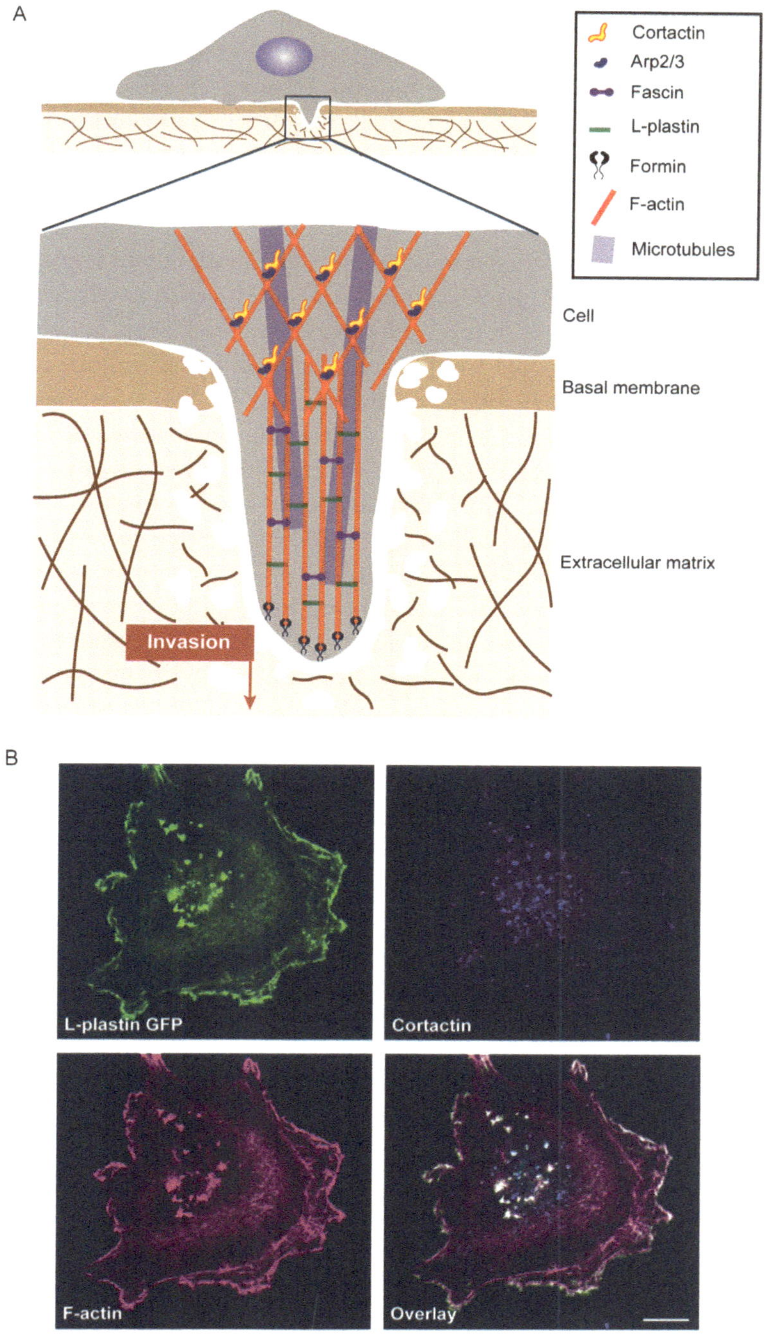

Fig. 2 See legend on next page.

the ERK/MAPK pathway, led to reduced formation of migratory structures concomitant with reduced L-plastin Ser5 phosphorylation. This study therefore suggested that Ser5 phosphorylated L-plastin is a molecular marker for deregulated ERK/MAPK pathway signaling in invasive breast carcinomas (Lommel et al., 2016).

MicroRNA-96 had previously been identified as a tumor suppressor by down-regulating ABCE1 expression at the post-transcriptional level (Pillar et al., 2018). In a follow-up study, the same group identified L-plastin as a further target of microRNA-96 and showed that L-plastin down-regulation reduced breast cancer progression. Notably, combined down-regulation of both target proteins ABCE1 and L-plastin had additive effects on reduction of *in vitro* breast cancer cell migration, invasion and proliferation as well as on inhibition of *in vivo* tumor growth and metastasis (Pillar et al., 2019).

In a highly interesting study of Riplinger and coworkers, the effect of L-plastin expression and Ser5 phosphorylation was investigated in PC3M metastatic prostate cancer cells and in MV3 melanoma cells (Riplinger et al., 2014). They found that L-plastin down-regulation in PC3M cells led to reduced cell migration, to a reduced tumor volume when the cells were injected subcutaneously into mice and to reduced metastasis formation when injected intracardially into mice. To assess the importance of L-plastin Ser5 phosphorylation, they neoexpressed L-plastin wild type and S5A phosphorylation variants in melanoma MV3 cells. The neoexpression of wild

Fig. 2—Cont'd L-plastin cooperates with other actin-binding proteins to form invadopodia. (A) Schematic representation of a mature invadopodium. These actin-based protrusions develop at the ventral surface of metastatic cancer cells, enabling matrix degradation by localized proteolysis. Cortactin associates with Arp2/3 complex for actin nucleation and formation of branched actin filaments at invadopodia initiation sites. Furthermore, cortactin regulates membrane trafficking for the recruitment of proteinases, such as matrix metalloproteases. Regarding the function of actin-bundling proteins, fascin guarantees rigidity and strength for protrusive force and structural stability, whereas L-plastin accounts for the flexibility needed during elongation. By driving the formation of linear (unbranched) actin filaments, formins play a role in invadopodia growth. Microtubules are thought to penetrate invadopodia at their mature state. (B) Confocal microscopy of MDA-MB-231 breast cancer cells showing the localization of L-plastin in invadopodia. MDA-MB-231 cells were transduced with L-plastin_GFP (green) and plated onto gelatin-coated coverslips. After 24h, cells were stained using anti-cortactin antibody (blue) and Alexa-633 phalloidin (magenta) to visualize F-actin. Single confocal slices of the ventral surface of a representative cell are shown. Invadopodia are characterized as cortactin and F-actin-containing punctae. Areas of cortactin, F-actin and L-plastin colocalization are seen in the overlay as white dot-like structures. Scale bar: 10 μm.

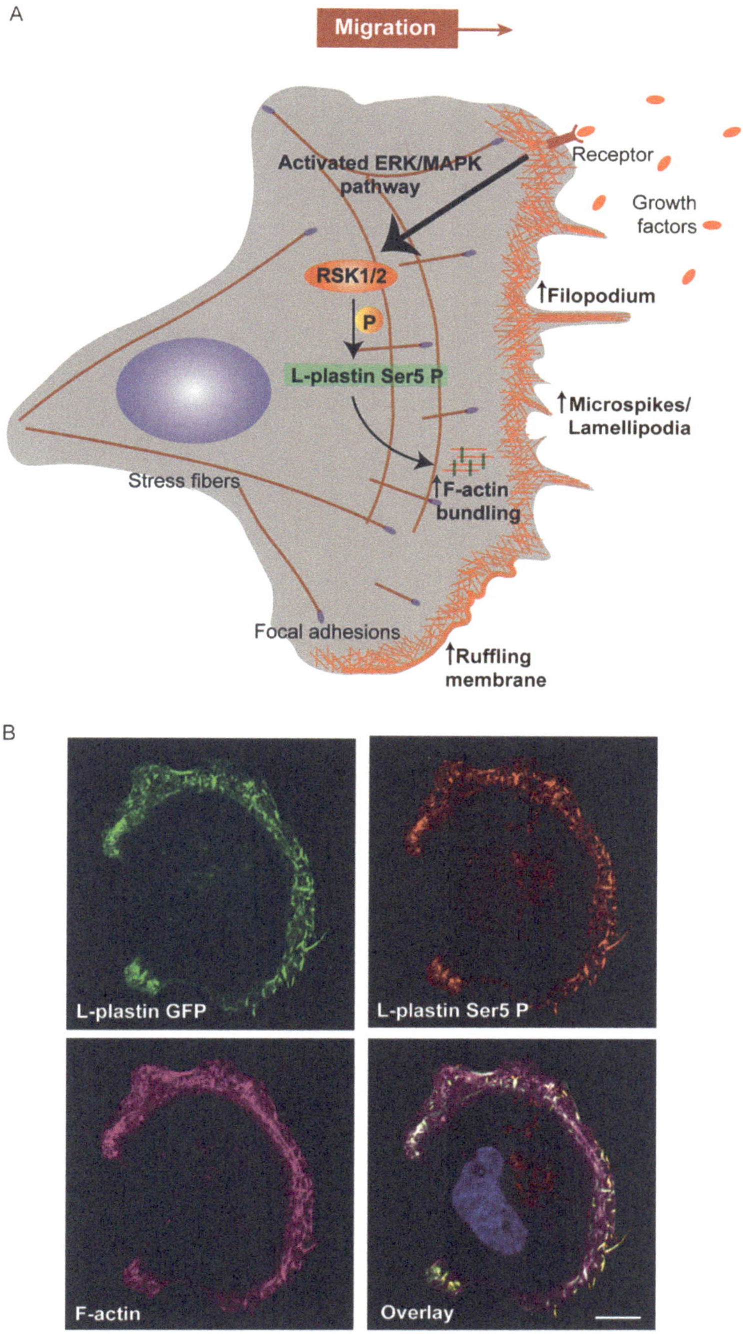

Fig. 3 See legend on next page.

type phosphorylatable L-plastin enhanced MV3 cell migration and tumor metastasis formation when injected intracardially into mice, whereas the non-phosphorylatable S5A variant had no effect. Thus, L-plastin expression promotes metastasis formation in a xenograft mouse model and, to be able to perform this metastasis-promoting activity, L-plastin has to be phosphorylated on residue Ser5 (Riplinger et al., 2014).

A study carried out in order to determine new factors able to distinguish multipotent somatic stem and progenitor cells (SSPCs) from differentiated cells, identified L-plastin as a novel discrimination factor. Indeed, shRNA-mediated silencing of L-plastin led to a striking decrease of the number and size of intestinal organoids (Schwalie et al., 2017).

Given that the L-plastin gene is down-regulated upon low-dose irradiation of human lymphoblastoid cells AHH-1 (Wang et al., 2006), another study has reported more recently that the production and exosomal secretion of L-plastin by cancer cells into the medium (conditioned medium) is also reduced in response to irradiation. As a result, indicator cells that were incubated with the conditioned medium of irradiated cells exhibited reduced cell proliferation and clonogenic activity, highlighting radiation-induced bystander effects. The authors identified L-plastin to be responsible for this radiation-triggered reduction of mitogenic/clonogenic activity. As an indicator of its activation state, L-plastin was shown to be Ser5 phosphorylated in the exosomes. Thus, this study suggests that tumor cell-released active L-plastin can be used to influence neighbor cells in the tumor microenvironment and, possibly, promote cancer progression. Irradiation of tumors could

Fig. 3—Cont'd Ser5 phosphorylated L-plastin is highly enriched in structures playing a role in cell migration. (A) Schematic model depicting the proposed role of L-plastin in response to different stimulations in human breast cancer cells. Growth factor stimulation induces rapid activation of the ERK/MAPK pathway that promotes the activation of the downstream ribosomal S6 kinases RSK1 and RSK2, which become able to directly phosphorylate L-plastin at the Ser5 site. This Ser5 phosphorylation enhances L-plastin-mediated actin-bundling at the leading edge of the cell. Notably, Ser5 phosphorylated L-plastin is highly enriched in structures that are involved in the migratory process, such as ruffling membranes, microspikes and filopodia-like structures. (B) Confocal microscopy of MDA-MB-231 breast cancer cells showing the localization of Ser5 phosphorylated L-plastin in migratory structures of a migrating cell. MDA-MB-231 cells were transduced with L-plastin_GFP (green) and plated onto fibronectin-coated coverslips. At ~90% confluence, a wound was scratched with a micropipette tip and cells were allowed to migrate for 1 h. Cells were stained using anti-L-plastin Ser5 P antibody (red), Alexa-633 phalloidin (magenta) to visualize F-actin and DAPI (blue) to visualize nuclei. Single confocal slices of the ventral surface of a representative cell are shown. Scale bar: 10 μm.

therefore target both the cancer cells and the tumor niche through inhibition of L-plastin production, explaining the efficiency of radiotherapy in tumor treatment (Freudenmann et al., 2019). Notably, γ-irradiation of L-plastin expressing PC3 prostate cancer cells also led to L-plastin thiol oxidation. Given that L-plastin thiol oxidation on Cys101 strongly impairs actin-based cellular processes such as cell migration, invasion as well as extracellular matrix degradation, L-plastin oxidation may contribute to the benefits of radiation therapy (Balta et al., 2019).

Furthermore, thiol oxidation of L-plastin residue Cys101 could also be induced by reactive oxygen species (ROS) treatment and regulate L-plastin activity by reducing its actin-bundling capacity. This process is reversible and reduction of L-plastin is mediated by the Thioredoxin1 (TRX1) system. Importantly, high levels of L-plastin oxidation occurred at spatially confined cellular protrusions, such as filopodia and invadopodia, and, as mentioned above, L-plastin oxidation led to impaired cell spreading and filopodia formation, migration, invasion and extracellular matrix degradation. Altogether these data led to the suggestion that L-plastin oxidation, which is spatially restricted to cellular protrusions, could serve as a molecular switch that translates the redox environment into altered peripheral actin dynamics (Balta et al., 2019).

7.2.2 Studies conducted using cancer tissues

Regarding L-plastin expression in cancer tissues, high levels of L-plastin expression have been reported for glandular epithelial cells in prostate carcinoma tissues as compared to normal tissues, where L-plastin staining was only observed in the fibromuscular stroma and not in the glandular epithelium (Zheng et al., 1997). Analyzing colorectal carcinoma tissue microarrays, Otsuka et al. provided evidence for a strong correlation between L-plastin expression with the progression of cancer staging. Indeed, L-plastin was preferentially expressed in stage III and stage IV tissues over stage I and stage II tissues, suggesting L-plastin expression level to be a candidate metastasis-related biomarker in colorectal cancer (Otsuka et al., 2001). According to a report of Ning et al., two germline variants of T- and L-plastin could be used to predict tumor recurrence as well as time to disease-relapse in stage II/III colon cancer patients. Their findings support a role for T- and L-plastin as prognostic factors based on gender and tumor stage and even suggested that selected patients might benefit from plastin-targeted therapy (Ning et al., 2014).

In splenic marginal zone lymphomas, L-plastin expression was significantly up-regulated as compared to other lymphoma types and high L-plastin protein levels could be found in the nuclei of these lymphoma cells (Trøen et al., 2004). The importance of L-plastin in tumor formation is highlighted by its up-regulation during lung tumorigenesis induced by carcinogen treatment of A/J mice, as quantified by mass spectrometry and Western blot analysis. Notably, L-plastin expression was reduced to a normal level when chemopreventive agents were administered in addition to the carcinogen treatment (Kassie et al., 2007). Therefore, the authors described L-plastin as a tumor-associated signature protein. Analysis of tissue microarrays of 251 bladder carcinomas highlighted a correlation of L-plastin, PTEN and Cx-26 expression with clinicopathologic features of bladder cancer such as tumor grade, stage and growth pattern. However none of these genes could be established as an independent predictor of patient survival (Harris et al., 2008). Identification of autoantibodies against L-plastin in the plasma of 21 out of 25 patients presenting with non-Hodgkin's lymphoma and its absence in 4 patients with Hodgkin's lymphoma suggests the presence of L-plastin autoantibody in the plasma of lymphoma patients as a new diagnostic biomarker for non-Hodgkin's lymphoma (Ueda et al., 2008). In an effort to find biomarkers specific for different subtypes of acute myeloid leukemia (AML), Luczak et al. focused on the AML subtypes M1 and M2, which are very difficult to differentiate. Using 2D electrophoresis followed by mass spectrometry to compare protein profiles of samples collected from AML-M1/M2 patients and from healthy volunteers, they found 25 proteins with differential profiles, among them L-plastin that accumulated only in the M2 group (Luczak et al., 2012).

Highly interesting results were provided by a proteomic study applying tissue matrix-assisted laser desorption/ionization – imaging mass spectrometry (MALDI-IMS). In this study, high expression levels of L-plastin were detected in the interface zones of ovarian cancers, which correspond to the region between tumors and normal tissues, and L-plastin was proposed as an interface marker in ovarian cancers (Kang et al., 2010).

In breast cancer, L-plastin was detected in a subset of ductal carcinomas devoid of hormone receptors. In normal breast tissues, only a staining of myoepithelial cells could be seen in 50% of the cases, whereas L-plastin could not be detected in the epithelial cells of the mammary gland. Notably, a correlation between L-plastin expression and tumor size, histological grade and lymph node status could not be established in the 29 breast tumor samples that were examined, although this absence of correlation might be due to the low number of tumor samples (Lapillonne et al., 2000). In another, proteome profiling-based study, L-plastin was found to be strongly overexpressed in breast tumors as compared to corresponding peripheral breast tissue samples or healthy mammary gland tissue samples. However, the comparison of this overexpression between non-metastatic and metastatic tumor tissues did not give conclusive results (Alldridge et al., 2008). When L-plastin expression level was compared between triple-negative breast carcinomas, devoid of HER2/neu receptors and the two hormone receptors for estrogen and progesterone ($ER^-/PR^-/HER2^-$), and HER2-positive/hormone receptor-negative carcinomas ($ER^-/PR^-/HER2^+$), a 1.43-fold up-regulation of L-plastin expression in triple-negative carcinomas was observed. Based on this result, the authors suggested that L-plastin might represent a potential candidate target for new therapeutic approaches for triple-negative breast cancer patients (Schulz et al., 2009). Additionally, using MERAV and cBioPortal tools, Pillar and coworkers showed that L-plastin expression levels are higher in breast tumors as compared to normal breast tissues and that they are correlated with tumor grade and lymph node micrometastases (Pillar et al., 2019).

Moreover, protein identification by mass spectrometry following 2D gel electrophoresis using renal cell carcinoma tumor tissues highlighted an up-regulation of L-plastin in all types of renal cell carcinoma, suggesting L-plastin to be, among others, a novel renal cell carcinoma biomarker (Kim et al., 2010). In a follow-up study, the same group analyzed the presence of three candidate markers for kidney tumors NNMT, L-plastin, NM23A in the plasma taken from patients with kidney cancer or from healthy donors and donors with benign tumors. As a result, they proposed the use of a 3-marker assay analyzing the plasma levels of NNMT/L-plastin/NM23A and allowing the early detection of renal cell carcinoma (Su Kim et al., 2013). In an effort to identify biomarkers that could be screened for in a non-invasive and simple way, several other studies were undertaken to determine potential biomarkers in the feces or the blood. Similar to renal cell carcinoma patients, L-plastin levels were also strongly increased in the plasma of colorectal cancer patients as compared to healthy volunteers. Importantly, this elevated L-plastin concentration in the colorectal cancer plasma was correlated to tumor differentiation, tumor size and lymph node metastasis as well as to the expression level of L-plastin in the corresponding colorectal tumor tissues. However, no correlation could be established between L-plastin expression and the extent of tumor invasion or distant metastasis (Li and Zhao, 2011). In another study, L-plastin was only present in the feces of colorectal cancer patients and absent in the feces of healthy volunteers. Notably, the report did not specify if the presence of L-plastin in the feces was due to its release by colon cancer cells or due to its inherent presence in hematopoietic cells of the blood that was detected in the colorectal cancer patient feces (Ang and Nice, 2010).

Up-regulation of L-plastin expression together with coronin-1C has been found in patients with metastatic laryngeal squamous cancer tissues *versus* paired adjacent normal tissues *via* proteomic profiling (Klobučar et al., 2016). Using qRT-PCR, Western blot analysis and immunohistochemistry, up-regulation of L-plastin expression was also detected in oral squamous cell carcinoma cell lines and tissues as compared with normal counterparts. Silencing of L-plastin induced reduced cell proliferation, invasion and migration and, notably, L-plastin expression in tissue samples correlated with the size of the primary tumor as well as regional lymph node metastasis (Koide et al., 2017). Altogether, these data suggest L-plastin as a useful biomarker for oral squamous cell carcinoma progression. Furthermore, in advanced head and neck squamous cell carcinoma an inverse correlation between miR-375 expression and the expression levels of the two invasion-associated proteins vimentin and L-plastin was observed. It was suggested that Runt-related transcription factor 1, which is a direct target candidate of miR-375, could regulate L-plastin and vimentin expression. The mechanism put forward by the authors suggested that miR-375 expression would lead to Runt-related transcription factor 1 inhibition and concomitant low L-plastin and vimentin expression levels, leading to decreased cell invasion (Jimenez et al., 2017). As described before, L-plastin expression is regulated through the key transcription factor AP-4 in prostate cancer cells. In line, a positive correlation could be shown between L-plastin expression and AP-4 expression in prostate cancer tissues. Moreover, L-plastin expression was found to be positively correlated with higher cancer stages as well as with lymph node metastasis (Chen et al., 2017).

However, as mentioned previously for the results obtained by Lapillonne, such a direct correlation between L-plastin expression and cancer is not always given (Lapillonne et al., 2000). Indeed, Klemke and coworkers did not see a link between L-plastin expression and the penetration depths or tumor stages of melanoma samples (Klemke et al., 2007). Similarly, although L-plastin expression seemed to be increased in the inflammatory environment of melanoma as compared to the benign nevi environment, no significant difference in L-plastin expression could be detected between tumor cells of melanoma samples and the cells of benign nevi samples (Strickler et al., 2014). As previously mentioned, Foran and coworkers found an up-regulation of L-plastin in metastatic cells as compared to the primary colon tumor cells (Foran et al., 2006). In contrast to these findings, L-plastin protein level was found to be more than 20-fold down-regulated in MDA-MB-231BR cells, which is a brain metastatic cell line derived from MDA-MB-231 cells, as compared to the parental MDA-MB-231 breast cancer cells (Dun et al., 2015).

Finally, another study has found contradicting results regarding differential expression levels of L-plastin in normal *versus* tumor tissues. This study showed a reduced L-plastin expression level in the stroma of nasopharyngeal carcinoma as compared to normal nasopharyngeal epithelial tissue. It is however noteworthy that L-plastin was highly expressed in lymphocytes of the normal stroma, which might account for this contradicting result (Li et al., 2009).

These controversies suggest that, in addition to L-plastin expression, L-plastin activity regulation by PTM such as Ser5 phosphorylation or reversible thiol oxidation of Cys101 should be taken into consideration for linking L-plastin to tumor development.

7.3 L-plastin phosphorylation: kinases and signaling pathways

L-plastin is the only plastin paralog to be phosphorylated. Whereas it is certain that residue Ser5 can be phosphorylated, controversies still exist regarding the phosphorylation of residue Ser7. In leukocytes, phosphorylation of L-plastin appeared to occur both on Ser5 and Ser7 and in fibroblasts, transfected L-plastin required the simultaneous presence of both residues for allowing phosphorylation to occur, suggesting the possibility of a phosphorylation zipper mechanism (Lin et al., 1998). Studying phosphorylation of transfected L-plastin phosphomutants in HeLa cells, Jones and coworkers found that the presence of Ser5 is a prerequisite for phosphorylation of L-plastin and might be necessary also for the phosphorylation of residue Ser7 (Jones et al., 1998). However, although T-plastin has an equivalent Ser7 residue, endogenous T-plastin in fibroblasts and transfected T-plastin in leukocytes is not phosphorylated (Lin et al., 1998).

Although L-plastin phosphorylation is in general triggered upon PMA stimulation, which is commonly used as a PKC activator, controversies also still exist regarding the identity of the kinase(s) responsible for L-plastin phosphorylation in leukocytes. Since PMA potently activated L-plastin phosphorylation in lymphokine-activated killer cells, a PKC-regulated kinase was suggested to be responsible for the phosphorylation (Frederick et al., 1996). In line, Lin et al. suggested that PKC and not CK2 is a candidate kinase for L-plastin phosphorylation (Lin et al., 1998). However, other studies reported the phosphorylation to be independent of PKC or PKA (Matsushima et al., 1987; Shiroo and Matsushima, 1990) and rather mediated by a serine kinase distinct from both PKC and PKA (Shiroo and Matsushima, 1989). In order to identify the kinases in charge of L-plastin Ser5 phosphorylation, Wang and Brown used kinase-specific activators and inhibitors. They showed that PKA was involved in immune complex-dependent integrin activation and L-plastin Ser5 phosphorylation with PI 3-kinase acting upstream of PKA. On the other hand, different kinases appeared to be responsible for this phosphorylation event in response to fMLP or PMA (Wang and Brown, 1999). Lipopolysaccharide stimulation of macrophages also led to Ser5 phosphorylation of L-plastin. This phosphorylation was sensitive to the PKA inhibitor H89 and insensitive to the CK2 inhibitor heparin. However, the PKA activator cAMP was not necessary for stimulation of Ser5 phosphorylation, leading to the conclusion that an as yet unidentified H89-sensitive protein kinase may be in charge of L-plastin Ser5 phosphorylation (Hagi et al., 2006). PKCβII-mediated Ser5 phosphorylation was reported in primary eosinophils upon GM-CSF stimulation and linked eosinophil chemotaxis and degranulation to integrin αMβ2-mediated signaling (Pazdrak et al., 2011). In a more recent study, Freeley and coworkers showed that L-plastin is involved in chemokine SDF-1α-triggered T cell polarization and migration. Indeed L-plastin was co-localized with F-actin at the leading edge upon SDF-1α stimulation and was required for polarization, lamellipodia formation and chemotaxis. Notably, L-plastin was phosphorylated on its residue Ser5 and this phosphorylation was dependent of PKCζ, which is a well-known regulator of cell polarity (Freeley et al., 2012).

An explanation for all these controversies regarding L-plastin phosphorylating kinases could be given by the hypothesis that the identity of the kinase responsible for L-plastin Ser5 phosphorylation depends on the cell type or even on the stimulus. In line, performing two-dimensional isoelectric focusing polyacrylamide gel electrophoresis and matrix-assisted laser desorption/ionization - time-of-flight mass spectrometry, Paclet et al. published in 2004 data demonstrating that two fMLP receptor subtypes exist in neutrophils and that fMLP stimulation of both subtypes leads to L-plastin phosphorylation. Notably, they showed that low-affinity fMLP receptors induced L-plastin phosphorylation through a pathway involving PI 3-kinase, PLD and PKC isotypes, whereas high-affinity fMLP receptors employed alternative, as yet unidentified pathways to trigger the same phosphorylation event. Thus, receptor subtypes for the same ligand in the same cells would lead to a similar endpoint through different signal transduction pathways (Paclet et al., 2004).

In a cancer context, PKA activation increased L-plastin Ser5 phosphorylation in HEK cells (Janji et al., 2006). Silencing and inhibition studies demonstrated that PKCδ was involved in PMA-induced actin cytoskeleton reorganization and L-plastin Ser5 phosphorylation in breast cancer MCF7 cells (Al Tanoury et al., 2010). More recently, a study combining whole-genome microarray analysis comparing non-invasive and invasive breast cancer cell lines, mathematical modeling of Western blot data and *in vitro* kinase assays, showed that kinases RSK1 and RSK2, located downstream of the ERK/MAPK pathway, are able to phosphorylate L-plastin on its residue Ser5 both *in vitro* and in breast cancer cells. RSK inhibition led to reduced formation of migratory structures concomitant with reduced L-plastin Ser5 phosphorylation (Lommel et al., 2016).

8. Conclusive remarks

In addition to L-plastin expression, L-plastin PTM has been shown to be important for the regulation of the immune response as well as for cancer progression. Thiol oxidation of L-plastin residue Cys101 might be a promising approach in cancer therapy, since this PTM inhibits the aggressiveness of cancer cells. Ser5 phosphorylated L-plastin is a candidate molecular marker for metastatic cancer cells and blocking L-plastin Ser5 phosphorylation in tumor cells is also a promising therapeutic strategy. So far, the ERK/MAPK pathway has been determined as (one of) the signal transduction pathway(s) leading to L-plastin Ser5 phosphorylation. However, cancer progression is normally the result of several deregulated and commonly interconnected signaling pathways. Therefore, further research is needed to thoroughly dissect the entire signal transduction network upstream of L-plastin Ser5 phosphorylation in order to pave the way for a potential clinical application of Ser5 phosphorylated L-plastin.

Acknowledgment

This work was supported by a grant from the Fondation Cancer, Luxembourg.

References

Adams, A.E., Shen, W., Lin, C.S., Leavitt, J., Matsudaira, P., 1995. Isoform-specific complementation of the yeast sac6 null mutation by human fimbrin. Mol. Cell. Biol. 15, 69–75.

Akbulut, H., Zhang, L., Tang, Y., Deisseroth, A., 2003. Cytotoxic effect of replication-competent adenoviral vectors carrying L-plastin promoter regulated E1A and cytosine deaminase genes in cancers of the breast, ovary and colon. Cancer Gene Ther. 10, 388–395.

Akbulut, H., Tang, Y., Maynard, J., Zhang, L., Pizzorno, G., Deisseroth, A., 2004. Vector targeting makes 5-fluorouracil chemotherapy less toxic and more effective in animal models of epithelial neoplasms. Clin. Cancer Res. 10, 7738–7746.

Al Tanoury, Z., Schaffner-Reckinger, E., Halavatyi, A., Hoffmann, C., Moes, M., Hadzic, E., Catillon, M., Yatskou, M., Friederich, E., 2010. Quantitative kinetic study of the actin-bundling protein L-plastin and of its impact on actin turn-over. PLoS One 5, e9210.

Alldridge, L., Metodieva, G., Greenwood, C., Al-Janabi, K., Thwaites, L., Sauven, P., Metodiev, M., 2008. Proteome profiling of breast tumors by gel electrophoresis and nanoscale electrospray ionization mass spectrometry. J. Proteome Res. 7, 1458–1469.

Anderson, N.L., Gemmell, M.A., Murao, S.I., Huberman, E., Coussens, P.M., 1985. Specific protein phosphorylation in human promyelocytic hl-60 leukemia cells susceptible or resistant to induction of cell differentiation by phorbol-12-my ristate-13-acetate. Cancer Res. 45, 4955–4962.

Ang, C.S., Nice, E.C., 2010. Targeted in-gel MRM: a hypothesis driven approach for colorectal cancer biomarker discovery in human feces. J. Proteome Res. 9, 4346–4355.

Arpin, M., Friederich, E., Algrain, M., Vernel, F., Louvard, D., 1994. Functional differences between L- and T-plastin isoforms. J. Cell Biol. 127, 1995–2008.

Babb, S.G., Matsudaira, P., Sato, M., Correia, I., Lim, S.S., 1997. Fimbrin in podosomes of monocyte-derived osteoclasts. Cell Motil. Cytoskeleton 37, 308–325.

Balta, E., Hardt, R., Liang, J., Kirchgessner, H., Orlik, C., Jahraus, B., Hillmer, S., Meuer, S., Hübner, K., Wabnitz, G.H., et al., 2019. Spatial oxidation of L-plastin downmodulates actin-based functions of tumor cells. Nat. Commun. 10, 4073.

Baumgartner, E.A., Compton, Z.J., Evans, S., Topczewski, J., LeClair, E.E., 2019. Identification of regulatory elements recapitulating early expression of L-plastin in the zebrafish enveloping layer and embryonic periderm. Gene Expr. Patterns 32, 53–66.

Berman, J.N., Kanki, J.P., Look, A.T., 2005. Zebrafish as a model for myelopoiesis during embryogenesis. Exp. Hematol. 33, 997–1006.

Bosseler, M., Marani, V., Broukou, A., Lequeux, A., Kaoma, T., Schlesser, V., François, J.H., Palissot, V., Berchem, G.J., Aouali, N., et al., 2018. Inhibition of HIF1α-dependent upregulation of phospho-L-plastin resensitizes multiple myeloma cells to frontline therapy. Int. J. Mol. Sci. 19, 1551.

Bostanci, N., Heywood, W., Mills, K., Parkar, M., Nibali, L., Donos, N., 2010. Application of label-free absolute quantitative proteomics in human gingival crevicular fluid by LC/MSE (Gingival Exudatome). J. Proteome Res. 9, 2191–2199.

Bretscher, A., 1981. Fimbrin is a cytoskeletal protein that crosslinks F-actin in vitro. Proc. Natl. Acad. Sci. U. S. A. 78, 6849–6853.

Bretscher, A., Weber, K., 1980. Fimbrin, a new microfilament-associated protein present in microvilli and other cell surface structures. J. Cell Biol. 86, 335–340.

Caseiro, A., Ferreira, R., Padrão, A., Quintaneiro, C., Pereira, A., Marinheiro, R., Vitorino, R., Amado, F., 2013. Salivary proteome and peptidome profiling in type 1 diabetes mellitus using a quantitative approach. J. Proteome Res. 12, 1700–1709.

Chafel, M.M., Shen, W., Matsudaira, P., 1995. Sequential expression and differential localization of I-, L-, and T-fimbrin during differentiation of the mouse intestine and yolk sac. Dev. Dyn. 203, 141–151.
Chaijan, S., Roytrakul, S., Mutirangura, A., Leelawat, K., 2014. Matrigel induces L-plastin expression and promotes L-plastin-dependent invasion in human cholangiocarcinoma cells. Oncol. Lett. 8, 993–1000.
Chaplin, D.D., Wedner, H.J., Parker, C.W., 1980. Protein phosphorylation in human peripheral blood lymphocytes: Mitogen-induced increases in protein phosphorylation in intact lymphocytes. J. Immunol. 124, 2390–2398.
Chellaiah, M.A., Ma, T., Majumdar, S., 2018a. L-plastin phosphorylation regulates the early phase of sealing ring formation by actin bundling process in mouse osteoclasts. Exp. Cell Res. 372, 73–82.
Chellaiah, M.A., Majumdar, S., Aljohani, H., 2018b. Peptidomimetic inhibitors of L-plastin reduce the resorptive activity of osteoclast but not the bone forming activity of osteoblasts in vitro. PLoS One 13, e0204209.
Chellaiah, M.A., Moorer, M.C., Majumdar, S., Aljohani, H., Morley, S.C., Yingling, V., Stains, J.P., 2020. L-Plastin deficiency produces increased trabecular bone due to attenuation of sealing ring formation and osteoclast dysfunction. Bone Res. 8, 3.
Chen, H., Mocsai, A., Zhang, H., Ding, R.X., Morisaki, J.H., White, M., Rothfork, J.M., Heiser, P., Colucci-Guyon, E., Lowell, C.A., et al., 2003. Role for plastin in host defense distinguishes integrin signaling from cell adhesion and spreading. Immunity 19, 95–104.
Chen, C., Cai, Q., He, W., Li, Z., Zhou, F., Liu, Z., Zhong, G., Chen, X., Zhao, Y., Dong, W., et al., 2016. An NKX3.1 binding site polymorphism in the l-plastin promoter leads to differential gene expression in human prostate cancer. Int. J. Cancer 138, 74–86.
Chen, C., Cai, Q., He, W., Lam, T.B., Lin, J., Zhao, Y., Chen, X., Gu, P., Huang, H., Xue, M., et al., 2017. AP4 modulated by the PI3K/AKT pathway promotes prostate cancer proliferation and metastasis of prostate cancer via upregulating L-plastin. Cell Death Dis. 8, e3060.
Chung, I., Schwartz, P.E., Crystal, R.G., Pizzorno, G., Leavitt, J., Deisseroth, A.B., 1999. Use of L-plastin promoter to develop an adenoviral system that confers transgene expression in ovarian cancer cells but not in normal mesothelial cells. Cancer Gene Ther. 6, 99–106.
Ciborowski, P., Kadiu, I., Rozek, W., Smith, L., Bernhardt, K., Fladseth, M., Ricardo-Dukelow, M., Gendelman, H.E., 2007. Investigating the human immunodeficiency virus type 1-infected monocyte-derived macrophage secretome. Virology 363, 198–209.
Correia, I., Chu, D., Chou, Y.H., Goldman, R.D., Matsudaira, P., 1999. Integrating the actin and vimentin cytoskeletons. Adhesion-dependent formation of fimbrin-vimentin complexes in macrophages. J. Cell Biol. 146, 831–842.
Crowhurst, M.O., Layton, J.E., Lieschke, G.J., 2002. Developmental biology of zebrafish myeloid cells. Int. J. Dev. Biol. 46, 483–492.
De Arruda, M.V., Watson, S., Lin, C.S., Leavitt, J., Matsudaira, P., 1990. Fimbrin is a homologue of the cytoplasmic phosphoprotein plastin and has domains homologous with calmodulin and actin gelation proteins. J. Cell Biol. 111, 1069–1079.
De Clercq, S., Boucherie, C., Vandekerckhove, J., Gettemans, J., Guillabert, A., 2013a. L-plastin nanobodies perturb matrix degradation, podosome formation, stability and lifetime in THP-1 macrophages. PLoS One 8, e78108.
De Clercq, S., Zwaenepoel, O., Martens, E., Vandekerckhove, J., Guillabert, A., Gettemans, J., 2013b. Nanobody-induced perturbation of LFA-1/L-plastin phosphorylation impairs MTOC docking, immune synapse formation and T cell activation. Cell. Mol. Life Sci. 70, 909–922.

Deady, L.E., Todd, E.M., Davis, C.G., Zhou, J.Y., Topcagic, N., Edelson, B.T., Ferkol, T.W., Cooper, M.A., Muenzer, J.T., Morley, S.C., 2014. L-plastin is essential for alveolar macrophage production and control of pulmonary pneumococcal infection. Infect. Immun. 82, 1982–1993.

Delanote, V., Van Impe, K., De Corte, V., Bruyneel, E., Vetter, G., Boucherie, C., Mareel, M., Vandekerckhove, J., Friederich, E., Gettemans, J., 2005. Molecular basis for dissimilar nuclear trafficking of the actin-bundling protein isoforms T- and L-plastin. Traffic 6, 335–345.

Delanote, V., Vanloo, B., Catillon, M., Friederich, E., Vandekerckhove, J., Gettemans, J., 2010. An alpaca single-domain antibody blocks filopodia formation by obstructing L-plastin-mediated F-actin bundling. FASEB J. 24, 105–118.

DeRosier, D., Mandelkow, E., Silliman, A., Tilney, L., Kane, R., 1977. Structure of actin-containing filaments from two types of non-muscle cells. J. Mol. Biol. 113, 679–695.

Dubey, M., Singh, A.K., Awasthi, D., Nagarkoti, S., Kumar, S., Ali, W., Chandra, T., Kumar, V., Barthwal, M.K., Jagavelu, K., et al., 2015. L-Plastin S-glutathionylation promotes reduced binding to β-actin and affects neutrophil functions. Free Radic. Biol. Med. 86, 1–15.

Dubovsky, J.A., Chappell, D.L., Harrington, B.K., Agrawal, K., Andritsos, L.A., Flynn, J.M., Jones, J.A., Paulaitis, M.E., Bolon, B., Johnson, A.J., et al., 2013. Lymphocyte cytosolic protein 1 is a chronic lymphocytic leukemia membrane-associated antigen critical to niche homing. Blood 122, 3308–3316.

Dun, M.D., Chalkley, R.J., Faulkner, S., Keene, S., Avery-Kiejda, K.A., Scott, R.J., Falkenby, L.G., Cairns, M.J., Larsen, M.R., Bradshaw, R.A., et al., 2015. Proteotranscriptomic profiling of 231-BR breast cancer cells: Identification of potential biomarkers and therapeutic targets for brain metastasis. Mol. Cell. Proteomics 14, 2316–2330.

Evans, S.W., Rennick, D., Farrar, W.L., 1986. Multilineage hematopoietic growth factor interleukin 3 and direct activators of protein kinase C stimulate phosphorylation of common substrates. Blood 68, 906–913.

Ezzell, R.M., Leung, J., Collins, K., Chafel, M.M., Cardozo, T.J., Matsudaira, P.T., 1992. Expression and localization of villin, fimbrin, and myosin I in differentiating mouse F9 teratocarcinoma cells. Dev. Biol. 151, 575–585.

Flock, Å., Bretscher, A., Weber, K., 1982. Immunohistochemical localization of several cytoskeletal proteins in inner ear sensory and supporting cells. Hear. Res. 7, 75–89.

Foran, E., McWilliam, P., Kelleher, D., Croke, D.T., Long, A., 2006. The leukocyte protein L-plastin induces proliferation, invasion and loss of E-cadherin expression in colon cancer cells. Int. J. Cancer 118, 2098–2104.

Forster, F., Paster, W., Supper, V., Schatzlmaier, P., Sunzenauer, S., Ostler, N., Saliba, A., Eckerstorfer, P., Britzen-Laurent, N., Schütz, G., et al., 2014. Guanylate binding protein 1–mediated interaction of t cell antigen receptor signaling with the cytoskeleton. J. Immunol. 192, 771–781.

Frederick, M.J., Rodriguez, L.V., Johnston, D.A., Darnay, B.G., Grimm, E.A., 1996. Characterization of the Mr 65,000 lymphokine-activated killer proteins phosphorylated after tumor target binding: evidence that pp65a and pp65b are phosphorylated forms of L-plastin. Cancer Res. 56, 138–144.

Freeley, M., O'Dowd, F., Paul, T., Kashanin, D., Davies, A., Kelleher, D., Long, A., 2012. L-plastin regulates polarization and migration in chemokine-stimulated human T lymphocytes. J. Immunol. 188, 6357–6370.

Freudenmann, L.K., Mayer, C., Rodemann, H.P., Dittmann, K., 2019. Reduced exosomal L-Plastin is responsible for radiation-induced bystander effect. Exp. Cell Res. 383, 111498.

Galkin, V.E., Orlova, A., Cherepanova, O., Lebart, M.C., Egelman, E.H., 2008. High-resolution cryo-EM structure of the F-actin-fimbrin/plastin ABD2 complex. Proc. Natl. Acad. Sci. U. S. A. 105, 1494–1498.
Glenney Jr., J.R., Kaulfus, P., Matsudaira, P., Weber, K., 1981. F-actin binding and bundling properties of fimbrin, a major cytoskeletal protein of microvillus core filaments. J. Biol. Chem. 256, 9283–9288.
Goldsmith, S.C., Pokala, N., Shen, W., Fedorov, A.A., Matsudaira, P., Almo, S.C., 1997. The structure of an actin-crosslinking domain from human fimbrin. Nat. Struct. Biol. 4, 708–712.
Goldstein, D., Djeu, J., Latter, G., Burbeck, S., Leavitt, J., 1985. Abundant synthesis of the transformation-induced protein of neoplastic human fibroblasts, plastin, in normal lymphocytes. Cancer Res. 45, 5643–5647.
González, E., Neuhaus, T., Kemper, M.J., Girardin, E., 2009. Proteomic analysis of mononuclear cells of patients with minimal-change nephrotic syndrome of childhood. Nephrol. Dial. Transplant. 24, 149–155.
Gruhlke, M.C.H., Antelmann, H., Bernhardt, J., Kloubert, V., Rink, L., Slusarenko, A.J., 2019. The human allicin-proteome: S-thioallylation of proteins by the garlic defence substance allicin and its biological effects. Free Radic. Biol. Med. 131, 144–153.
Hagi, A., Hirata, H., Shinomiya, H., 2006. Analysis of a bacterial lipopolysaccharide-activated serine kinase that phosphorylates p65/L-plastin in macrophages. Microbiol. Immunol. 50, 331–335.
Hagiwara, M., Shinomiya, H., Kashihara, M., Kobayashi, K., Tadokoro, T., Yamamoto, Y., 2011. Interaction of activated Rab5 with actin-bundling proteins, L- and T-plastin and its relevance to endocytic functions in mammalian cells. Biochem. Biophys. Res. Commun. 407, 615–619.
Hanein, D., Matsudaira, P., DeRosier, D.J., 1997. Evidence for a conformational change in actin induced by fimbrin (N375) binding. J. Cell Biol. 139, 387–396.
Hanein, D., Volkmann, N., Goldsmith, S., Michon, A.M., Lehman, W., Craig, R., DeRosier, D., Almo, S., Matsudaira, P., 1998. An atomic model of fimbrin binding to F-actin and its implications for filament crosslinking and regulation. Nat. Struct. Biol. 5, 787–792.
Harris, L.D., De La Cerda, J., Tuziak, T., Rosen, D., Xiao, L., Shen, Y., Sabichi, A.L., Czerniak, B., Grossman, H.B., 2008. Analysis of the expression of biomarkers in urinary bladder cancer using a tissue microarray. Mol. Carcinog. 47, 678–685.
Hayashi, J., Kihara, M., Kato, H., Nishimura, T., 2015. A proteomic profile of synoviocyte lesions microdissected from formalin-fixed paraffin-embedded synovial tissues of rheumatoid arthritis. Clin. Proteomics 12, 20.
Henning, S.W., Meuer, S.C., Samstag, Y., 1994. Serine phosphorylation of a 67-kDa protein in human T lymphocytes represents an accessory receptor-mediated signaling event. J. Immunol. 152, 4808–4815.
Hornbeck, P., Paul, W.E., 1986. Anti-immunoglobulin and phorbol ester induce phosphorylation of proteins associated with the plasma membrane and cytoskeleton in murine B lymphocytes. J. Biol. Chem. 261, 14817–14824.
Inaguma, S., Riku, M., Ito, H., Tsunoda, T., Ikeda, H., Kasai, K., 2015. GLI1 orchestrates CXCR4/CXCR7 signaling to enhance migration and metastasis of breast cancer cells. Oncotarget 6, 33648–33657.
Ishida, H., Jensen, K.V., Woodman, A.G., Hyndman, M.E., Vogel, H.J., 2017. The calcium-dependent switch helix of l-plastin regulates actin bundling. Sci. Rep. 7, 40662.
Janji, B., Giganti, A., De Corte, V., Catillon, M., Bruyneel, E., Lentz, D., Plastino, J., Gettemans, J., Friederich, E., 2006. Phosphorylation on Ser5 increases the F-actin-binding activity of L-plastin and promotes its targeting to sites of actin assembly in cells. J Cell Sci 119, 1947–1960.

Janji, B., Vallar, L., Al-Tanoury, Z., Bernardin, F., Vetter, G., Schaffner-Reckinger, E., Berchem, G., Friederich, E., Chouaib, S., 2010. The actin filament cross-linker L-plastin confers resistance to TNF-alpha in MCF-7 breast cancer cells in a phosphorylation-dependent manner. J. Cell. Mol. Med. 14, 1264–1275.

Jimenez, L., Lim, J., Burds, B., Harris, T.M., Ow, T.J., Kawachi, N., Belbin, T.J., Angeletti, R., Prystowsky, M.B., Childs, G., et al., 2017. miR-375 regulates invasion-related proteins vimentin and L-plastin. Am. J. Pathol. 187, 1523–1536.

Jones, S.L., Brown, E.J., 1996. FcgammaRII-mediated adhesion and phagocytosis induce L-plastin phosphorylation in human neutrophils. J. Biol. Chem. 271, 14623–14630.

Jones, S.L., Wang, J., Turck, C.W., Brown, E.J., 1998. A role for the actin-bundling protein L-plastin in the regulation of leukocyte integrin function. Proc. Natl. Acad. Sci. U. S. A. 95, 9331–9336.

Kang, S., Shim, H.S., Lee, J.S., Kim, D.S., Kim, H.Y., Hong, S.H., Kim, P.S., Yoon, J.H., Cho, N.H., 2010. Molecular proteomics imaging of tumor interfaces by mass spectrometry. J. Proteome Res. 9, 1157–1164.

Kassie, F., Anderson, L.B., Scherber, R., Yu, N., Lahti, D., Upadhyaya, P., Hecht, S.S., 2007. Indole-3-carbinol inhibits 4-(methylnitrosamino)-1-(3-pyridyl)-1-butanone plus benzo(a)pyrene-induced lung tumorigenesis in A/J mice and modulates carcinogen-induced alterations in protein levels. Cancer Res. 67, 6502–6511.

Kell, M.J., Riccio, R.E., Baumgartner, E.A., Compton, Z.J., Pecorin, P.J., Mitchell, T.A., Topczewski, J., LeClair, E.E., 2018. Targeted deletion of the zebrafish actin-bundling protein L-plastin (lcp1). PLoS One 13, e0190353.

Khoontawad, J., Intuyod, K., Rucksaken, R., Hongsrichan, N., Pairojkul, C., Pinlaor, P., Boonmars, T., Wongkham, C., Jones, A., Plieskatt, J., et al., 2018. Discovering proteins for chemoprevention and chemotherapy by curcumin in liver fluke infection-induced bile duct cancer. PLoS One 13, e0207405.

Kim, D.S., Choi, Y.P., Kang, S., Gao, M.Q., Kim, B., Park, H.R., Choi, Y.D., Lim, J.B., Na, H.J., Kim, H.K., et al., 2010. Panel of candidate biomarkers for renal cell carcinoma. J. Proteome Res. 9, 3710–3719.

Klein, M.G., Shi, W., Ramagopal, U., Tseng, Y., Wirtz, D., Kovar, D.R., Staiger, C.J., Almo, S.C., 2004. Structure of the actin crosslinking core of fimbrin. Structure 12, 999–1013.

Klemke, M., Rafael, M.T., Wabnitz, G.H., Weschenfelder, T., Konstandin, M.H., Garbi, N., Autschbach, F., Hartschuh, W., Samstag, Y., 2007. Phosphorylation of ectopically expressed L-plastin enhances invasiveness of human melanoma cells. Int. J. Cancer 120, 2590–2599.

Klobučar, M., Sedić, M., Gehrig, P., Grossmann, J., Bilić, M., Kovač-Bilić, L., Pavelić, K., Kraljević Pavelić, S., 2016. Basement membrane protein ladinin-1 and the MIF-CD44-β 1 integrin signaling axis are implicated in laryngeal cancer metastasis. Biochim. Biophys. Acta, Mol. Basis Dis. 1862, 1938–1954.

Koide, N., Kasamatsu, A., Endo-Sakamoto, Y., Ishida, S., Shimizu, T., Kimura, Y., Miyamoto, I., Yoshimura, S., Shiiba, M., Tanzawa, H., et al., 2017. Evidence for critical role of lymphocyte cytosolic protein 1 in oral cancer. Sci. Rep. 7, 43379.

Kovar, D.R., Staiger, C.J., Weaver, E.A., McCurdy, D.W., 2000. AtFim1 is an actin filament crosslinking protein from Arabidopsis thaliana. Plant J. 24, 625–636.

Kübler, E., Riezman, H., 1993. Actin and fimbrin are required for the internalization step of endocytosis in yeast. EMBO J. 12, 2855–2862.

Langereis, J.D., Prinsen, B.H.C.M.T., De Sain-van Der Velden, M.G.M., Coppens, C.J.C., Koenderman, L., Ulfman, L.H., 2009. A 2D-DIGE approach to identify proteins involved in inside-out control of integrins. J. Proteome Res. 8, 3824–3833.

Lapillonne, A., Coue, O., Friederich, E., Nicolas, A., Del Maestro, L., Louvard, D., Robine, S., Sastre-Garau, X., 2000. Expression patterns of L-plastin isoform in normal and carcinomatous breast tissues. Anticancer Res 20, 3177–3182.

Le Goff, E., Vallentin, A., Harmand, P.O., Aldrian-Herrada, G., Rebiere, B., Roy, C., Benyamin, Y., Lebart, M.C., 2010. Characterization of L-plastin interaction with beta integrin and its regulation by micro-calpain. Cytoskeleton 67, 286–296.

Leavitt, J., Chen, Z.P., Lockwood, C.J., Schatz, F., 1994. Regulation of synthesis of the transformation-induced protein, leukocyte plastin,by ovarian steroid hormones. Cancer Res. 54, 3447–3454.

Lebart, M.C., Hubert, F., Boiteau, C., Venteo, S., Roustan, C., Benyamin, Y., 2004. Biochemical characterization of the L-plastin-actin interaction shows a resemblance with that of alpha-actinin and allows a distinction to be made between the two actin-binding domains of the molecule. Biochemistry 43, 2428–2437.

Li, J., Zhao, R., 2011. Expression and clinical significance of L-plastin in colorectal carcinoma. J. Gastrointest. Surg. 15, 1982–1988.

Li, M.X., Xiao, Z.Q., Liu, Y.F., Chen, Y.H., Li, C., Zhang, P.F., Li, M.Y., Li, F., Peng, F., Duan, C.J., et al., 2009. Quantitative proteomic analysis of differential proteins in the stroma of nasopharyngeal carcinoma and normal nasopharyngeal epithelial tissue. J. Cell. Biochem. 106, 570–579.

Lin, C.S., Aebersold, R.H., Kent, S.B., Varma, M., Leavitt, J., 1988. Molecular cloning and characterization of plastin, a human leukocyte protein expressed in transformed human fibroblasts. Mol. Cell. Biol. 8, 4659–4668.

Lin, C.S., Aebersold, R.H., Leavitt, J., 1990. Correction of the N-terminal sequences of the human plastin isoforms by using anchored polymerase chain reaction: identification of a potential calcium-binding domain. Mol. Cell. Biol. 10, 1818–1821.

Lin, C.S., Chen, Z.P., Park, T., Ghosh, K., Leavitt, J., 1993a. Characterization of the human L-plastin gene promoter in normal and neoplastic cells. J. Biol. Chem. 268, 2793–2801.

Lin, C.S., Park, T., Chen, Z.P., Leavitt, J., 1993b. Human plastin genes. Comparative gene structure, chromosome location, and differential expression in normal and neoplastic cells. J. Biol. Chem. 268, 2781–2792.

Lin, C.S., Shen, W., Chen, Z.P., Tu, Y.H., Matsudaira, P., 1994. Identification of I-plastin, a human fimbrin isoform expressed in intestine and kidney. Mol. Cell. Biol. 14, 2457–2467.

Lin, C.S., Lau, A., Lue, T.F., 1998. Analysis and mapping of plastin phosphorylation. DNA Cell Biol. 17, 1041–1046.

Lin, C.S., Lau, A., Yeh, C.C., Chang, C.H., Lue, T.F., 2000. Upregulation of L-plastin gene by testosterone in breast and prostate cancer cells: identification of three cooperative androgen receptor-binding sequences. DNA Cell Biol. 19, 1–7.

Liu, F., Shinomiya, H., Kirikae, T., Hirata, H., Asano, Y., 2004. Characterization of murine grancalcin specifically expressed in leukocytes and its possible role in host defense against bacterial infection. Biosci. Biotechnol. Biochem. 68, 894–902.

Liu, Y., Ye, T., Maynard, J., Akbulut, H., Deisseroth, A., 2006. Engineering conditionally replication-competent adenoviral vectors carrying the cytosine deaminase gene increases the infectivity and therapeutic effect for breast cancer gene therapy. Cancer Gene Ther. 13, 346–356.

Lollike, K., Johnsen, A.H., Durussel, I., Borregaard, N., Cox, J.A., 2001. Biochemical characterization of the penta-EF-hand protein grancalcin and identification of L-plastin as a binding partner. J. Biol. Chem. 276, 17762–17769.

Lommel, M.J., Trairatphisan, P., Gabler, K., Laurini, C., Muller, A., Kaoma, T., Vallar, L., Sauter, T., Schaffner-Reckinger, E., 2016. L-plastin Ser5 phosphorylation in breast cancer cells and in vitro is mediated by RSK downstream of the ERK/MAPK pathway. FASEB J. 30, 1218–1233.

Luczak, M., Kaźmierczak, M., Handschuh, L., Lewandowski, K., Komarnicki, M., Figlerowicz, M., 2012. Comparative proteome analysis of acute myeloid leukemia with and without maturation. J. Proteomics 75, 5734–5748.

Ma, T., Sadashivaiah, K., Madayiputhiya, N., Chellaiah, M.A., 2010. Regulation of sealing ring formation by L-plastin and cortactin in osteoclasts. J. Biol. Chem. 285, 29911–29924.

Ma, G., Gezer, D., Herrmann, O., Feldberg, K., Schemionek, M., Jawhar, M., Reiter, A., Brümmendorf, T.H., Koschmieder, S., Chatain, N., 2020. LCP1 triggers mTORC2/AKT activity and is pharmacologically targeted by enzastaurin in hypereosinophilia. Mol. Carcinog. 59, 87–103.

Marchisio, P.C., Cirillo, D., Naldini, L., Primavera, M.V., Teti, A., Zambonin-Zallone, A., 1984. Cell-substratum interaction of cultured avian osteoclasts is mediated by specific adhesion structures. J. Cell Biol. 99, 1696–1705.

Matsudaira, P.T., Burgess, D.R., 1979. Identification and organization of the components in the isolated microvillus cytoskeleton. J. Cell Biol. 83, 667–673.

Matsudaira, P., Mandelkow, E., Renner, W., Hesterberg, L.K., Weber, K., 1983. Role of fimbrin and villin in determining the interfilament distances of actin bundles. Nature 301, 209–214.

Matsushima, K., Kobayashi, Y., Copeland, T.D., Akahoshi, T., Oppenheim, J.J., 1987. Phosphorylation of a cytosolic 65-kDa protein induced by interleukin 1 in glucocorticoid pretreated normal human peripheral blood mononuclear leukocytes. J. Immunol. 139, 3367–3374.

Matsushima, K., Shiroo, M., Kung, H.F., Copeland, T.D., 1988. Purification and characterization of a cytosolic 65-kilodalton phosphoprotein in human leukocytes whose phosphorylation is augmented by stimulation with interleukin 1. Biochemistry 27, 3765–3770.

Messier, J.M., Shaw, L.M., Chafel, M., Matsudaira, P., Mercurio, A.M., 1993. Fimbrin localized to an insoluble cytoskeletal fraction is constitutively phosphorylated on its headpiece domain in adherent macrophages. Cell Motil. Cytoskeleton 25, 223–233.

Miyakawa, T., Shinomiya, H., Yumoto, F., Miyauchi, Y., Tanaka, H., Ojima, T., Kato, Y.S., Tanokura, M., 2012. Different Ca2+-sensitivities between the EF-hands of T- and L-plastins. Biochem. Biophys. Res. Commun. 429, 137–141.

Morley, S.C., Wang, C., Lo, W.L., Lio, C.W., Zinselmeyer, B.H., Miller, M.J., Brown, E.J., Allen, P.M., 2010. The actin-bundling protein L-plastin dissociates CCR7 proximal signaling from CCR7-induced motility. J. Immunol. 184, 3628–3638.

Namba, Y., Ito, M., Zu, Y., Shigesada, K., Maruyama, K., 1992. Human T cell L-plastin bundles actin filaments in a calcium-dependent manner. J. Biochem. 112, 503–507.

Nara, M., Morii, H., Shimizu, T., Shinomiya, H., Furuta, Y., Miyazono, K., Miyakawa, T., Tanokura, M., 2019. Characterization of the Ca2+ -coordination structures of L- and T-plastins in combination with their synthetic peptide analogs by FTIR spectroscopy. Sci. Rep. 9, 4217.

Ning, Y., Gerger, A., Zhang, W., Hanna, D.L., Yang, D., Winder, T., Wakatsuki, T., Labonte, M.J., Stintzing, S., Volz, N., et al., 2014. Plastin polymorphisms predict gender and stage-specific colon cancer recurrence after adjuvant chemotherapy. Mol. Cancer Ther. 13, 528–539.

Nishida, Y., Aida, K., Kihara, M., Kobayashi, T., 2014. Antibody-validated proteins in inflamed islets of fulminant type 1 diabetes profiled by laser-capture microdissection followed by mass spectrometry. PLoS One 9, e107664.

Ohsawa, K., Imai, Y., Sasaki, Y., Kohsaka, S., 2004. Microglia/macrophage-specific protein Iba1 binds to fimbrin and enhances its actin-bundling activity. J. Neurochem. 88, 844–856.

Otsuka, M., Kato, M., Yoshikawa, T., Chen, H., Brown, E.J., Masuho, Y., Omata, M., Seki, N., 2001. Differential expression of the L-plastin gene in human colorectal cancer progression and metastasis. Biochem. Biophys. Res. Commun. 289, 876–881.

Otto, J.J., Kane, R.E., Bryan, J., 1979. Formation of filopodia in coelomocytes: localization of fascin, a 58,000 Dalton actin cross-linking protein. Cell 17, 285–293.

Pacaud, M., Derancourt, J., 1993. Purification and further characterization of macrophage 70-kDa protein, a calcium-regulated, actin-binding protein identical to L-plastin. Biochemistry 32, 3448–3455.

Pacaud, M., Harricane, M.C., 1987. Calcium control of macrophage cytoplasmic gelation: evidence for the involvement of the 70,000 Mr actin-bundling protein. J. Cell Sci. 88, 81–94.

Paclet, M.H., Davis, C., Kotsonis, P., Godovac-Zimmermann, J., Segal, A.W., Dekker, L.V., 2004. N-Formyl peptide receptor subtypes in human neutrophils activate L-plastin phosphorylation through different signal transduction intermediates. Biochem. J. 377, 469–477.

Park, T., Chen, Z.P., Leavitt, J., 1994. Activation of the leukocyte plastin gene occurs in most human cancer cells. Cancer Res. 54, 1775–1781.

Patterson, A., Harris, A.L., 1999. Molecular chemotherapy for breast cancer. Drugs Aging 14, 75–90.

Pazdrak, K., Young, T.W., Straub, C., Stafford, S., Kurosky, A., 2011. Priming of eosinophils by GM-CSF is mediated by protein kinase CβII-phosphorylated L-plastin. J. Immunol. 186, 6485–6496.

Peng, X.Y., Won, J.H., Rutherford, T., Fujii, T., Zelterman, D., Pizzorno, G., Sapi, E., Leavitt, J., Kacinski, B., Crystal, R., et al., 2001. The use of the L-plastin promoter for adenoviral-mediated, tumor-specific gene expression in ovarian and bladder cancer cell lines. Cancer Res. 61, 4405–4413.

Pillar, N., Polsky, A.L., Weissglas-Volkov, D., Shomron, N., 2018. Comparison of breast cancer metastasis models reveals a possible mechanism of tumor aggressiveness. Cell Death Dis. 9, 1040.

Pillar, N., Polsky, A.L., Shomron, N., 2019. Dual inhibition of ABCE1 and LCP1 by microRNA-96 results in an additive effect in breast cancer mouse model. Oncotarget 10, 2086–2094.

Pollard, T.D., 2016. Actin and actin-binding proteins. Cold Spring Harb. Perspect. Biol. 8, a018226.

Pollard, T.D., Borisy, G.G., 2003. Cellular motility driven by assembly and disassembly of actin filaments. Cell 112, 453–465.

Ralser, M., Nonhoff, U., Albrecht, M., Lengauer, T., Wanker, E.E., Lehrach, H., Krobitsch, S., 2005. Ataxin-2 and huntingtin interact with endophilin-A complexes to function in plastin-associated pathways. Hum. Mol. Genet. 14, 2893–2909.

Reuter, S., Reiermann, S., Wörner, R., Schröter, R., Edemir, B., Buck, F., Henning, S., Peter-Katalinic, J., Vollenbröker, B., Amann, K., et al., 2010. IF/TA-related metabolic changes—proteome analysis of rat renal allografts. Nephrol. Dial. Transplant. 25, 2492–2501.

Riplinger, S.M., Wabnitz, G.H., Kirchgessner, H., Jahraus, B., Lasitschka, F., Schulte, B., van der Pluijm, G., van der Horst, G., Hammerling, G.J., Nakchbandi, I., et al., 2014. Metastasis of prostate cancer and melanoma cells in a preclinical in vivo mouse model is enhanced by L-plastin expression and phosphorylation. Mol. Cancer 13, 10.

Schenk, L.K., Möller-Kerutt, A., Klosowski, R., Wolters, D., Schaffner-Reckinger, E., Weide, T., Pavenstädt, H., Vollenbröker, B., 2017. Angiotensin II regulates phosphorylation of actin-associated proteins in human podocytes. FASEB J. 31, 5019–5035.

Schulz, D.M., Böllner, C., Thomas, G., Atkinson, M., Esposito, I., Höfler, H., Aubele, M., 2009. identification of differentially expressed proteins in triple-negative breast carcinomas using DIGE and mass spectrometry. J. Proteome Res. 8, 3430–3438.

Schwalie, P.C., Ordóñez-Morán, P., Huelsken, J., Deplancke, B., 2017. Cross-tissue identification of somatic stem and progenitor cells using a single-cell RNA-sequencing derived gene signature. Stem Cells 35, 2390–2402.
Schwebach, C.L., Agrawal, R., Lindert, S., Kudryashova, E., Kudryashov, D.S., 2017. The roles of actin-binding domains 1 and 2 in the calcium-dependent regulation of actin filament bundling by human plastins. J. Mol. Biol. 429, 2490–2508.
Shibata, M., Yamakawa, Y., Ohoka, T., Mizuno, S., Suzuki, K., 1993. Characterization of a 64-kd protein phosphorylated during chemotactic activation with IL-8 and fMLP of human polymorphonuclear leukocytes. II. Purification and amino acid analysis of phosphorylated 64-kd protein. J. Leukoc. Biol. 54, 10–16.
Shin, J., Dan, K., Han, D., Kim, J.W., Kim, K.K., Koh, Y., Shin, D.Y., Hong, J., Yoon, S.S., Park, S., et al., 2019. Plasma-based protein biomarkers can predict the risk of acute graft-versus-host disease and non-relapse mortality in patients undergoing allogeneic hematopoietic stem cell transplantation. Blood Cells Mol Dis 74, 5–12.
Shinomiya, H., Hirata, H., Nakano, M., 1991. Purification and characterization of the 65-kDa protein phosphorylated in murine macrophages by stimulation with bacterial lipopolysaccharide. J. Immunol. 146, 3617–3625.
Shinomiya, H., Hirata, H., Saito, S., Yagisawa, H., Nakano, M., 1994. Identification of the 65-kDa phosphoprotein in murine macrophages as a novel protein: homology with human L-plastin. Biochem. Biophys. Res. Commun. 202, 1631–1638.
Shinomiya, H., Hagi, A., Fukuzumi, M., Mizobuchi, M., Hirata, H., Utsumi, S., 1995. Complete primary structure and phosphorylation site of the 65-kDa macrophage protein phosphorylated by stimulation with bacterial lipopolysaccharide. J. Immunol. 154, 3471–3478.
Shinomiya, H., Shinjo, M., Fengzhi, L., Asano, Y., Kihara, H., 2007. Conformational analysis of the leukocyte-specific EF-hand protein p65/L-plastin by X-ray scattering in solution. Biophys. Chem. 131, 36–42.
Shiroo, M., Matsushima, K., 1989. A protein kinase distinct from PKC and PKA is activated by IL 1 or TNF in human peripheral blood mononuclear cells. In: FASEB Journal. 3, p. A671.
Shiroo, M., Matsushima, K., 1990. Enhanced phosphorylation of 65 and 74 kDa proteins by tumor necrosis factor and interleukin-1 in human peripheral blood mononuclear cells. Cytokine 2, 13–20.
Si, M., Goodluck, H., Zeng, C., Pan, S., Todd, E.M., Morley, S.C., Qin, X., Mohan, S., Xing, W., 2018. *LRRK1* regulation of actin assembly in osteoclasts involves serine 5 phosphorylation of L-plastin. J. Cell. Biochem. 119, 10351–10357.
Skopeliti, M., Kratzer, U., Altenberend, F., Panayotou, G., Kalbacher, H., Stevanovic, S., Voelter, W., Tsitsilonis, O.E., 2007. Proteomic exploitation on prothymosin α-induced mononuclear cell activation. Proteomics 7, 1814–1824.
Strickler, A.G., Vasquez, J.G., Yates, N., Ho, J., 2014. Potential diagnostic significance of HSP90, ACS/TMS1, and L-plastin in the identification of melanoma. Melanoma Res. 24, 535–544.
Su Kim, D., Choi, Y.D., Moon, M., Kang, S., Lim, J.B., Kim, K.M., Park, K.M., Cho, N.H., 2013. Composite three-marker assay for early detection of kidney cancer. Cancer Epidemiol. Biomarkers Prev. 22, 390–398.
Tiedemann, K., Sadvakassova, G., Mikolajewicz, N., Juhas, M., Sabirova, Z., Tabariès, S., Gettemans, J., Siegel, P.M., Komarova, S.V., 2019. Exosomal release of l-plastin by breast cancer cells facilitates metastatic bone osteolysis. Transl. Oncol. 12, 462–474.
Tit-Oon, P., Chokchaichamnankit, D., Khongmane, A., Sawangaretrakul, P., Svasti, J., Srisomsap, C., 2014. Comparative secretome analysis of cholangiocarcinoma cell line in three-dimensional culture. Int. J. Oncol. 45, 2108–2116.

Todd, E.M., Deady, L.E., Morley, S.C., 2011. The actin-bundling protein L-plastin is essential for marginal zone B cell development. J. Immunol. 187, 3015–3025.
Todd, E.M., Deady, L.E., Morley, S.C., 2013. Intrinsic T- and B-cell defects impair T-cell-dependent antibody responses in mice lacking the actin-bundling protein L-plastin. Eur. J. Immunol. 43, 1735–1744.
Todd, E.M., Zhou, J.Y., Szasz, T.P., Deady, L.E., D'Angelo, J.A., Cheung, M.D., Kim, A.H.J., Morley, S.C., 2016. Alveolar macrophage development in mice requires L-plastin for cellular localization in alveoli. Blood 128, 2785–2796.
Toro, E.J., Zuo, J., Ostrov, D.A., Catalfamo, D., Bradaschia-Correa, V., Arana-Chavez, V., Caridad, A.R., Neubert, J.K., Wronski, T.J., Wallet, S.M., et al., 2012. Enoxacin directly inhibits osteoclastogenesis without inducing apoptosis. J. Biol. Chem. 287, 17894–17904.
Trøen, G., Nygaard, V., Jenssen, T.K., Ikonomou, I.M., Tierens, A., Matutes, E., Gruszka-Westwood, A., Catovsky, D., Myklebost, O., Lauritzsen, G., et al., 2004. Constitutive expression of the AP-1 transcription factors c-jun, junD, junB, and c-fos and the marginal zone B-cell transcription factor notch2 in splenic marginal zone lymphoma. J. Mol. Diagn. 6, 297–307.
Ueda, K., Nakanishi, T., Shimizu, A., Takubo, T., Matsuura, N., 2008. Identification of L-plastin autoantibody in plasma of patients with non-Hodgkin's lymphoma using a proteomics-based analysis. Ann. Clin. Biochem. 45, 65–69.
Van Audenhove, I., Denert, M., Boucherie, C., Pieters, L., Cornelissen, M., Gettemans, J., 2016. Fascin rigidity and L-plastin flexibility cooperate in cancer cell invadopodia and Filopodia. J. Biol. Chem. 291, 9148–9160.
Volkmann, N., DeRosier, D., Matsudaira, P., Hanein, D., 2001. An atomic model of actin filaments cross-linked by fimbrin and its implications for bundle assembly and function. J. Cell Biol. 153, 947–956.
Wabnitz, G.H., Kocher, T., Lohneis, P., Stober, C., Konstandin, M.H., Funk, B., Sester, U., Wilm, M., Klemke, M., Samstag, Y., 2007. Costimulation induced phosphorylation of L-plastin facilitates surface transport of the T cell activation molecules CD69 and CD25. Eur. J. Immunol. 37, 649–662.
Wabnitz, G.H., Lohneis, P., Kirchgessner, H., Jahraus, B., Gottwald, S., Konstandin, M., Klemke, M., Samstag, Y., 2010. Sustained LFA-1 cluster formation in the immune synapse requires the combined activities of L-plastin and calmodulin. Eur. J. Immunol. 40, 2437–2449.
Wabnitz, G.H., Michalke, F., Stober, C., Kirchgessner, H., Jahraus, B., van den Boomen, D.J., Samstag, Y., 2011. L-plastin phosphorylation: a novel target for the immunosuppressive drug dexamethasone in primary human T cells. Eur. J. Immunol. 41, 3157–3169.
Wabnitz, G., Balta, E., Samstag, Y., 2017. L-plastin regulates the stability of the immune synapse of naive and effector T-cells. Adv. Biol. Regul. 63, 107–114.
Wang, J., Brown, E.J., 1999. Immune complex-induced integrin activation and L-plastin phosphorylation require protein kinase A. J. Biol. Chem. 274, 24349–24356.
Wang, J., Chen, H., Brown, E.J., 2001. L-plastin peptide activation of alpha(v)beta(3)-mediated adhesion requires integrin conformational change and actin filament disassembly. J. Biol. Chem. 276, 14474–14481.
Wang, H.P., Long, X.H., Sun, Z.Z., Rigaud, O., Xu, Q.Z., Huang, Y.C., Sui, J.L., Bai, B., Zhou, P.K., 2006. Identification of differentially transcribed genes in human lymphoblastoid cells irradiated with 0.5 Gy of γ-ray and the involvement of low dose radiation inducible CHD6 gene in cell proliferation and radiosensitivity. Int. J. Radiat. Biol. 82, 181–190.
Wang, C., Morley, S.C., Donermeyer, D., Peng, I., Lee, W.P., Devoss, J., Danilenko, D.M., Lin, Z., Zhang, J., Zhou, J., et al., 2011. Actin-bundling protein L-plastin regulates T cell activation. J. Immunol. 185, 7487–7497.

Wang, L., Fan, J., Francis, J.M., Georghiou, G., Hergert, S., Li, S., Gambe, R., Zhou, C.W., Yang, C., Xiao, S., et al., 2017. Integrated single-cell genetic and transcriptional analysis suggests novel drivers of chronic lymphocytic leukemia. Genome Res. 27, 1300–1311.

Xu, X., Wang, X., Todd, E.M., Jaeger, E.R., Vella, J.L., Mooren, O.L., Feng, Y., Hu, J., Cooper, J.A., Morley, S.C., et al., 2016. Mst1 kinase regulates the actin-bundling protein L-plastin to promote t cell migration. J. Immunol. 197, 1683–1691.

Zhang, L., Akbulut, H., Tang, Y., Peng, X., Pizzorno, G., Sapi, E., Manegold, S., Deisseroth, A., 2002. Adenoviral vectors with E1A regulated by tumor-specific promoters are selectively cytolytic for breast cancer and melanoma. Mol. Ther. 6, 386–393.

Zhang, B., Yu, Q., Wang, Y., Xiao, C., Li, J., Huo, D., Zhang, D., Jia, C., Li, M., 2016. The Candida albicans fimbrin Sac6 regulates oxidative stress response (OSR) and morphogenesis at the transcriptional level. Biochim. Biophys. Acta, Mol. Cell Res. 1863, 2255–2266.

Zheng, J., Rudra-Ganguly, N., Miller, G.J., Moffatt, K.A., Cote, R.J., Roy-Burman, P., 1997. Steroid hormone induction and expression patterns of L-plastin in normal and carcinomatous prostate tissues. Am. J. Pathol. 150, 2009–2018.

Zheng, J., Rudra-Ganguly, N., Powell, W.C., Roy-Burman, P., 1999. Suppression of prostate carcinoma cell invasion by expression of antisense L-plastin gene. Am. J. Pathol. 155, 115–122.

Zhou, J.Y., Szasz, T.P., Stewart-Hutchinson, P.J., Sivapalan, J., Todd, E.M., Deady, L.E., Cooper, J.A., Onken, M.D., Morley, S.C., 2016. L-Plastin promotes podosome longevity and supports macrophage motility. Mol. Immunol. 78, 79–88.

Zu, Y., Hanaoka, M., Namba, Y., Shigesada, K., Nishida, E., Kubota, I., Kohno, M., 1990a. 65-kilodalton protein phosphorylated by interleukin 2 stimulation bears two putative actin-binding sites and two calcium-binding sites. Biochemistry 29, 8319–8324.

Zu, Y., Kohno, M., Kubota, I., Nishida, E., Hanaoka, M., Namba, Y., 1990b. Characterization of interleukin 2 stimulated 65-kilodalton phosphoprotein in human T cells. Biochemistry 29, 1055–1062.

CHAPTER FIVE

Profilin choreographs actin and microtubules in cells and cancer

Morgan L. Pimm[a], Jessica Hotaling[a], Jessica L. Henty-Ridilla[a,b,*]

[a]Department of Cell and Developmental Biology, State University of New York (SUNY) Upstate Medical University, Syracuse, NY, United States
[b]Department of Biochemistry and Molecular Biology, State University of New York (SUNY) Upstate Medical University, Syracuse, NY, United States
*Corresponding author: e-mail address: ridillaj@upstate.edu

Contents

Abstract

Actin and microtubules play essential roles in aberrant cell processes that define and converge in cancer including: signaling, morphology, motility, and division. Actin and microtubules do not directly interact, however shared regulators coordinate these polymers. While many of the individual proteins important for regulating and choreographing actin and microtubule behaviors have been identified, the way these molecules collaborate or fail in normal or disease contexts is not fully understood. Decades of research focus on Profilin as a signaling molecule, lipid-binding protein, and canonical regulator of actin assembly. Recent reports demonstrate that Profilin also

International Review of Cell and Molecular Biology, Volume 355
ISSN 1937-6448
https://doi.org/10.1016/bs.ircmb.2020.05.005

regulates microtubule dynamics and polymerization. Thus, Profilin can coordinate both actin and microtubule polymer systems. Here we reconsider the biochemical and cellular roles for Profilin with a focus on the essential cytoskeletal-based cell processes that go awry in cancer. We also explore how the use of model organisms has helped to elucidate mechanisms that underlie the regulatory essence of Profilin in vivo and in the context of disease.

1. Introduction: Cellular infrastructure constructed from cytoskeletal building blocks

Actin and microtubules bestow useful properties to cells including: paths for transport, shape, and the forces that propel and steer movement (Fig. 1). The actin and microtubule cytoskeletons are traditionally studied as separate entities, restricted to specific cellular regions, and regulated by different suites of binding proteins unique for each polymer. Mounting genetic and pharmacological evidence suggests a rich degree of functional crosstalk and physical interactions between both systems, ultimately manifesting in the normal function of fundamental cellular processes, e.g., motility, adhesion, intracellular transport, wound healing, phagocytosis, and cell division (Coles and Bradke, 2015; Dogterom and Koenderink, 2019; Etienne-Manneville, 2004; Li and Gundersen, 2008; Rodriguez et al., 2003). Many proteins likely to be involved in actin-microtubule interactions have already been identified and characterized with regard to either actin or microtubules alone, owing to pioneering work in genetically tractable model organisms (Chang and Martin, 2009; Coles and Bradke, 2015; Dogterom and Koenderink, 2019; Etienne-Manneville, 2004; Gardiner and Marc, 2011; Prokop et al., 2013; Roeles and Tsiavaliaris, 2019; Slater et al., 2017). However, the detailed mechanisms that underlie cytoskeletal synergy and that go wrong in cancer and disease are just starting be investigated (Akhmanova and Steinmetz, 2010; Coles and Bradke, 2015; Dogterom and Koenderink, 2019; Mitchison and Kirschner, 1984a, 1984b; Rodriguez et al., 2003; Salmon et al., 2002). Here we focus on Profilin as a regulator of actin and microtubules in biochemical assays, and cancer-relevant cell processes and signaling pathways.

Actin filaments and microtubules share many properties. All cells contain both polymers in subunit-based and polarized filament forms (Fig. 1, inset). New assembly of either polymer is kinetically unfavorable and requires a template to organize polymer growth. For actin: de novo assembly

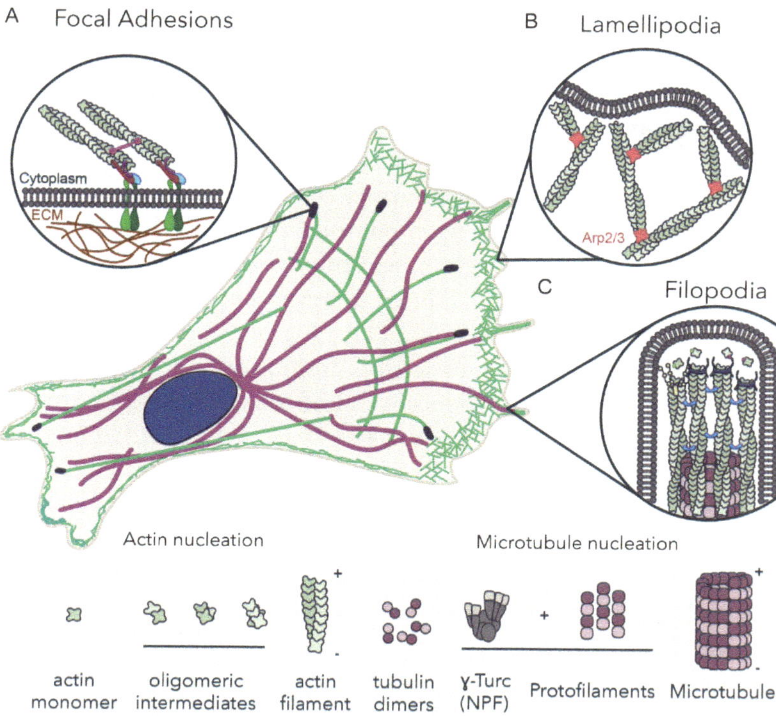

Fig. 1 Cellular actin and microtubule architectures. Cartoon of actin (green) and microtubule (purple) structures in a motile cell. (A) Diagram of actin contact with focal adhesion proteins including: Integrins (green), Vinculin (purple), Talin (red), and Paxillin (blue). Actin filaments are bundled by α-actinin (magenta). (B) Leading edge lamellipodial actin branched actin structures are generated by the Arp2/3 complex. (C) Formins (navy) and Ena/VASP (yellow) elongate straight filaments bundled by Fascin proteins (cyan) in filopodial structures. Inset: Diagram of nucleation steps for actin and microtubules. NPF, nucleation promoting factor. +, the faster growing end of actin or microtubules. −, the slower growing end of actin or microtubules.

requires a seed of three to four monomers that can further polymerize into helical two-stranded filaments with a diameter ~7 nm (Barshop et al., 1983; Cooper et al., 1983; Courtemanche, 2018; Oda et al., 2016; Sept and McCammon, 2001; Skruber et al., 2018; Svitkina et al., 1997). For microtubules: spontaneous microtubule assembly can occur in vitro with high concentrations of tubulin and time (Caudron et al., 2000; Desai and Mitchison, 1997; Fygenson et al., 1995; Job et al., 2003; Nogales, 2001; Roostalu and Surrey, 2017). In cells the critical concentration required for microtubule assembly far exceeds the amount of available free tubulin (estimated ~20 μM) (Fygenson et al., 1995; Voter and Erickson, 1984; Wieczorek et al., 2015). Thus, cellular microtubule assembly requires a stable template to ultimately form a ~25 nm cylinder held together through the lateral contact of ~13 parallel protofilaments (this number varies depending on cell source and experiment) (Chaaban and Brouhard, 2017; Kollman et al., 2010; Roostalu and Surrey, 2017; Wieczorek et al., 2015). Protofilaments are formed from αβ-tubulin heterodimers that intrinsically self-assemble in a head-to-tail fashion and impart structural polarity to the microtubule, with an exposed β-subunit pointing outward (Alushin et al., 2014; Löwe et al., 2001; Mitchison, 1993; Nogales, 2001; Nogales et al., 1998). Cells employ a plethora nucleation promoting factors to overcome these barriers at specific times and locations (for actin: the Arp2/3 complex, Formins, Ena/Vasp, Spire; for microtubules: ɣ-Turc, XMAP215) (Chesarone et al., 2010; Courtemanche, 2018; Kollman et al., 2010; Kovar et al., 2006; Krause et al., 2003; Machesky et al., 1999; Moritz et al., 2000; Mullins and Pollard, 1999; Mullins et al., 1998; Oakley et al., 2015; Pollard, 2007; Popov et al., 2002; Pruyne et al., 2002; Quinlan et al., 2005; Sagot et al., 2002; Thawani et al., 2018). Shared actin and microtubule nucleation proteins can also link dynamic cytoskeletal behaviors in cells and reconstituted systems (Chang, 2000; Colin et al., 2018; Elie et al., 2015; Gaillard et al., 2011; Henty-Ridilla et al., 2016; Inoue et al., 2019; Kita et al., 2019; Lewkowicz et al., 2008; Plessner et al., 2019; Prezel et al., 2018; Szikora et al., 2017). For example, Profilin can block actin filament assembly rendering building blocks sterically or nucleotide assembly incompetent (De La Cruz et al., 2000; Skruber et al., 2018). Profilin can also stabilize parameters of microtubule growth at concentrations below the normal concentration required for microtubule assembly (Henty-Ridilla et al., 2017).

Many cellular actin and microtubule structures like filopodia or the mitotic spindle apparatus are not long lived. Actin and microtubules each display periods of growth (although on different time scales) and rapid disassembly. Dismantling either cytoskeletal polymer is part of a recycling process required for polymer growth to occur again, elsewhere. Many disassembly factors recognize the nucleotide state of actin (ATP, ADP-Pi, or ADP) or microtubules (GTP, GTP-Pi, or GDP) for their function, which can result in a targeted destruction of "aged" regions of polymer (Berro et al., 2010; Brieher, 2013; Manandhar et al., 2018; Margolis, 1981; Pollard and Borisy, 2003). The role of actin or microtubule disassembly factors in coordinating both cytoskeletal polymers have not been extensively studied. However, Profilin promotes actin filament disassembly by sequestering actin monomers and sterically blocking actin assembly (Carlier et al., 1993; Pantaloni and Carlier, 1993). Compared to actin filaments, microtubules have an additional intrinsic disassembly/recycling property (i.e., dynamic instability) that results in the co-existence of growing and shrinking microtubules (Burbank and Mitchison, 2006; Erickson and O'Brien, 1992; Mitchison and Kirschner, 1984a,b; Zhang et al., 2015). Catastrophe events, denoted by the stochastic switch from periods of microtubule growth to rapid depolymerization, may or may not be rescued. Thus, dynamic instability may provide an example of how physical linkages between actin and microtubules disengage. In microtubule dynamics assays Profilin did not have a significant effect on these parameters of microtubule disassembly (Henty-Ridilla et al., 2017). However, because many parameters of actin and microtubule dynamics are concentration dependent, control of either subunit pool through assembly-disassembly dynamics may ultimately influence the coordinated behaviors manifested during diverse cell processes.

In cancers, the unchecked dynamics (nucleation, assembly, disassembly, and crosstalk) of either actin or microtubules results in abnormal cell division, aberrant cell morphology or size, and invasive cell migration (Bae et al., 2009; Ding et al., 2014; Hall, 2009; Mouneimne et al., 2012; Roy and Jacobson, 2004; Tang et al., 2015). Many tumor suppressors (APC, N- and E-cadherin, neurofibromin) interact with both microtubules and actin and have been proposed to disrupt epithelial-mesenchymal transitions (EMT), epithelial positioning, and many aspects of cell migration and division (Hernandez and Tirnauer, 2010; Juanes et al., 2017, 2019). Thus, Profilin is poised as a convergence point linking the cytoskeleton and the signaling pathways that regulate cancer onset, progression, and severity in several tissues. Below we explore the biochemical and cellular facets of Profilin in coordinating actin and microtubules in normal cells, cancer, and lessons learned from unique model organisms.

2. The Profilin origin story

Profilin was first isolated and crystalized from calf spleen extracts as an unrecognized protein present in a 1:1 complex with monomeric actin (Fig. 2A) (Carlsson et al., 1976a). This low molecular weight contaminant was later characterized as the first actin monomer binding protein—remarkable because the hypothesis that a large pool of unpolymerized actin could exist in muscle or non-muscle cells was an emerging idea at the time (Carlsson et al., 1976b, 1977; Tilney, 1976). The association of this

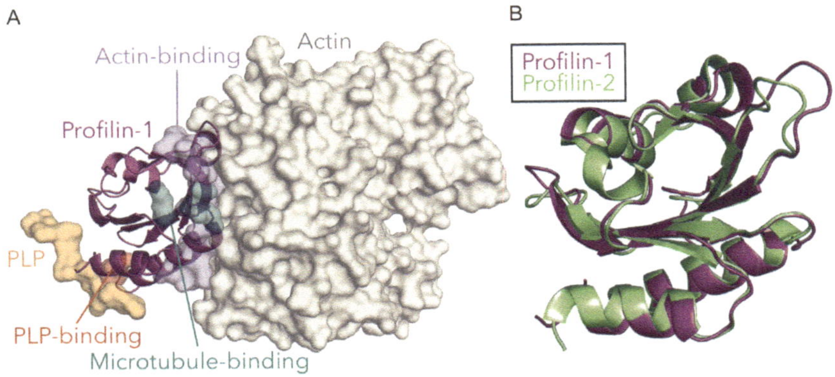

C

	1	2a	2b	3	4
Profilin-1	100	75	78	61	45
Profilin-2a	75	100	95	58	33
Profilin-2b	78	95	100	61	34
Profilin-3	61	58	61	100	38
Profilin-4	45	33	34	38	100

Fig. 2 Structural Features of Profilin. (A) View of Profilin-1 (purple) interacting with actin (gray) and with the poly-L-proline region of VASP (yellow). Important Profilin residues for interactions with actin (blue), microtubules (teal), and poly-L-proline (PLP) peptides (orange) are highlighted. Structures modeled using PDBIDs 2PAVE and 2BTF (Ferron et al., 2007; Schutt et al., 1993). (B) Profilin-1 (purple) and Profilin-2 (green) structures are very similar (RMSD = 0.3), as evident from the aligned structures. Structures modeled using PDBIDs 1FIK and 1D1J (Nodelman et al., 1999). (C) Table of amino acid similarity from BLASTp alignments for Human Profilin isoforms.

unknown protein was sufficient to fully explain the presence of a monomeric actin state in various cell extracts (Carlsson et al., 1977). Profilin was proposed to be a rapidly reversible storage solution, as sequestered actin monomers could be rapidly interconverted between different polymerization states to promote new actin filament formation to adapt to changing cell requirements (Carlsson et al., 1976b; Tilney, 1976). To date, no eukaryotic cells have been described that do not contain actin or Profilin, and both proteins are among the most abundant proteins on Earth (Beck et al., 2011; Dominguez and Holmes, 2011).

Profilins are small (12–15 kDa), well conserved, and highly abundant (Carlsson et al., 1976b; Pollard et al., 2000; Witke, 2004; Witke et al., 1998). They are essential in all forms of life from *Archaea* to eukaryotes, and even some viruses (Akıl and Robinson, 2018; Zaremba-Niedzwiedzka et al., 2017). Profilins exist as a single gene in many organisms (yeast, *Drosophila*, *Acanthamoeba*, certain viruses, *Toxoplasma*) and as several isoforms in others (*Arabidopsis*, *Chlamydomonas*, humans, worms, *Dictyostelium*). The greatest number and diversity of Profilin isoforms comes from *Maize* (10; and 2 additional non-annotated sequences); however, even more may be present in higher ploidy phytozome genomes (Bao et al., 2011). Most Profilins have highly conserved actin-, poly-*L*-proline (PLP)-, and lipid-binding regions, which impart the potential to interact with a variety of additional cytoskeletal regulating factors (Dominguez and Holmes, 2011; Haarer and Brown, 1990; Kaiser et al., 1999; Kovar et al., 2000; Krishnan and Moens, 2009; Thorn et al., 1997; Vinson et al., 1998). The role of Profilin as a major regulator of actin assembly is broadly conserved in each of these systems (Akıl and Robinson, 2018; Di Nardo et al., 2000; Dominguez and Holmes, 2011; Witke, 2004; Zaremba-Niedzwiedzka et al., 2017).

There are four Profilin isoforms present in humans with ~33–95% similarity in amino acids (Fig. 2B–C). Depending on cell type, the estimate and measured amount of Profilin in mammalian cells is between 50 and 200 μM, which is generally considered sufficient or even excessive to the concentration of actin monomers in non-muscle cells (Funk et al., 2019; Kaiser et al., 1999; Lämmermann and Sixt, 2009; Vargas et al., 2017). Profilin-1 and Profilin-2 are ubiquitously expressed and the most abundant Profilin homologs, whereas Profilin-3 and Profilin-4 proteins and transcripts have only been detected in the testes and kidneys (Behnen et al., 2009; Fagerberg et al., 2014; Mouneimne et al., 2012; Witke, 2004; Witke et al., 1998). Profilin-1 is the best characterized isoform, which is a ligand for over 128 binding partners and plays roles in actin assembly, nuclear transport and DNA repair, phase-separation, and as a tumor suppressor in diverse cancers (Burke et al., 2014; Henty-Ridilla and Goode, 2015; Hurst and Welch, 2011; Percipalle and Vartiainen, 2019; Posey et al., 2018; Rotty et al., 2015; Stüven et al., 2003; Suarez et al., 2015; Suarez and Kovar, 2016; Wittenmayer et al., 2004).

Finally, while much focus (>2000 research articles) has illuminated the role of Profilin isoforms with regard to cellular processes that require proper actin dynamics for function, evidence suggests Profilin isoforms can regulate properties of microtubule dynamics through direct and indirect mechanisms in cells and reconstituted assays (see below) (Henty-Ridilla et al., 2017; Holzinger et al., 2000; Nejedla et al., 2016; Witke et al., 1998). Both Profilin-1 and Profilin-2 have been localized to microtubule structures in cells. Profilin-2 has been found on mitotic spindles and asters during mitosis, and Profilin-1 remains present on the sides of microtubules in mouse melanoma cells following cytoplasmic extraction (Di Nardo et al., 2000; Nejedla et al., 2016). Together these findings suggest an intriguing and emerging role for Profilin isoforms in regulating actin-microtubule crosstalk mechanisms (Henty-Ridilla et al., 2017; Holzinger et al., 2000; Pinto-Costa and Sousa, 2019; Wu et al., 2012). Such interactions and isoform differences may ultimately dictate a protective or malignant function in many pathologies including allergies, cardiovascular disease, Amyotrophic Lateral Sclerosis (ALS), Huntington's Disease, Alzheimer's Disease, diabetes, and lung, breast, and colorectal cancers (Ali et al., 2016; Alvarado et al., 2014; Bae et al., 2009; Boopathy et al., 2015; Burnett et al., 2008; Caglayan et al., 2010; Chakraborty et al., 2014; Coumans et al., 2014; Hauser et al., 2010; Henty-Ridilla et al., 2017; Horrevoets, 2007; Janke et al., 2000; Kim et al., 2015; Kooij et al., 2016; Lee et al., 2005; Roy and Jacobson, 2004; Santos and Van Ree, 2011; Shao et al., 2008; Tang et al., 2015; Wittenmayer et al., 2004; Wu et al., 2012; Zoidakis et al., 2012).

3. Biochemistry of Profilin in actin and microtubule dynamics

Competitive interactions underpin all biological processes ranging from individual molecules to ecological scales. Profilin is a champion of navigating the biochemical competition that ultimately regulates actin and microtubule dynamics, lipid and receptor interactions at the membrane, and diverse aspects of cancer signaling. Attempts to engineer a completely functional tagged version of Profilin to assess interactions between >100 cellular binding partners or actin monomers have been extremely challenging (Funk et al., 2019; Melak et al., 2017; Plastino and Blanchoin, 2018; Skruber et al., 2018, 2020; Vitriol et al., 2015). Conventional fluorescent tags (~30 kDa) are twice as large as Profilin (~15 kDa). Tagging Profilin on the C-terminal end disrupts PLP-binding and using the N-terminus or an exposed loop can reduce Profilin-actin binding interactions (Nejedla et al., 2017; Wittenmayer et al., 2000). Direct labeling approaches for detecting single molecules of Profilin require four to five mutations and can result in aggregation (Henty-Ridilla et al., 2017; Vinson et al., 1998). As a consequence, most of what we know about Profilin has been determined from biochemical investigations of purified proteins, rather than cellular observations.

Cellular actin exists in several forms that are generated and maintained by different classes of regulatory proteins: straight filaments, bundles, branched networks, and globular monomers (Blanchoin et al., 2014; Michelot and Drubin, 2011; Svitkina, 2018; Svitkina and Borisy, 1999). Based solely on biochemical principles (the critical concentration for assembly is 0.1 μM), and estimates of the total cellular concentration of actin (~20–200 μM) and Profilin (20–200 μM), actin in cells predominately exists as polymerized filaments (Dominguez and Holmes, 2011; Funk et al., 2019; Gordon et al., 1977; Lodish et al., 2000; Pollard, 1986; Pollard et al., 2000). Actin filaments are constantly remodeled, assembled, destroyed, and recycled to reflect changing cell requirements in response diverse stimuli (Henty-Ridilla et al., 2013; Pollard, 2016). To achieve such actin dynamics, cells maintain a substantial stash of monomeric actin (~50–90% of the total actin depending on cell source), buffered by sequestering proteins that include Profilin and Thymosin-β4 (Tβ4) (Carlsson et al., 1977; Dominguez and Holmes, 2011; Pollard et al., 2000; Safer et al., 1991; Skruber et al., 2018). It has been tempting to speculate that most monomeric actin in cells exists in a 1:1 complex with Profilin because the concentration is commonly reported as near equimolar with actin, Profilin effectively outcompetes Tβ4 for binding actin monomers, and structurally inhibits new actin filament polymerization (Blanchoin et al., 2014; Carlier et al., 1996; Carlsson et al., 1976b; Cooper et al., 1984; Dominguez and Holmes, 2011; Funk et al., 2019; Goldschmidt-Clermont et al., 1992; Kaiser et al., 1999; Vinson et al., 1998). Although the activities of Profilin sum to an attractive explanation for the requirement of nucleation factors to stimulate actin assembly, the balance between free actin monomers and those bound to Profilin, Tβ4, or other cellular ligands is not well characterized and extremely challenging to determine (Aroush et al., 2017; Henty-Ridilla et al., 2017; Plastino and Blanchoin, 2018; Skruber et al., 2018). Further, the availability of cellular actin monomers also fluctuates due to mechanisms of actin filament turnover and the presence of suites of disassembly proteins that synergize with Profilin to convert actin filaments into a polymerizable or sequestered monomeric state (Aroush et al., 2017; Blanchoin et al., 2000; Blanchoin and Pollard, 1998; Brieher, 2013; Bubb et al., 2003; Didry et al., 1998; Pollard, 2016; Pollard and Borisy, 2003; Pollard et al., 2000; Yarmola et al., 2001).

Profilin has a profound effect on actin polymerization in other ways besides sequestering actin including recharging actin monomers with ATP for new filament assembly, and interactions with signaling lipids and proline-rich proteins (Chou and Pollard, 2019; Machesky and Pollard, 1993; Merino et al., 2018; Schlüter et al., 1997). Only two proteins are known to catalyze nucleotide exchange on actin monomers, Cyclase-Associated Protein (CAP) and Profilin. Historically Profilin has been depicted as the key driver of actin monomer recycling because it is able to bind the "aged" ADP-bound actin monomers that are released from depolymerizing actin filaments (Bertling et al., 2007; Goldschmidt-Clermont et al., 1992; Kotila et al., 2018; Mockrin and Korn, 1980; Vinson et al., 1998; Wolven et al., 2000). New evidence suggests that CAP can do this more efficiently than Profilin in mammalian systems, and there are organisms where Profilin is not known to catalyze this nucleotide exchange (Kotila et al., 2018; Ono, 2013). Profilin also has strikingly different affinities for monomers and for the growing ends of polymerizing actin filaments, and this property permits the efficient dissociation of Profilin from the ends of growing filaments (Courtemanche and Pollard, 2013; Funk et al., 2019; Jégou et al., 2011). Profilins have high affinity for binding PIP lipids which are conveniently located in the cellular membranes that polymerizing actin filaments push against to generate forces, create protrusions, and drive endocytosis (Dürre et al., 2018; Lu and Pollard, 2001). Interactions between PIP lipids and Profilin facilitate the release of Profilin-bound actin monomers for actin assembly (Lassing and Lindberg, 1985; Lu and Pollard, 2001; Ostrander et al., 1999; Schlett, 2017). All Profilins are known to bind some form of PIP lipid alone or in complex with actin monomers, with the exact PIP binding site defined by ~11 positively charged amino acids (Lu and Pollard, 2001; Sohn et al., 1995; Vinson et al., 1993). It is difficult to compare the exact affinities of Profilin for various PIP lipids across studies because different experimental methods (i.e., PIP compositions, ionic strength buffers, and wash steps) have been used and each of these variables drastically influences PIP binding (Shirey et al., 2017).

Profilin can reliably recognize and jointly bind, actin monomers and proline-rich motifs (i.e., PLP) present in many cytoskeletal regulatory proteins including: Formins, Ena/VASP, WASP/VCA-domain activators of the Arp2/3 complex, and Drebrin (Chang et al., 1997; Evangelista, 1997; Evangelista et al., 2002; Ferron et al., 2007; Gertler et al., 1996; Higgs and Pollard, 1999; Mammoto et al., 1998; Miki et al., 1998; Reinhard et al., 1995; Rodal et al., 2003; Suetsugu et al., 1998). The ability to simultaneously bind actin monomers and PLP motifs endows Profilin with strong regulatory powers over actin polymerization and the actin monomer pool. Profilin-Formin interactions are mediated by PLP tracks present in all Formin proteins (Paul et al., 2008). Profilin interactions with specific Formin PLP tracks can strongly enhance the elongation and nucleation phases of actin assembly with micromolar to millimolar affinities (Chang et al., 1997; Courtemanche, 2018; Funk et al., 2019; Horan et al., 2018; Paul and Pollard, 2008; Perelroizen et al., 1994; Petrella et al., 1996; Sagot et al., 2002; Sherer et al., 2018; Watanabe et al., 1997; Zweifel and Courtemanche, 2020). The number and lengths of PLP tracks vary with each Formin, but each track competes to bind Profilin-bound actin monomers which ultimately increases the probability and speed actin monomers will be added to the growing actin filament in the correct orientation (Courtemanche and Pollard, 2012; Horan et al., 2018; Sherer et al., 2018; Zweifel and Courtemanche, 2020). Profilin-bound actin monomers hinder branched actin filament assembly mediated by the Arp2/3 complex which prefers unbound actin monomers for assembly (Burke et al., 2014; Mullins et al., 1998; Rodal et al., 2003; Rotty et al., 2015; Skruber et al., 2018, 2020; Suarez et al., 2015; Suarez and Kovar, 2016). Contrary to this mechanism, actin assembly mediated by Formin proteins requires Profilin and is significantly enhanced with Profilin-bound actin monomers (Burke et al., 2014; Chang et al., 1997; Evangelista et al., 2002; Funk et al., 2019; Henty-Ridilla and Goode, 2015; Kovar et al., 2003; Neidt et al., 2009; Romero et al., 2004; Rotty et al., 2015; Skruber et al., 2018, 2020; Suarez et al., 2015; Suarez and Kovar, 2016). Thus, competition between different actin nucleation systems has led the popular idea that Profilin tunes specific forms of actin assembly depending on the concentration of *active* nucleation proteins present (Rotty et al., 2015; Skruber et al., 2020; Suarez et al., 2015). Profilin-Formin isoform pairs in worms can further tune these activities (Neidt et al., 2009), which may have important implications in systems with higher numbers of Formin and Profilin isoforms present.

While much attention has focused on the role of Profilin in regulating actin dynamics, Profilin is also capable of regulating microtubule polymers and actin-microtubule crosstalk. In one of the first comprehensive studies comparing Profilin isoforms, tubulin and microtubule-associated proteins were first identified as ligands of Profilin-1 and Profilin-2 from affinity chromatography of mouse brain extracts (Witke et al., 1998). Profilin directly binds to microtubule sides ($K_D = \sim 11\,\mu M$) through specific amino acids in sites adjacent to the actin-binding surface on Profilin, and this microtubule binding activity is sensitive to the presence of actin monomers when both cytoskeletal elements are present in equal concentrations (Henty-Ridilla et al., 2017). In cells, Profilin resides on spindle and astral microtubules during mitosis and influences microtubule dynamics (Di Nardo et al., 2000; Henty-Ridilla et al., 2017; Nejedla et al., 2016). Some microtubule

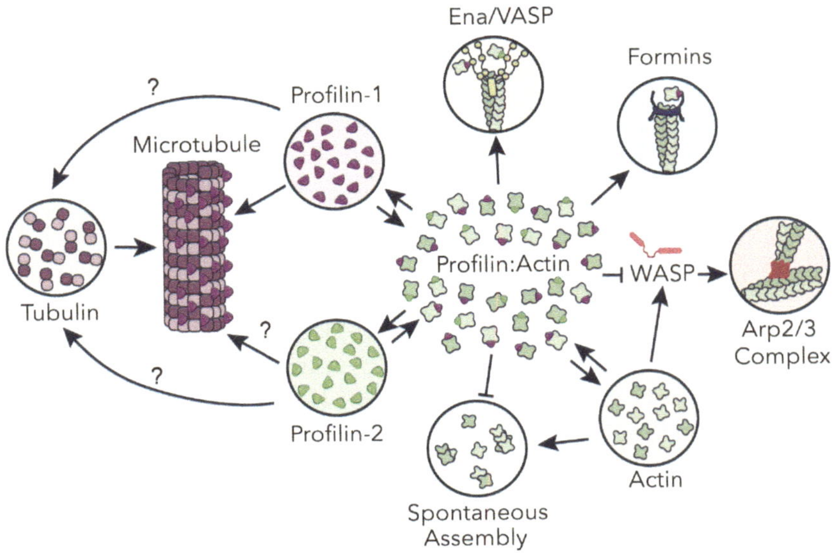

Fig. 3 Competition for Profilin Dictates the Types of Cellular Cytoskeletal Structures Formed. Cartoon model for the distribution of Profilin to actin, microtubules, or regulatory ligands (Formins, Ena/VASP, the Arp2/3 complex). Based on biochemical principles, free Profilin pools likely exist in cells. Direct interactions between isoforms of Profilin and tubulin are hypothesized but not yet directly confirmed (Henty-Ridilla et al., 2017; Nejedla et al., 2016; Pinto-Costa and Sousa, 2019; Witke et al., 1998).

effects may be indirectly mediated through interactions between Profilin and Formin proteins that can also bind to microtubules (Bender et al., 2014; Nejedla et al., 2016; Pinto-Costa and Sousa, 2019; Szikora et al., 2017). At present there is not a simple assay to assess whether endogenous Profilin influences microtubule dynamics through direct mechanisms in cells. However, based on biochemical observations, cellular concentrations, estimates of the size of the Profilin-bound actin monomer pool, and relevant protein affinities, it is very likely that a pool of free "unbound" Profilin exists in the cytoplasm of mammalian cells and is available to bind microtubules and additional ligands at physiological concentrations (Fig. 3) (Henty-Ridilla et al., 2017; Henty-Ridilla and Goode, 2015; Plastino and Blanchoin, 2018).

4. Role of Profilin isoforms in cancer

Humans have four Profilin isoforms, with Profilin-1 commonly accepted as is the most ubiquitous and abundant isoform in almost all tissues and cell types (Fig. 4A) (Behnen et al., 2009; Fagerberg et al., 2014;

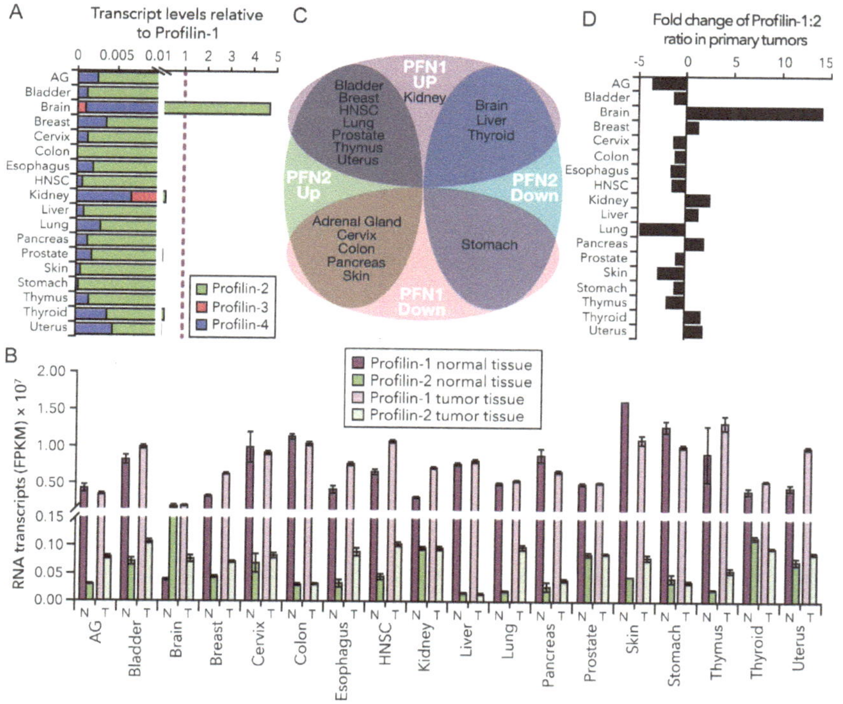

Fig. 4 Profilin Transcripts in Normal and Tumor Tissues. (A) Transcript levels for Profilin-2, Profilin-3, and Profilin-4 relative to the most ubiquitous expressed isoform, Profilin-1. We normalized the means of all available RNAseq transcript data in "Projects" that contained normal tissue samples from cancer patients ($n = 1$–215) currently available in the National Cancer Institute Genomic Data Commons (https://portal.gdc.cancer.gov; Grossman et al., 2016). The purple dotted line represents Profilin-1 levels in each tissue shown. AG, adrenal gland. HNSC, head neck squamous cells. (B) Mean RNA transcripts of Profilin-1 and Profilin-2 obtained from the database in (A) for normal and tumor tissues in fragments per kilobase exon model per million mapped reads (FPKM), ($n = 1$–1191). Error bars, SEM. (C) A Venn diagram summarizing the changes found in (B) for Profilin-1 (PFN1) and Profilin-2 (PFN2) transcripts in tumors compared to normal tissues. (D) Fold change in the Profilin-1:2 ratio in primary tumors. All data was downloaded and analyzed from https://portal.gdc.cancer.gov on 10 April 2020 (data release 23.0). All transcript data was sorted by tissue source across "Project" databases with the exception of HNSC which was is presented a mix of tissue sources.

Mouneimne et al., 2012; Witke, 2004; Witke et al., 1998). Thus, the majority of cellular and biochemical studies have focused on the activities of Profilin-1. Profilin-3 transcripts are virtually absent from all tissues except kidneys where transcripts are 83-fold less abundant than Profilin-1 (Fig. 4A). Profilin-4 transcripts are more abundant than Profilin-3 across tissues except kidneys, but are still much less abundant than either Profilin-1 or

Profilin-2 isoforms (Fig. 4A). The only known location where Profilin-1 is not the most predominate isoform is in neuronal-derived tissues and cells. Here, Profilin-2 proteins and transcripts have been measured ~5-fold more abundant than Profilin-1, although the exact mechanisms that underlie this distinct distribution are still not fully elucidated (Fig. 4A) (Gareus et al., 2006; Mouneimne et al., 2012; Witke et al., 1998). There are two alternatively spliced versions of Profilin-2 (designated 2a and 2b) differing by nine amino acids in the C-terminal region and an extended patch of aromatic residues (Gieselmann et al., 2008; Lambrechts et al., 1997; Nodelman et al., 1999). Both splice variants of Profilin-2 have similar affinities for actin but differ in binding other ligands (Nodelman et al., 1999; Witke et al., 1998). Profilin-2a is the predominant form, whereas Profilin-2b is restricted to very limited tissues (Lambrechts et al., 2006). While Profilin-1 and Profilin-2 have similar effects on many biochemical properties pertaining to actin dynamics including nucleotide exchange and binding phosphatidylinositol (PIP) lipids, Profilin-2 has a five-fold lower binding affinity for actin monomers and has higher affinity for EVL and VASP PLP-containing ligands (Gieselmann et al., 2008; Mouneimne et al., 2012). Thus far, Profilin-2 has been studied as a regulator of actin dynamics and in PLP ligand binding assays.

Several studies have suggested the intriguing and controversial idea that perturbations to Profilin-1 and Profilin-2 have opposing phenotypes in several cancers (Baraniskin et al., 2012; Cui et al., 2016; Janke et al., 2000; Mouneimne et al., 2012; Wittenmayer et al., 2004; Zhang et al., 2018; Zoidakis et al., 2012). In short, elevated levels of Profilin-1 are correlated with a tumor suppressive effect and reduced metastasis in breast, lung, colorectal, bladder, esophageal, and thyroid cancers, whereas elevated levels of Profilin-2 produce higher metastatic behaviors (Janke et al., 2000; Jiang et al., 2017; Mouneimne et al., 2012; Wittenmayer et al., 2004; Zhang et al., 2018; Zou et al., 2007). There are inconsistencies to this dichotomy, however, and no satisfying explanation reconciling the differences between these studies has been reached beyond differences arising from tissue-derived or cell line specific phenotypes and sample sizes used.

To clarify whether Profilin isoform levels are correlated with tumor development and/or metastasis across diverse cancers, we analyzed all RNAseq transcript data in "Projects" that contained normal tissue and primary tumor samples from cancer patients currently available in the National Cancer Institute Genomic Data Commons (https://portal.gdc.cancer.gov; Grossman et al., 2016) (Fig. 4). In most of the tissues examined (i.e., bladder, breast, esophagus, head and neck squamous cell (HNSC), lung, prostate, thymus, and uterus), both Profilin-1 and Profiln-2 transcript levels were elevated in primary tumors compared to normal tissues (Fig. 4B and C). The second most common trend observed was that primary tumor transcripts displayed decreased Profilin-1 transcripts and elevated Profilin-2 transcripts compared to normal tissues (i.e., adrenal gland, cervix, colon, pancreas, and skin) (Fig. 4B and C). The only measurement where both Profilin isoforms decreased was in stomach primary tumors (Fig. 4B and C). We did not observe any transcript-based trends between Profilin isoforms for metastatic and reoccurring tumors, although this data was much less abundant and not available for all the tissues investigated above. The most spectacular observation was in brain tumors—in normal tissue Profilin-2 transcripts are ~5-fold more abundant than Profilin-1; however, in primary tumors Profilin-1 transcripts outnumber Profilin-2 by 2.7-fold (Fig. 4B and C). Combined this results in an over 12-fold relative increase in Profilin-1 transcripts in brain tumors!

These observations and previous studies may raise the exciting possibility that cellular ratios of Profilin-1 and Profilin-2 underlie the behavioral differences between cancers (Fig. 4D). In general terms the ratio of Profilin-1 to Profilin-2 transcripts in primary tumors decreases (Fig. 4D), but this is not simply the result of a change in Profilin-1 transcripts (i.e., Profilin-1 transcripts can be reduced, Profilin-2 transcripts can become elevated, or combination of both may occur) (Fig. 4B and C). We also explored whether patient survival was connected to changes in Profilin transcripts or ratios from this data, but no specific correlation was observed. However, this observation has several confounding factors that should be considered including: the similarities and differences in patient treatment plans, age, gender, and progression at diagnosis. Finally, while transcript levels are relatively easy to obtain or measure, they do not always reflect the amount of protein present and available for cellular activities, particularly those associated with regulating the cytoskeleton. Future studies will likely have to measure these parameters in specific cell types and circumstances.

5. Profilin in cell division

Cancer in the most basic sense is disordered or uncontrolled cell division, instigating changes in the rate cells divide, the activity of cell cycle regulators and signals, or inhibition of normal cell maintenance/death. In a cancer-free context, the details underlying the dynamics of the mitotic spindle have been the subject of intense scrutiny for literally hundreds of years (McIntosh and Hays, 2016). Changes in cell architecture supported by the actin and microtubule cytoskeletons is a normal requirement to progress through the cell cycle. Some of the most dramatic cytoskeletal reorganizations are triggered by Cyclin complexes which alter the dynamics of motor proteins organizing the mitotic spindle, Rho GTPases, and proteins that facilitate actin-microtubule interactions (Blangy et al., 1995; Böttcher et al., 2009; Jiang et al., 2015; Kita et al., 2019; Miller, 2011; Plessner et al., 2019; Pollard and Wu, 2010; Ubersax et al., 2003; Yamashiro et al., 1991). Much information that clarifies the roles for microtubules and many signaling factors is known, however functions of actin (and regulatory partners) in cell division have been more elusive and are just starting to emerge. For example, many tools used for visualizing cytoskeletal proteins in mitotic spindles were considered technically limiting and challenging for actin filaments, despite plentiful descriptions of its presence associated with the mitotic spindle (Cande et al., 1977; Gawadi, 1971; Herman and Pollard, 1979; Sanger, 1975). While the classic and most characterized roles for actin in cell division pertain to generating the forces required for cytokinesis, additional studies demonstrate that actin is important for positioning the spindle and spindle pole separation (Miller, 2011; Pelham and Chang, 2002; Rosenblatt et al., 2004; Théry et al., 2005; Toyoshima and Nishida, 2007; Watanabe et al., 2008), extensively reviewed: (Pollard and O'Shaughnessy, 2019). Recent observations show that there are dynamic populations of actin and actin-microtubule-associated structures localized to the mitotic spindle, and these structures reorganize as cells advance through the mitotic phase (Kita et al., 2019; Plessner et al., 2019). Actin nucleation proteins, microtubule-associated proteins, and Profilin are proposed mediators of these dynamics, but the detailed mechanisms of how they may go wrong in cancer are not clear (Henty-Ridilla et al., 2016, 2017; Ishizaki et al., 2001; Kita et al., 2019; Nejedla et al., 2016; Plessner et al., 2019; Roth-Johnson et al., 2014; Wade, 2007).

The role for Profilin in the cell cycle has been difficult to discern, complicated by its affinity for a plethora of signaling molecules and actin and microtubule regulation proteins. Classic genetics has demonstrated that Profilin-1 is essential for generating the cytokinetic ring, cell survival, and division in many organisms (Chang et al., 1997; Kandasamy et al., 2002; Kovar et al., 2003; Polet et al., 2006; Severson et al., 2002; Vidali et al., 2007; Witke et al., 2001). Further, in systems where multiple isoforms of Profilin are available, the loss of the most ubiquitous Profilin cannot be fully rescued for cell cycle effects with the other isoforms (Polet et al., 2006;

Witke et al., 2001). Specialized mouse cells lacking Profilin-1 displayed morphological defects and aberrant actin filament distributions but were able to complete mitosis albeit on a slower timescale than normal cells (Böttcher et al., 2009). These phenotypes were unable to be rescued with Profilin point mutants deficient for binding actin or Formin proteins. These results may indicate cell dependent differences in Profilin function but may suggest the involvement of actin-independent functions of Profilin.

In addition to organizing microtubules, centrosomes organize a network of actin filaments generated by the Arp2/3 complex. Intriguingly, increasing density or crosslinking of actin filaments correlates with a reduction in microtubules in vitro and at centrosomes in cells (Colin et al., 2018; Farina et al., 2016, 2019; Inoue et al., 2019; Ricketts et al., 2019). Hence, the centrosome is a coordinator of actin and microtubules, and in light of biochemical studies elucidating the role of Profilin with actin nucleation promoting factors and microtubules, this relationship may be indirectly regulated through Profilin (Burke et al., 2014; Funk et al., 2019; Rotty et al., 2015; Skruber et al., 2020; Suarez et al., 2015). In sum, the complete functions of Profilin in cell division remain unclear. Some roles are likely directly related to Arp2/3- or Formin-mediated actin filament assembly or in regulating microtubule dynamics, and some may go beyond including generating the forces for cytokinesis (Chang et al., 1997; Kita et al., 2019; Kovar et al., 2003; Nejedla et al., 2016; Plessner et al., 2019; Severson et al., 2002).

6. Profilin in cell motility and metastasis

The migration of cells is a complex biological process that requires the reorganization of actin, microtubules, membrane receptors, lipids, and the cell-matrix. The loss of proper cell migration has profound effects on neuronal pathfinding, development, wound healing, and overactive migration is a classic hallmark of metastasis and ultimately responsible for distributing tumorigenic cells to sites in the body where they do not normally belong. Cells initiate movements by extending lamellar membrane protrusions driven by physical forces produced by assembling actin filaments that push on the membrane surface. Traction is produced from directionally elongated focal adhesion sites and actin stress fibers to propel the cell forward. Meanwhile contractile forces produced by actin-myosin stress fibers retract the back of the cell as it advances onward (Pollard and Borisy, 2003). Less is known about microtubules or crosstalk between actin and microtubules in this process, however pharmacological evidence demonstrates that perturbing either system alters motile behaviors (Coles and Bradke, 2015;

Dogterom and Koenderink, 2019; Etienne-Manneville, 2004; Rodriguez et al., 2003). There are also numerous connections between microtubules and focal adhesion complexes as well as integrin-based extracellular matrix adhesions (Borisy et al., 2016; Bouchet and Akhmanova, 2017; Bouchet et al., 2016; Dziezanowski et al., 1980; Euteneuer and Schliwa, 1984; Kaverina and Straube, 2011; Kaverina et al., 1998; Rodionov et al., 1998; Wittenmayer et al., 2004; Wittmann and Waterman-Storer, 2001).

Motility can be recapitulated on a bead in vitro or in genetically tractable motile organisms like *Listeria* from a core set of proteins including actin, the Arp2/3 complex, Capping Protein, Cofilin, and Profilin (Loisel et al., 1999; Pantaloni et al., 2001; Pollard and Borisy, 2003; Theriot et al., 1992, 1994; Tilney et al., 1992; Wiesner et al., 2003). Actin filaments at the leading edge are mostly formed by the Arp2/3 complex with the faster growing end oriented outward (Pollard and Borisy, 2003; Rouiller et al., 2008; Small, 1988; Small and Celis, 1978; Svitkina, 2018; Svitkina and Borisy, 1999; Svitkina et al., 2003; Symons and Mitchison, 1991). Growing filaments are capped relatively quickly at short lengths and are therefore mechanically suited to generate/sustain sufficient forces to propel the cell forward (Akin et al., 2008; Blanchoin et al., 2000; Mogilner and Oster, 1996). Biochemical evidence demonstrates that the presence of Profilin-bound actin inhibits branched actin assembly, that Profilin is required for nucleotide exchange to assemble new actin filaments, and that amounts of free Profilin can compete for the faster growing end of actin filaments with Capping Proteins, Formins, and other ligands (Bubb et al., 2003; Cooper et al., 1984; Dos Remedios et al., 2003; Kaiser et al., 1999; Mockrin and Korn, 1980; Mullins et al., 1998; Pernier et al., 2016; Rotty et al., 2015; Skruber et al., 2020; Suarez et al., 2015; Vinson et al., 1998). There are still many questions that underlie the behavior of actin and microtubule networks at the leading edge of crawling cells. How do cells assemble and rearrange cytoskeletal polymers so quickly? Why do cells expend so much energy incessantly constructing and disassembling these proteins? What detailed roles do the five minimal proteins perform in cells, can they be visualized, and how do they go awry in disease?

7. Cell signals converging on Profilin

Many interconnected signaling pathways and feedback loops contribute to cell homeostasis and respond in disease, particularly in cancer. This labyrinth of signals commonly challenges the development of therapeutics, especially when target molecules exhibit high sensitivity to a diversity of ligands spanning pathways on very rapid timescales. Several signaling pathways converge on Profilin in cancer (Fig. 5), however whether or not these pathways use Profilin as a regulator of the cytoskeleton are not always clear. In addition, specific modifications to Profilin (usually phosphorylation) directly impact actin dynamics, but the identity and timing of signals and kinases that regulate these modifications in disease remain unknown.

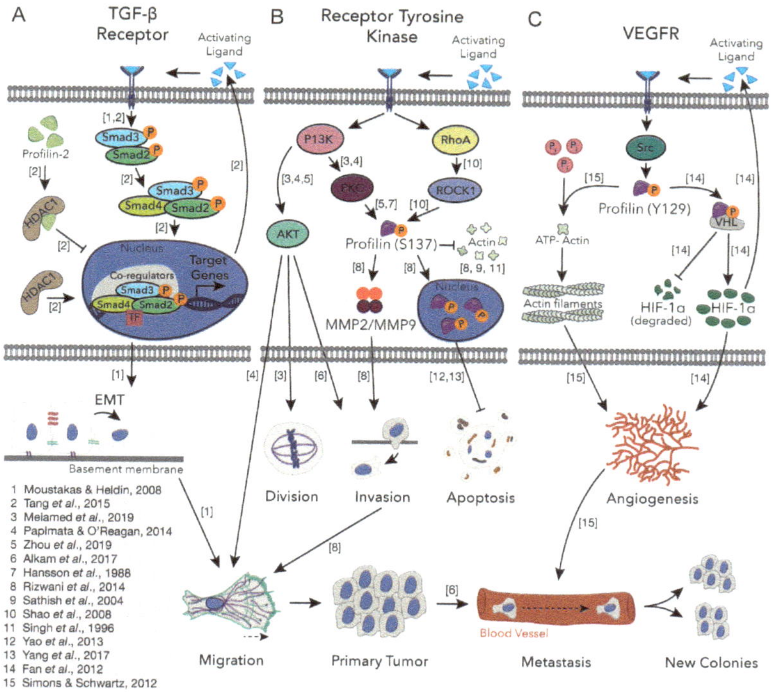

Fig. 5 Signaling Pathways that Converge on Profilin. (A) Upon stimulation by an activating ligand the TGFβ signaling pathway activates Smads which promote the epithelial to mesenchymal transition (EMT). Profilin-2 has an inhibitory effect on this pathway by preventing HDAC1 signals to the nucleus. TF, transcription factor. (B) Upon activation by diverse receptor tyrosine kinases the P13K/TOR/AKT pathway ultimately stimulates cell proliferation, invasion, and migration. Phosphorylation of Profilin-1 (S137) loses the ability to bind actin monomers and becomes translocated to the nucleus and inhibits cell death pathways. (C) Activation of the VEGFR pathway leads to the phosphorylation of Profilin-1 (Y129), stimulating angiogenesis. Phosphorylated Profilin-1 (Y129) enhances nucleotide exchange on actin monomers, ultimately increasing overall actin filament polymerization. Numbers present throughout the figure are references (purple box) for particular signaling steps.

The Transforming Growth Factor Beta (TGFβ) pathway is essential in development, regulating cell growth and differentiation, apoptosis, and is a common place where signals go astray in diverse cancers, ultimately driving the epithelial to mesenchymal transition (EMT) of cancer cells and permitting invasive migratory behaviors (Moustakas and Heldin, 2008). TGFβ ligands bind cell receptors which recruit and phosphorylate signal transducing transcription factors (SMADs) to mediate downstream responses. While Profilin-1 has no reported effect on TGFβ signals, increased Profilin-2 protein correlates with SMAD2/3 signals, reducing Profilin-2 or SMAD2/3 levels correlated with tumor suppression in mice, and an early spike in TGFβ activity in a luciferase assay was reduced (Tang et al., 2015). Additional analysis revealed a downstream cytoplasmic interaction between Profilin-2 with HDAC1 that further reinforces SMAD nuclear activities by inhibiting HDAC1 (Fig. 5A) (Tang et al., 2015). The P13K/AKT/mTOR (Phosphatidylinositol 3-kinase/Protein Kinase B/mammalian Target of Rapamycin) intracellular pathway is one of the most commonly mutated in cancer (Melamed et al., 2019; Paplomata and O'Regan, 2014). The overexpression of Profilin-2 in head and neck cancer cell lines increased cell proliferation, and these cells displayed higher levels of phosphorylation for AKT and downstream effectors including β-catenin (Zhou et al., 2019). Therefore, Profilins are an important link in the web of cancer signaling pathways that require the cytoskeleton for function.

Protein phosphorylation is widely used to regulate biological functions. Phosphorylation is the only known category of post-translational modification of Profilin, and it can occur at several amino acid sites to regulate actin-based activities (Alkam et al., 2017). In cells, Profilin-1 can be targeted for phosphorylation by Protein Kinase C (PKC) at S137 downstream of P13K signals, and also by Rho-associated Kinase-1 (ROCK1) downstream of GTP-signals, and is linked to promoting metastasis and invasion in breast cancer cells (Fig. 5B) (Hansson et al., 1988; Rizwani et al., 2014; Sathish et al., 2004; Shao et al., 2008; Singh et al., 1996; Yang et al., 2017; Yao et al., 2014). In biochemical assays phosphomimetic mutations at this site indicate that actin monomers do not bind this modification of Profilin as efficiently compared to wild-type versions (Shao et al., 2008). Profilin-1 can also be phosphorylated at Y129, which is present in the actin binding site of the protein and unsurprisingly reduces the binding capacity of Profilin-1 for monomeric actin (Fan et al., 2012). In cells this modification occurs by Src kinase initiated through a Vascular Endothelial Growth Factor

Receptor Kinase 2 (VEGFR2) signaling cascade (Fan et al., 2012; Simons and Schwartz, 2012). The Y129 phosphorylation has also been linked to the progression of glioblastoma by forming a complex with the Von Hippel-Lindau protein that prevents the degradation of hypoxia induced factor 1 alpha (HIF-1α), ultimately driving the vascularization of tumors and cancer progression (Fig. 5C) (Fan et al., 2014). The impacts of phosphorylation on Profilin-2 have not been as extensively investigated. Using an in silico approach, 14 potential phosphorylation sites on the Profilin-2a protein have been identified, seven of which were biochemically characterized: Y29, S71, S76, Y78, S129, Y133, and S137 (Walter et al., 2020). Phosphorylation of Profilin-2 at S71, S76, or S129 disrupted actin-binding activities, and intriguingly phosphorylation of S76 was able to stimulate the elongation phase of actin polymerization (Walter et al., 2020). Lastly, to use post-translational modifications as effective molecular switches, the cellular balance between the phosphorylated and dephosphorylated states of Profilin must be maintained. To date Protein Phosphatase-1 (PP1) is the only known phosphatase to effectively dephosphorylate Profilin-1, specific to amino acid S137 (Shao and Diamond, 2012). Whether PP1 can perform this role at other phosphorylation sites in Profilin-1 or Profilin-2 or if other kinases contribute to these functions has not been fully explored.

8. Profilin in immune system responses

How do cancer cells avoid detection or eradication by the immune system? Can targeting the host immune responses contribute to better treatment outcomes? The presence of inflammatory immune cells in human tumors and innate (receptor-ligand interactions) and adaptive immune responses (phagocytic macrophages) contribute to the progression of cancer by inducing immunosuppression, stimulating cancer proliferation and metastasis (Palucka and Coussens, 2016). Reports have investigated the contribution of Profilin in the innate and adaptive immune responses elicited by different microbes, but these responses in the context of cancer or cancer recovery have not been explored. In the human adaptive immune system, actin and microtubules are essential for migration, phagocytosis, cell secretion, and cell-cell interactions (Mostowy and Shenoy, 2015; Pfajfer et al., 2018; Wickramarachchi et al., 2010). Although only disruptions to actin dynamics were investigated, cells without Profilin-1 or Profilin-2 fail to perform actin-microtubule mediated phagocytosis in macrophages

(Coppolino et al., 2002; Kim et al., 2012). In Cytotoxic T Lymphocytes (CTLs) Profilin negatively regulates the exocytosis of lytic granules, which may suggest an enhanced ability to both eliminate tumor cells and increase the migration and invasion of these "helpful" immune cells (Schoppmeyer et al., 2017). In other immune cells (e.g., dendritic cells, neutrophils) Profilin-1 protein levels were higher than cancer cell lines (HT1080 and B16F10) and Profilin-2 was only detected in the dendritic cell line used (Funk et al., 2019). Presumably these findings suggest a role for Profilin in ameboid migratory behaviors controlled by the actin and microtubule cytoskeletons but the exact mechanisms have not been fully elucidated (Lämmermann and Sixt, 2009).

9. Targeting Profilin as a cancer therapeutic

The timing of the cell cycle, the morphology of cells and tissues, and the direction and speed of cellular movements are essential processes regulated by the broad actions of cellular actin and microtubule dynamics. As a consequence, compounds (synthetic and natural) that disrupt cytoskeletal dynamics are among the most widely utilized chemotherapeutics available. These properties also cause treatments to be extremely toxic to patients. With the critical nature of microtubules in cells, drugs that target microtubule dynamics are some of the most effective therapeutics available (Mukhtar et al., 2014). For example, one of the first compounds that targeted the cytoskeleton to treat cancer was Taxol, which stabilizes microtubules and effectively arrests cell division in breast, ovarian, lung, prostate, blood, and many other cancers (Fife et al., 2014; Weaver, 2014). Existing pharmaceutical agents target actin dynamics (i.e., latrunculins, cytochalasins, jasplakinolides), however these compounds are indiscriminately toxic to numerous organs, cardiac, and muscle function in addition to cancerous tumors (Bonello et al., 2009). The development of compounds that target actin and microtubule regulatory proteins are even more rare, but come with the advantage of a targeting specific features of cytoskeleton dynamics (i.e., cytoskeletal assembly, disassembly, stabilization, turnover, or motor protein dynamics) or potentially actin-microtubule crosstalk. Compounds targeting the microtubule-associated proteins Tau or the kinesin Eg5 are excellent at inducing mitotic arrest and limiting tumor proliferation but frequently fail as a clinical monotherapy due to their acute specificity (Chan et al., 2010; Engelke et al., 2016; Hancock, 2014; Milic et al., 2018; Pan et al., 2017; Smith et al., 2013; Sturgill et al., 2016). Other small-molecule screens targeting actin assembly identified inhibitors for the Arp2/3 complex, N-WASP, Tropomyosins, and Formins, although recent evidence questions the specificity of some of these molecules in cells (Bolger-Munro et al., 2019; Hetrick et al., 2013; Isogai et al., 2015; Nolen et al., 2009; Peterson et al., 2004; Rizvi et al., 2009; Sellers et al., 2020; Stehn et al., 2013).

Profilin may represent an effective therapeutic target to fight diverse cancers due to its roles in cytoskeletal regulation and position in cancer signaling cascades. Intriguingly hyperactive Profilin-1 in signaling pathways can lead to precocious apoptosis and resistance to several chemotherapeutics, while silencing Profilin-1 can reduce tumor growth in vivo (Frantzi et al., 2016; Zou et al., 2010). Small molecule screens have revealed compounds that mitigate breast cancer-induced changes in Profilin expression and migration and two small molecules that prevent the interaction of Profilin with actin monomers have been identified (C1 and C2) (Gau et al., 2018; Joy et al., 2014). In biochemical assays C1 and C2 obstruct Profilin-actin binding at high concentrations (50–100 μM), supporting more total actin polymerization than control assays conducted in their absence (Gau et al., 2018). In cells these compounds slowed endothelial cell migration, proliferation, and inhibited angiogenesis (Gau et al., 2018). Do these small molecules also target the actin binding affinity of other Profilin isoforms and actin? Further biochemical and cellular characterization of C1 and C2 with regard to cancer signaling, phosphorylation state, lipid-binding, or microtubule effects may provide valuable mechanistic insights for using these molecules to treat diseases.

10. Tales from diverse model systems

Many mechanisms underlying protein function are conserved across evolutionary scales. The natural course of disease can sometimes take a lifetime to manifest (evolutionary lifetime is shorter in some organisms). Model organisms can quickly develop a disease or its symptoms allowing researchers to study links between genetic factors, aberrant protein functions and cellular processes on a much shorter time frame. Studying the diversity of protein homologs could lead us down unexpected paths illuminating new therapeutics or elucidating new molecular connections that can be exploited with new treatment approaches. Profilins are evolutionarily conserved in all forms of life where it has been investigated (including *Archaea*, bacteria, viruses, and eukaryotes) and this provides an exceptional opportunity to take advantage of model organisms to study its role as a regulator of actin and microtubule dynamics in disease.

There is much to learn about cancer not only by observing the differences between normal biology and how normal biology goes wrong, but also from how organisms use the same or similar biology in unique ways. Developmental regimes in the *Drosophila*, zebrafish, and *C. elegans* model systems are similar to cancer progression and metastasis in that they require the same tools for execution: cytoskeletal dynamics, cell migration, and cell division. The simplest mechanism explaining the connection of the development of each these model organisms with cancer is that that the loss or misregulation of Profilin is linked to failures in actin assembly and microtubule dynamics. Intriguingly these dynamics are not restored by other Profilin isoforms in vertebrates and many eukaryotes (Cooley et al., 1992; Kovar et al., 2000; Lai et al., 2008; Müssar et al., 2015; Polet et al., 2006; Reeve et al., 2005; Verheyen and Cooley, 1994; Witke et al., 2001). Profilin is critical for maintaining the mesh of actin bundles keeping *Drosophila* nurse cells (in oocytes) intact (Ghiglione et al., 2018; Verheyen and Cooley, 1994). In addition to reduced viability, flies with reduced Profilin levels possess a weakened microtubule spindle apparatus, a less contractile actomyosin cytokinetic ring, and over-proliferative somatic cells (Giansanti et al., 1998; Giansanti and Fuller, 2012; Shields et al., 2014; Verheyen and Cooley, 1994). Zebrafish and *C. elegans* require Profilin for the growth of neuronal cells, neuronal maturation, myelination, and muscle development (Ehler, 2018; Kooij et al., 2016; Kwak et al., 2013; LeCorgne et al., 2018; Majesky, 2007; Meyers, 2018; Murk et al., 2009; Polet et al., 2006; Roth et al., 1999; Yuan et al., 2018). Profilin from *C. elegans* is required for anterior-posterior establishment, DNA positioning and abscission during mitosis, and to optimally regulate Formin-mediated actin polymerization through specific Formin-Profilin isoform pairs (Davies et al., 2018; Neidt et al., 2009; Panzica et al., 2017; Severson et al., 2002). Zebrafish undergo extensive cell migration phases during development that require key regulators of actin for normal execution. Profilin is required for the completion of gastrulation, endothelial cell proliferation, neural cord development, and establishing heart progenitor cell lineage (Ding et al., 2006; Lai et al., 2008; Yuan et al., 2018). Thus, Profilin-mediated development has historically revealed fresh perspectives for the underlying mechanisms that explain how Profilin goes rogue in cancer.

Several viral Profilin homologs effectively bind mammalian actin monomers but with weaker affinity and actin nucleotide exchange (Blasco et al., 1991; Butler-Cole et al., 2007; Machesky et al., 1994; Moreau et al., 2017, 2020). Neither *Vaccinia* nor *Ectromelia* homologs bind PLP regions

(Butler-Cole et al., 2007; Machesky et al., 1994). Intriguingly, *Vaccinia* Profilin binds phosphoinositide (PIP) lipids with higher affinity than human Profilin-1, and Profilin from *Ectromelia* does not bind PIP lipids but directly interacts with other actin regulatory proteins like tropomyosin for function (Butler-Cole et al., 2007; Machesky et al., 1994). *Apicomplexa* are extremely susceptible to actin-polymerizing and depolymerizing agents (Baum et al., 2006; Gordon and Sibley, 2005). *Toxoplasma gondii* Profilin binds and sequesters actin monomers, and loss of *Toxoplasma* Profilin prevents parasite replication and host invasion by disrupting host and parasite actin dynamics (Plattner et al., 2008; Skillman et al., 2012). Genetic, immunological, structural, and cell biological studies have further demonstrated *Toxoplasma* Profilin is important in pathogen-host interactions initiated through interleukins and toll-like receptors from both the host and parasite (Denkers, 2010; Kucera et al., 2010; Plattner et al., 2008; Yarovinsky et al., 2005). Similar observations converging on Profilin and actin dynamics in innate immunity have been investigated in chytrid fungi and plants (Babik et al., 2014; Cao et al., 2016; Qiao et al., 2019; Sun et al., 2018). Thus, actin regulation by Profilin proteins has been important in the "evolutionary arms race" between hosts and diverse microbes.

The role of cilia in cancer signaling and cell cycle regulation with regard to the microtubule cytoskeleton has been studied extensively (Fabbri et al., 2020; Goetz and Anderson, 2010; Golemis et al., 2018; Higgins et al., 2019). Emerging evidence demonstrating the involvement of actin in ciliary formation and development from the tractable model system *Chlamydomonas reinhardtii* demonstrate important new forms of ciliary regulation that challenge long-held ideas suggesting actin or actin-microtubule crosstalk is necessary for normal ciliary assembly, motility, and signaling. *Chlamydomonas* is an excellent model system for studying the duality of the actin and microtubule cytoskeletons in disease. It features two easily accessible cilia that behave and are regulated by mechanisms virtually identical to mammalian forms and has yielded important insights into human diseases and developmental disorders including primary ciliary dyskinesia (PCD), polycystic kidney disease (PKD), *situs inversus*, and numerous ciliopathies (Harris, 2001; Pazour and Witman, 2009; Wase et al., 2019). *Chlamydomonas* actin structures depend on specific actin architectures and localizations to execute diverse cell processes (Christensen et al., 2019; Craig et al., 2019; Detmers, 1985; Detmers et al., 1983; Harper et al., 1992; Jack et al., 2019; Kovar et al., 2001; Onishi et al., 2016; Piperno and Luck, 1979; Wilson et al., 1997). This includes cilia which were historically studied with microtubules as the predominant cytoskeletal polymer in ciliary assembly (Avasthi et al., 2014; Jack et al., 2019; Kovar et al., 2001; Tai et al., 1999). *Chlamydomonas* Profilin is found throughout the organism including ciliary structures (Kovar et al., 2001). *Chlamydomonas* Profilin binds actin, can inhibit aspects of Arp2/3-mediated actin assembly and enhance actin assembly through Formins. However, this Profilin is unique—it does not recycle nucleotides on actin monomers, it very potently inhibits spontaneous filament nucleation, it caps the fast-growing ends of actin filaments 5- to 10-fold more efficiently than other homologs, and seems to protect a specific actin isoform (IDA5) from degradation (Christensen et al., 2019; Courtemanche and Pollard, 2013; Kovar et al., 2001; Onishi et al., 2016; Pernier et al., 2016). Competitive interactions between Profilin and diverse actin assembly factors may ultimately dictate timing and dimensions of assembled actin in these cells. In addition, a single Profilin regulates two very different actin isoforms (IDA5 and NAP1). The occupation of actin filament ends by Profilin may limit the role for Profilin on ciliary microtubules or liberate shared actin-microtubule regulators to orchestrate linked cytoskeletal behaviors. Thus, *Chlamydomonas*, is uniquely situated to elucidate foundation mechanisms concerning the role of Profilin in actin assembly and as a facilitator of actin-microtubule crosstalk.

Yeast model systems are genetic powerhouses that have been indispensable in developing a "parts list" and interactome for many complex pathways and cellular processes. Many foundational studies dissecting Profilin-mediated cytoskeletal dynamics come from yeast, including the discoveries that: Profilin facilitates nucleotide exchange with Srv2/CAP (Amberg et al., 1995; Lila and Drubin, 1997; Ono, 2013; Wolven et al., 2000); Profilin interacts with essential actin assembly factors during cell division (Chang et al., 1997); Profilin synergizes with Formin proteins to promote actin polymerization (Pruyne et al., 2002; Sagot et al., 2002); and that Profilin dictates actin structure (straight or branched filaments) by regulating the actin monomer pool (Suarez and Kovar, 2016). The ability to precisely engineer yeast coupled with a rapid lifecycle are unparalleled for dissecting mechanisms in vivo. In one compelling example, the interaction and mechanism of how Profilin binds to PLP motifs was dissected by introducing 87-point mutations into yeast Profilin (Lu and Pollard, 2001). Similar approaches have been used to quickly assess the viability and role of disease-specific Profilin variants in human disease (Figley et al., 2014). Further, yeast-based technologies have been utilized in drug discovery screens, to produce anti-cancer drugs, and personalized cancer therapies (Ferreira et al., 2019).

Profilins and many of their biochemical activities are conserved across evolution—features that provide an exceptional opportunity to employ model organisms in studying the roles of Profilin as a regulator of actin and microtubule dynamics in disease. A major drawback of conventional mammalian systems in such endeavors is the inability to resolve the fine details of cytoskeletal dynamics in vivo by the standard microscopy techniques used in many screens (i.e., interactor, small-molecule therapeutics, localization) and even fewer in living organisms. Traditional single-molecule attempts to visualize individual actin filaments require injecting fluorescently tagged actin polymers or though FRAP/photoconversion methods that track fiducial marks on preformed actin filaments or bundles (Dovas et al., 2011; McGrath et al., 1998; Wang, 1984; Waterman-Storer and Salmon, 1998). Plant model systems (*Arabidopsis* and *Physcomitrella*) may be the only model systems where dynamics of individual actin filaments have been measured in cells due to the presence of sparse cytoskeletal arrays that afford high-resolution on fast scales (ms) (Augustine et al., 2011; Staiger et al., 2009). Other model systems are approaching this resolution for actin and microtubules through creative combinations of traditional approaches (i.e., microinjection, FRAP, photoactivation), super-resolution imaging modalities, genetically-encodable fluorescently-stable single-molecule tags/tools, and gene-editing technologies (Aumeier et al., 2016; Fritzsche et al., 2017; Funk et al., 2019; Huang et al., 2008; Rust et al., 2006; Skruber et al., 2020; Tas et al., 2017; Vignaud et al., 2020; Vitriol et al., 2015).

11. Open questions and future directions

Profilin is a much more complicated and elegant molecule than suggested by its defining role as a sequestering protein. Profilin was first identified in the 1970s and since then thousands of publications have described and defined its mechanisms regulating the actin cytoskeleton. Recent studies suggest that Profilin still has secrets to share regarding how tubulin, microtubules, and cytoskeletal proteins are regulated, how Profilin is uniquely positioned to choreograph the cytoskeleton during essential cell processes, and the snafus that cause diseases including cancer, neurodegeneration, cardiovascular decline, allergies, and many more. We have still not discerned complete molecular mechanisms that connect Profilin to these processes. When it comes to the role of Profilin the details are truly important.

Clearly actin regulation is an essential function of Profilin, but what other roles do Profilin proteins fulfill? Some versions of Profilin contain extended structural regions, and although these regions do not alter Profilin-actin binding, point mutations there correlate with a loss of parasite motility and force generation (Kursula et al., 2008; Moreau et al., 2020; Nodelman et al., 1999; Qiao et al., 2019; Sun et al., 2018). Some structurally distinct eukaryotic Profilins form oligomers (dimers, trimers, and tetramers) that can reduce actin assembly mediated by Formins by reducing the affinity of Profilin for PLP and also obstructing Profilin-actin binding sites along Formin homology domains (Qiao et al., 2019; Sun et al., 2018). Profilin oligomers are also present in mammalian systems and are suggested to play important roles in disease (specifically neurodegeneration and immune responses), although the detailed mechanisms that underlie the physiological function and formation of these higher-order Profilin configurations are far from clear (Babich et al., 1996; Korupolu et al., 2009; Mares-Mejía et al., 2016; Posey et al., 2018).

Does a free Profilin pool exist in cells? Studies have consistently reported similar concentrations of actin and Profilin in diverse cell types and species. While these calculations are appealing to explain Profilin function with regard to actin monomers, they often fail to account for the presence of actin filaments or the affinity of Profilin for other ligands (i.e., Ena/VASP, Formins, activators of the Arp2/3 complex, microtubules). Thus, with the assumption that 50–90% of cellular actin is polymerized into filaments (Funk et al., 2019; Pollard et al., 2000), at least half the total amount of Profilin may be available for functions beyond binding actin monomers or remaining unbound as a free Profilin pool. Does the presence of Profilin isoforms further complicate these interactions? Human Profilin isoforms 1–3 each bind actin, PIP lipids, and PLP albeit with different affinities, and whether the isoforms can be used interchangeably for these processes is unclear (Behnen et al., 2009; Lambrechts et al., 2006; Michaelsen et al., 2010). Profilin-2 binds PLP residues more strongly, but actin less efficiently, than Profilin-1 (Lu and Pollard, 2001; Vinson et al., 1998). Do these properties translate into more or less proficient Formin-mediated assembly? These interactions might slow Formin-mediated actin polymerization because this version of Profilin doesn't bind actin well. On the other hand, these circumstances might aid Formin proteins in finding cytoplasmic actin monomers and more efficiently releasing Profilin from the polymerizing actin filament. Additionally, neuronal tissues are known to have more Profilin-2 than Profilin-1 and more tubulin than other tissues in the body (Witke et al., 1998). Although not explicitly tested, does this suggest that Profilin-2 is a better regulator of microtubule dynamics than

Profilin-1? If so, this may further elucidate some of the complex details that underlie how cellular actin and microtubule dynamics are linked. Intriguingly, the testes-specific Profilin-4 isoform does not bind actin or PLP stretches, and mouse studies have demonstrated that each additional isoform of Profilin does not fully rescue the effects of Profilin-1. This suggests that there may be distinct tissue-specific roles for Profilin that go beyond actin assembly and define even more interactions that compete for cellular Profilin (Behnen et al., 2009; Polet et al., 2006; Witke et al., 1998).

Do we really understand the role of Profilin in cancer? Some have tried to bin specific Profilin isoforms as indicators of cancer prognosis with Profilin-1 behaving as a tumor suppressor and Profilin-2 suggesting malignancy, but there are multiple lines of evidence to contradict both of these statements. Data available from one of the most comprehensive databases quantifying RNA transcripts in cancers suggests that Profilin-1 and Profilin-2 RNA transcripts become elevated in most (but not all) cancers (Fig. 4). Is Profilin a good target for chemotherapies or to reinforce for better patient recovery? Regardless, to develop an effective pharmaceutical target we need to understand the exact timescales and mechanisms of Profilin function that are disrupted in disease and whether inhibiting one of them (i.e., actin binding) is enough. What if the role of Profilin goes beyond regulating cytoskeletal dynamics? Some reports suggest Profilin may play mysterious roles unrelated to the cytoskeleton in the nucleus (i.e., nuclear import/export and signaling), however if and how these studies link to cancer has not been elucidated (Holzinger et al., 2000; Lederer et al., 2005; Söderberg et al., 2012; Stüven et al., 2003). Alternatively, perhaps Profilin-facilitated post-translational modification of actin goes askew in cancer, ultimately leading to changes to specific forms of actin assembly and cell migratory behaviors (e.g., more actin N-terminal-acetylation decreases cell migration, and the formation of filopodia and lamellipodial protrusions) (Rebowski et al., 2020). Model organisms are powerful tools in this regard and can help to elucidate new and underexplored roles of Profilin in these and cancer-relevant cell processes.

Finally, an increasing number of recent studies link specific cancers directly to proteins that form or regulate the formation of biomolecular condensates (Chen et al., 2019; Kamagata et al., 2020). Although Profilin does not phase-separate on its own, it is found in many cellular condensates and can regulate their size and dynamics (Ghosh et al., 2019; Molliex et al., 2015; Posey et al., 2018). To date no studies have explored whether

Profilin contributes to the pathological role of biomolecular condensates in cancer, however understanding exactly how Profilin regulates these processes could be a valuable asset in the development of cancer pharmaceutical agents and therapies.

Acknowledgments

We are especially grateful to Marc R. Ridilla (Repair Biotechnologies), Amanda Young (SUNY Upstate), Naomi Courtemanche (University of Minnesota), Prachee Avasthi (University of Kansas Medical Center), Jonathan Wendel (Iowa State University), Fatima Cvrčková (Univerzita Karlova) for their helpful suggestions and/or for reading parts of this composition. This work was supported by the Michael E. Connolly Endowment for Lung Cancer Research and NIH R35 GM133485-01 to J.L.H-R.

Author contributions

J.L.H.-R. downloaded and analyzed the transcript data from the National Cancer Institute Genomic Data Commons (https://portal.gdc.cancer.gov; Grossman et al., 2016). M.L.P. and J.L.H-R designed the figures. J.H., M.L.P., and J.L.H-R wrote portions of this work.

Conflicts of interest

The authors declare that they have no conflicts of interest.

References

Akhmanova, A., Steinmetz, M.O., 2010. Microtubule +TIPs at a glance. J. Cell Sci. 123, 3415–3419.

Akıl, C., Robinson, R.C., 2018. Genomes of Asgard archaea encode profilins that regulate actin. Nature 562 (7727), 439–443.

Akin, S., Can, G., Durna, Z., Aydiner, A., 2008. The quality of life and self-efficacy of Turkish breast cancer patients undergoing chemotherapy. Eur. J. Oncol. Nurs. 12 (5), 449–456.

Ali, M., Heyob, K., Jacob, N.K., Rogers, L.K., 2016. Alterative expression and localization of profilin-1/VASPpS157 and Cofilin-1/VASPpS239 regulates metastatic growth and is modified by DHA supplementation. Mol. Cancer Ther. 15 (9), 2220–2231.

Alkam, D., Feldman, E.Z., Singh, A., Kiaei, M., 2017. Profilin1 biology and its mutation, actin(g) in disease. Cell. Mol. Life Sci. 74 (6), 967–981.

Alushin, G.M., Lander, G.C., Kellogg, E.H., Zhang, R., Baker, D., Nogales, E., 2014. High-resolution microtubule structures reveal the structural transitions in αβ-tubulin upon GTP hydrolysis. Cell 157 (5), 1117–1129.

Alvarado, M.I., Jimeno, L., De La Torre, F., Boissy, P., Rivas, B., Lázaro, M.J., Barber, D., 2014. Profilin as a severe food allergen in allergic patients overexposed to grass pollen. Allergy 69 (12), 1610–1616.

Amberg, D.C., Basart, E., Botstein, D., 1995. Defining protein interactions with yeast actin in vivo. Nat. Struct. Biol. 2, 28–34.

Aroush, D.R.-B., Ofer, N., Abu-Shah, E., Allard, J., Krichevsky, O., Mogilner, A., Keren, K., 2017. Actin turnover in lamellipodial fragments. Curr. Biol. 27 (19), 2963–2973.e14.

Augustine, R.C., Pattavina, K.A., Tüzel, E., Vidali, L., Bezanilla, M., 2011. Actin interacting protein1 and actin depolymerizing factor drive rapid actin dynamics in *Physcomitrella patens*. Plant Cell 23, 3696–3710.

Aumeier, C., Schaedel, L., Gaillard, J., John, K., Blanchoin, L., Théry, M., 2016. Self-repair promotes microtubule rescue. Nat. Cell Biol. 18 (10), 1054–1064.

Avasthi, P., Onishi, M., Karpiak, J., Yamamoto, R., Mackinder, L., Jonikas, M.C., Sale, W.S., Shoichet, B., Pringle, J.R., Marshall, W.F., 2014. Actin is required for IFT regulation in *Chlamydomonas reinhardtii*. Curr. Biol. 24 (17), 2025–2032.

Babich, M., Foti, L.R.P., Sykaluk, L.L., Clark, C.R., 1996. Profilin forms tetramers that bind to G-actin. Biochem. Biophys. Res. Commun. 218, 125–131.

Babik, W., Dudek, K., Fijarczyk, A., Pabijan, M., Stuglik, M., Szkotak, R., Zieliński, P., 2014. Constraint and adaptation in newt toll-like receptor genes. Genome Biol. Evol. 7 (1), 81–95.

Bae, Y.H., Ding, Z., Zou, L., Wells, A., Gertler, F., Roy, P., 2009. Loss of profilin-1 expression enhances breast cancer cell motility by Ena/VASP proteins. J. Cell. Physiol. 219 (2), 354–364.

Bao, Y., Hu, G., Flagel, L.E., Salmon, A., Bezanilla, M., Paterson, A.H., Wang, Z., Wendel, J.F., 2011. Parallel up-regulation of the profilin gene family following independent domestication of diploid and allopolyploid cotton (*Gossypium*). Proc. Natl. Acad. Sci. 108 (52), 21152–21157.

Baraniskin, A., Birkenkamp-Demtroder, K., Maghnouj, A., Zöllner, H., Munding, J., Klein-Scory, S., Reinacher-Schick, A., Schwarte-Waldhoff, I., Schmiegel, W., Hahn, S.A., 2012. MiR-30a-5p suppresses tumor growth in colon carcinoma by targeting DTL. Carcinogenesis 33 (4), 732–739.

Barshop, B.A., Wrenn, R.F., Frieden, C., 1983. Analysis of numerical methods of computer simulation of kinetic processes: development of KINSIM—a flexible, portable system. Anal. Biochem. 72, 248–254.

Baum, J., Papenfuss, A.T., Baum, B., Speed, T.P., Cowman, A.F., 2006. Regulation of apicomplexan actin-based motility. Nat. Rev. Microbiol. 4 (8), 621–628.

Beck, M., Schmidt, A., Malmstroem, J., Claassen, M., Ori, A., Szymborska, A., Herzog, F., Rinner, O., Ellenberg, J., Aebersold, R., 2011. The quantitative proteome of a human cell line. Mol. Syst. Biol. 7, 549.

Behnen, M., Murk, K., Kursula, P., Cappallo-Obermann, H., Rothkegel, M., Kierszenbaum, A.L., Kirchhoff, C., 2009. Testis-expressed profilins 3 and 4 show distinct functional characteristics and localize in the acroplaxome-manchette complex in spermatids. BMC Cell Biol. 10, 34.

Bender, M., Stritt, S., Nurden, P., van Eeuwijk, J.M.M., Zieger, B., Kentouche, K., Schulze, H., Morbach, H., Stegner, D., Heinze, K.G., Heinze, K., Dütting, S., Gupta, S., Witke, W., Falet, H., Fischer, A., Hartwig, J.H., Nieswandt, B., 2014. Megakaryocyte-specific Profilin1-deficiency alters microtubule stability and causes a Wiskott-Aldrich syndrome-like platelet defect. Nat. Commun. 5, 4746.

Berro, J., Sirotkin, V., Pollard, T.D., 2010. Mathematical modeling of endocytic actin patch kinetics in fission yeast: disassembly requires release of actin filament fragments. Mol. Biol. Cell 21, 2905–2915.

Bertling, E., Quintero-Monzon, O., Mattila, P.K., Goode, B.L., Lappalainen, P., 2007. Mechanism and biological role of profilin-Srv2/CAP interaction. J. Cell Sci. 120, 1225–1234.

Blanchoin, L., Pollard, T.D., 1998. Interaction of actin monomers with *Acanthamoeba* actophorin (ADF/cofilin) and profilin. J. Biol. Chem. 273, 25106–25111.

Blanchoin, L., Pollard, T.D., Mullins, R.D., 2000. Interactions of ADF/cofilin, Arp2/3 complex, capping protein and profilin in remodeling of branched actin filament networks. Curr. Biol. 10, 1273–1282.

Blanchoin, L., Boujemaa-Paterski, R., Sykes, C., Plastino, J., 2014. Actin dynamics, architecture, and mechanics in cell motility. Physiol. Rev. 94 (1), 235–263.
Blangy, A., Lane, H.A., d'Hérin, P., Harper, M., Kress, M., Nigg, E.A., 1995. Phosphorylation by p34cdc2 regulates spindle association of human Eg5, a kinesin-related motor essential for bipolar spindle formation in vivo. Cell 83 (7), 1159–1169.
Blasco, R., Cole, N.B., Moss, B., 1991. Sequence analysis, expression, and deletion of a vaccinia virus gene encoding a homolog of profilin, a eukaryotic actin-binding protein. J. Virol. 65, 4598–4608.
Bolger-Munro, M., Choi, K., Scurll, J.M., Abraham, L., Chappell, R.S., Sheen, D., Dang-Lawson, M., Wu, X., Priatel, J.J., Coombs, D., Hammer, J.A., Gold, M.R., 2019. Arp2/3 complex-driven spatial patterning of the BCR enhances immune synapse formation, BCR signaling and B cell activation. eLife 8, e44574.
Bonello, L., Camoin-Jau, L., Armero, S., Com, O., Arques, S., Burignat-Bonello, C., Giacomoni, M.-P., Bonello, R., Collet, F., Rossi, P., Barragan, P., Dignat-George, F., Paganelli, F., 2009. Tailored clopidogrel loading dose according to platelet reactivity monitoring to prevent acute and subacute stent thrombosis. Am. J. Cardiol. 103 (1), 5–10.
Boopathy, S., Silvas, T.V., Tischbein, M., Jansen, S., Shandilya, S.M., Zitzewitz, J.A., Landers, J.E., Goode, B.L., Schiffer, C.A., Bosco, D.A., 2015. Structural basis for mutation-induced destabilization of profilin 1 in ALS. Proc. Natl. Acad. Sci. 112 (26), 7984–7989.
Borisy, G., Heald, R., Howard, J., Janke, C., Musacchio, A., Nogales, E., 2016. Microtubules: 50 years on from the discovery of tubulin. Nat. Rev. Mol. Cell Biol. 17 (5), 322–328.
Böttcher, R.T., Wiesner, S., Braun, A., Wimmer, R., Berna, A., Elad, N., Medalia, O., Pfeifer, A., Aszódi, A., Costell, M., Fässler, R., 2009. Profilin 1 is required for abscission during late cytokinesis of chondrocytes. EMBO J. 28 (8), 1157–1169.
Bouchet, B.P., Akhmanova, A., 2017. Microtubules in 3D cell motility. J. Cell Sci. 130 (1), 39–50.
Bouchet, B.P., Gough, R.E., Ammon, Y.-C., van de Willige, D., Post, H., Jacquemet, G., Altelaar, A.M., Heck, A.J., Goult, B.T., Akhmanova, A., 2016. Talin-KANK1 interaction controls the recruitment of cortical microtubule stabilizing complexes to focal adhesions. Elife 5, e18124.
Brieher, W., 2013. Mechanisms of actin disassembly. Mol. Biol. Cell 24 (15), 2299–2302.
Bubb, M.R., Yarmola, E.G., Gibson, B.G., Southwick, F.S., 2003. Depolymerization of actin filaments by profilin: effects of profilin on capping protein function. J. Biol. Chem. 278, 24629–24635.
Burbank, K.S., Mitchison, T.J., 2006. Microtubule dynamic instability. Curr. Biol. 16 (14), R516–R517.
Burke, T.A., Christensen, J.R., Barone, E., Suarez, C., Sirotkin, V., Kovar, D.R., 2014. Homeostatic actin cytoskeleton networks are regulated by assembly factor competition for monomers. Curr. Biol. 24 (5), 579–585.
Burnett, B.G., Andrews, J., Ranganathan, S., Fischbeck, K.H., Di Prospero, N.A., 2008. Expression of expanded polyglutamine targets profilin for degradation and alters actin dynamics. Neurobiol. Dis. 30 (3), 365–374.
Butler-Cole, C., Wagner, M.J., Da Silva, M., Brown, G.D., Burke, R.D., Upton, C., 2007. An ectromelia virus profilin homolog interacts with cellular tropomyosin and viral A-type inclusion protein. Virol. J. 4, 76.
Caglayan, E., Romeo, G.R., Kappert, K., Odenthal, M., Südkamp, M., Body, S.C., Shernan, S.K., Hackbusch, D., Vantler, M., Kazlauskas, A., Rosenkranz, S., 2010. Profilin-1 is expressed in human atherosclerotic plaques and induces atherogenic effects on vascular smooth muscle cells. PLoS One 5 (10), e13608.

Cande, W.Z., Lazarides, E., McIntosh, J.R., 1977. A comparison of the distribution of actin and tubulin in the mammalian mitotic spindle as seen by indirect immunofluorescence. J. Cell Biol. 72 (3), 552–567.

Cao, L., Henty-Ridilla, J.L., Blanchoin, L., Staiger, C.J., 2016. Profilin-dependent nucleation and assembly of actin filaments controls cell elongation in *Arabidopsis*. Plant Physiol. 170 (1), 220–233.

Carlier, M.-F., Jean, C., Rieger, K.J., Lenfant, M., Pantaloni, D., 1993. Modulation of the interaction between g-actin and Thymosin-ß4 by ATP/ADP ratio: possible implication in the regulation of actin dynamics. Proc. Natl. Acad. Sci. 90, 5034–5038.

Carlier, M.F., Didry, D., Erk, I., Lepault, J., Vantroys, M.L., Vandekerchkove, J., Perelroizen, I., Yin, H., Doi, Y.K., Pantaloni, D., 1996. Tß4 is not a simple g-actin sequestering protein and interacts with F-actin at high concentration. J. Biol. Chem. 271, 9231–9239.

Carlsson, L., Nyström, L.-E., Lindberg, U., 1976a. Crystallization of a non-muscle actin. J. Mol. Biol. 105, 353–366.

Carlsson, L., Nyström, L.E., Sundkvist, I., Markey, F., Lindberg, U., 1976b. Profilin, a low-molecular weight protein controlling actin polymerisability. In: Contractile systems in non-muscle tissues. Elsevier North-Holland Biomedical Press, Amsterdam, pp. 39–49.

Carlsson, L., Nyström, L.E., Sundkvist, I., Markey, F., Lindberg, U., 1977. Actin polymerizability is influenced by profilin, a low molecular weight protein in non-muscle cells. J. Mol. Biol. 115 (3), 465–483.

Caudron, N., Valiron, O., Usson, Y., Valiron, P., Job, D., 2000. A reassessment of the factors affecting microtubule assembly and disassembly in vitro. J. Mol. Biol. 297 (1), 211–220.

Chaaban, S., Brouhard, G.J., 2017. A microtubule bestiary: structural diversity in tubulin polymers. Mol. Biol. Cell 28 (22), 2924–2931.

Chakraborty, J., Pandey, M., Navneet, A.K., Appukuttan, T.A., Varghese, M., Sreetama, S.C., Rajamma, U., Mohanakumar, K.P., 2014. Profilin-2 increased expression and its altered interaction with β-actin in the striatum of 3-nitropropionic acid-induced Huntington's disease in rats. Neuroscience 281, 216–228.

Chan, K.Y., Matthews, K.R., Ersfeld, K., 2010. Functional characterisation and drug target validation of a mitotic Kinesin-13 in *Trypanosoma brucei*. PLoS Pathog. 6 (8), e1001050.

Chang, F., 2000. Microtubule and actin-dependent movement of the formin cdc12p in fission yeast. Microsc. Res. Tech. 49, 161–167.

Chang, F., Martin, S.G., 2009. Shaping fission yeast with microtubules. CSH Perspect. Biol. 1 (1), a001347.

Chang, F., Drubin, D., Nurse, P., 1997. cdc12p, a protein required for cytokinesis in fission yeast, is a component of the cell division ring and interacts with profilin. J. Cell Biol. 137, 169–182.

Chen, R.R., Yung, M.M.H., Xuan, Y., Zhan, S., Leung, L.L., Liang, R.R., Leung, T.H.Y., Yang, H., Xu, D., Sharma, R., Chan, K.K.L., Ngu, S.-F., Ngan, H.Y.S., Chan, D.W., 2019. Targeting of lipid metabolism with a metabolic inhibitor cocktail eradicates peritoneal metastases in ovarian cancer cells. Commun. Biol. 2 (1), 281.

Chesarone, M.A., DuPage, A.G., Goode, B.L., 2010. Unleashing formins to remodel the actin and microtubule cytoskeletons. Nat. Rev. Mol. Cell Biol. 11 (1), 62–74.

Chou, S.Z., Pollard, T.D., 2019. Mechanism of actin polymerization revealed by cryo-EM structures of actin filaments with three different bound nucleotides. Proc. Natl. Acad. Sci. 116 (10), 4265–4274.

Christensen, J.R., Craig, E.W., Glista, M.J., Mueller, D.M., Li, Y., Sees, J.A., Huang, S., Suarez, C., Mets, L.J., Kovar, D.R., Avasthi, P., 2019. *Chlamydomonas reinhardtii* formin FOR1 and profilin PRF1 are optimized FOR acute rapid actin filament assembly. Mol. Biol. Cell 30 (26), 3123–3135.

Coles, C.H., Bradke, F., 2015. Coordinating neuronal actin-microtubule dynamics. Curr. Biol. 25 (15), R677–R691.

Colin, A., Singaravelu, P., Théry, M., Blanchoin, L., Gueroui, Z., 2018. Actin-network architecture regulates microtubule dynamics. Curr. Biol. 28 (16), 2647–2656.e4.

Cooley, L., Verheyen, E., Ayers, K., 1992. *Chickadee* encodes a profilin required for intercellular cytoplasm transport during *Drosophila* oogenesis. Cell 69 (1), 173–184.

Cooper, J.A., Walker, S.B., Pollard, T.D., 1983. Pyrene actin: documentation of the validity of a sensitive assay for actin polymerization. J. Muscle Res. Cell Motil. 4, 253–262.

Cooper, J.A., Blum, J.D., Pollard, T.D., 1984. Acanthamoeba castellanii capping protein: properties, mechanism of action, immunologic cross-reactivity, and localization. J. Cell Biol. 99, 217–225.

Coppolino, M.G., Dierckman, R., Loijens, J., Collins, R.F., Pouladi, M., Jongstra-Bilen, J., Schreiber, A.D., Trimble, W.S., Anderson, R., Grinstein, S., 2002. Inhibition of phosphatidylinositol-4-phosphate 5-kinase I alpha impairs localized actin remodeling and suppresses phagocytosis. J. Biol. Chem. 277, 43849–43857.

Coumans, J.V.F., Gau, D., Poljak, A., Wasinger, V., Roy, P., Moens, P.D.J., 2014. Profilin-1 overexpression in MDA-MB-231 breast cancer cells is associated with alterations in proteomics biomarkers of cell proliferation, survival, and motility as revealed by global proteomics analyses. Omics 18 (12), 778–791.

Courtemanche, N., 2018. Mechanisms of formin-mediated actin assembly and dynamics. Biophys. Rev. 10 (6), 1553–1569.

Courtemanche, N., Pollard, T.D., 2012. Determinants of formin homology 1 (FH1) domain function in actin filament elongation by formins. J. Biol. Chem. 287 (10), 7812–7820.

Courtemanche, N., Pollard, T.D., 2013. Interaction of profilin with the barbed end of actin filaments. Biochemistry 52 (37), 6456–6466.

Craig, E.W., Mueller, D.M., Bigge, B.M., Schaffer, M., Engel, B.D., Avasthi, P., 2019. The elusive actin cytoskeleton of a green alga expressing both conventional and divergent actins. Mol. Biol. Cell 30 (22), 2827–2837.

Cui, X., Zhang, S., Xu, Y., Dang, H., Liu, C., Wang, L., Yang, L., Hu, J., Liang, W., Jiang, J., Li, N., Li, Y., Chen, Y., Li, F., 2016. PFN2, a novel marker of unfavorable prognosis, is a potential therapeutic target involved in esophageal squamous cell carcinoma. J. Transl. Med. 14 (1), 137.

Davies, T., Kim, H.X., Romano Spica, N., Lesea-Pringle, B.J., Dumont, J., Shirasu-Hiza, M., Canman, J.C., 2018. Cell-intrinsic and -extrinsic mechanisms promote cell-type-specific cytokinetic diversity. Elife 7, e36204.

De La Cruz, E.M., Ostap, E.M., Brundage, R.A., Reddy, K.S., Sweeney, H.L., Safer, D., 2000. Thymosin-β_4 changes the conformation and dynamics of actin monomers. Biophys. J. 78, 2516–2527.

Denkers, E.Y., 2010. Toll-like receptor initiated host defense against *Toxoplasma gondii*. J. Biomed. Biotechnol. 2010, 1–7.

Desai, A., Mitchison, T.J., 1997. Microtubule polymerization dynamics. Annu. Rev. Cell Dev. Biol. 13, 83–117.

Detmers, P.A., 1985. Elongation of cytoplasmic processes during gametic mating: models for actin-based motility. Can. J. Biochem. Cell Biol. 63, 599–607.

Detmers, P.A., Goodenough, U.W., Condeelis, J., 1983. Elongation of the fertilization tubule in Chlamydomonas: new observations on the core microfilaments and the effect of transient intracellular signals in their structural integrity. J. Cell Biol. 97, 522–532.

Di Nardo, A., Gareus, R., Kwiatkowski, D., Witke, W., 2000. Alternative splicing of the mouse profilin II gene generates functionally different profilin isoforms. J. Cell Sci. 113, 3795–3803.

Didry, D., Carlier, M.-F., Pantaloni, D., 1998. Synergy between actin depolymerizing factor/cofilin and profilin in increasing actin filament turnover. J. Biol. Chem. 273, 25602–25611.

Ding, Z., Lambrechts, A., Parepally, M., Roy, P., 2006. Silencing profilin-1 inhibits endothelial cell proliferation, migration and cord morphogenesis. J. Cell Sci. 119, 4127–4137.
Ding, Z., Joy, M., Bhargava, R., Gunsaulus, M., Lakshman, N., Miron-Mendoza, M., Petroll, M., Condeelis, J., Wells, A., Roy, P., 2014. Profilin-1 downregulation has contrasting effects on early vs late steps of breast cancer metastasis. Oncogene 33 (16), 2065–2074.
Dogterom, M., Koenderink, G.H., 2019. Actin-microtubule crosstalk in cell biology. Nat. Rev. Mol. Cell Biol. 20 (1), 38–54.
Dominguez, R., Holmes, K.C., 2011. Actin structure and function. Annu. Rev. Biophys. 40, 169–186.
Dos Remedios, C.G., Chhabra, D., Kekic, M., Dedova, I.V., Tsubakihara, M., Berry, D.A., Nosworthy, N.J., 2003. Actin binding proteins: regulation of cytoskeletal microfilaments. Physiol. Rev. 83, 433–473.
Dovas, A., Gligorijevic, B., Chen, X., Entenberg, D., Condeelis, J., Cox, D., 2011. Visualization of actin polymerization in invasive structures of macrophages and carcinoma cells using photoconvertible β-actin-Dendra2 fusion proteins. PLoS One 6 (2), e16485.
Dürre, K., Keber, F.C., Bleicher, P., Brauns, F., Cyron, C.J., Faix, J., Bausch, A.R., 2018. Capping protein-controlled actin polymerization shapes lipid membranes. Nat. Commun. 9 (1), 1630.
Dziezanowski, M.A., DeStefano, M.J., Rabinovitch, M., 1980. Effect of antitubulins on spontaneous and chemotactic migration of neutrophils under agarose. J. Cell Sci. 42, 379–388.
Ehler, E., 2018. Actin-associated proteins and cardiomyopathy—the 'unknown' beyond troponin and tropomyosin. Biophys. Rev. 10 (4), 1121–1128.
Elie, A., Prezel, E., Guérin, C., Denarier, E., Ramirez-Rios, S., Serre, L., Andrieux, A., Fourest-Lieuvin, A., Blanchoin, L., Arnal, I., 2015. Tau co-organizes dynamic microtubule and actin networks. Sci. Rep. 5, 9964.
Engelke, M.F., Winding, M., Yue, Y., Shastry, S., Teloni, F., Reddy, S., Blasius, T.L., Soppina, P., Hancock, W.O., Gelfand, V.I., Verhey, K.J., 2016. Engineered kinesin motor proteins amenable to small-molecule inhibition. Nat. Commun. 7 (1), 11159.
Erickson, H.P., O'Brien, E.T., 1992. Microtubule dynamic instability and GTP hydrolysis. Annu. Rev. Biophys. Biomol. Struct. 21 (1), 145–166.
Etienne-Manneville, S., 2004. Actin and microtubules in cell motility: which one is in control? Traffic 5 (7), 470–477.
Euteneuer, U., Schliwa, M., 1984. Persistent, directional motility of cells and cytoplasmic fragments in the absence of microtubules. Nature 310 (5972), 58–61.
Evangelista, M., 1997. Bni1p, a yeast Formin linking Cdc42p and the actin cytoskeleton during polarized morphogenesis. Science 276 (5309), 118–122.
Evangelista, M., Pruyne, D., Amberg, D.C., Boone, C., Bretscher, A., 2002. Formins direct Arp2/3-independent actin filament assembly to polarize cell growth in yeast. Nat. Cell Biol. 4 (3), 260–269.
Fabbri, M., Wiemann, J., Manucci, F., Briggs, D.E.G., 2020. Three-dimensional soft tissue preservation revealed in the skin of a non-avian dinosaur. Palaeontology 63 (2), 185–193.
Fagerberg, L., Hallström, B.M., Oksvold, P., Kampf, C., Djureinovic, D., Odeberg, J., Habuka, M., Tahmasebpoor, S., Danielsson, A., Edlund, K., Asplund, A., Sjöstedt, E., Lundberg, E., Szigyarto, C.A.-K., Skogs, M., Takanen, J.O., Berling, H., Tegel, H., Mulder, J., … Uhlén, M., 2014. Analysis of the human tissue-specific expression by genome-wide integration of transcriptomics and antibody-based proteomics. Mol. Cell. Proteomics 13 (2), 397–406.
Fan, Y., Arif, A., Gong, Y., Jia, J., Eswarappa, S.M., Willard, B., Horowitz, A., Graham, L.M., Penn, M.S., Fox, P.L., 2012. Stimulus-dependent phosphorylation of profilin-1 in angiogenesis. Nat. Cell Biol. 14 (10), 1046–1056.

Fan, Y., Potdar, A.A., Gong, Y., Eswarappa, S.M., Donnola, S., Lathia, J.D., Hambardzumyan, D., Rich, J.N., Fox, P.L., 2014. Profilin-1 phosphorylation directs angiocrine expression and glioblastoma progression through HIF-1α accumulation. Nat. Cell Biol. 16 (5), 445–456.

Farina, F., Gaillard, J., Guérin, C., Couté, Y., Sillibourne, J., Blanchoin, L., Théry, M., 2016. The centrosome is an actin-organizing centre. Nat. Cell Biol. 18 (1), 65–75.

Farina, F., Ramkumar, N., Brown, L., Samandar Eweis, D., Anstatt, J., Waring, T., Bithell, J., Scita, G., Thery, M., Blanchoin, L., Zech, T., Baum, B., 2019. Local actin nucleation tunes centrosomal microtubule nucleation during passage through mitosis. EMBO J. 38 (11), e99843.

Ferreira, R., Limeta, A., Nielsen, J., 2019. Tackling cancer with yeast-based technologies. Trends Biotech. 37 (6), 592–603.

Ferron, F., Rebowski, G., Lee, S.H., Dominguez, R., 2007. Structural basis for the recruitment of profilin-actin complexes during filament elongation by Ena/VASP. EMBO J. 26 (21), 4597–4606.

Fife, C.M., McCarroll, J.A., Kavallaris, M., 2014. Movers and shakers: cell cytoskeleton in cancer metastasis. Brit. J. Pharm. 171 (24), 5507–5523.

Figley, M.D., Bieri, G., Kolaitis, R.-M., Taylor, J.P., Gitler, A.D., 2014. Profilin 1 associates with stress granules and ALS-linked mutations alter stress granule dynamics. J. Neurosci. 34 (24), 8083–8097.

Frantzi, M., van Kessel, K.E., Zwarthoff, E.C., Marquez, M., Rava, M., Malats, N., Merseburger, A.S., Katafigiotis, I., Stravodimos, K., Mullen, W., Zoidakis, J., Makridakis, M., Pejchinovski, M., Critselis, E., Lichtinghagen, R., Brand, K., Dakna, M., Roubelakis, M.G., Theodorescu, D., Vlahou, A., Mischak, H., Anagnou, N.P., 2016. Development and validation of urine-based peptide biomarker panels for detecting bladder cancer in a multi-center study. Clin. Cancer Res. 22 (16), 4077–4086.

Fritzsche, M., Fernandes, R.A., Chang, V.T., Colin-York, H., Clausen, M.P., Felce, J.H., Galiani, S., Erlenkämper, C., Santos, A.M., Heddleston, J.M., Pedroza-Pacheco, I., Waithe, D., de la Serna, J.B., Lagerholm, B.C., Liu, T., Chew, T.-L., Betzig, E., Davis, S.J., Eggeling, C., 2017. Cytoskeletal actin dynamics shape a ramifying actin network underpinning immunological synapse formation. Sci. Adv. 3 (6), e1603032.

Funk, J., Merino, F., Venkova, L., Heydenreich, L., Kierfeld, J., Vargas, P., Raunser, S., Piel, M., Bieling, P., 2019. Profilin and formin constitute a pacemaker system for robust actin filament growth. Elife 8, e50963.

Fygenson, D.K., Flyvbjerg, H., Sneppen, K., Libchaber, A., Leibler, S., 1995. Spontaneous nucleation of microtubules. Phys. Rev. E. 51 (5), 5058–5063.

Gaillard, J., Ramabhadran, V., Neumanne, E., Gurel, P., Blanchoin, L., Vantard, M., Higgs, H.N., 2011. Differential interactions of the formins INF2, mDia1, and mDia2 with microtubules. Mol. Biol. Cell 22 (23), 4575–4587.

Gardiner, J., Marc, J., 2011. *Arabidopsis thaliana*, a plant model organism for the neuronal microtubule cytoskeleton? J. Exp. Bot. 62 (1), 89–97.

Gareus, R., Di Nardo, A., Rybin, V., Witke, W., 2006. Mouse profilin 2 regulates endocytosis and competes with SH3 ligand binding to dynamin 1. J. Biol. Chem. 281 (5), 2803–2811.

Gau, D., Lewis, T., McDermott, L., Wipf, P., Koes, D., Roy, P., 2018. Structure-based virtual screening identifies a small-molecule inhibitor of the profilin 1-actin interaction. J. Biol. Chem. 293 (7), 2606–2616.

Gawadi, N., 1971. The spindle at metaphase. Nature 232 (5305), 61–62.

Gertler, F.B., Niebuhr, K., Reinhard, M., Wehland, J., Soriano, P., 1996. Mena, a relative of VASP and *Drosophila* enabled, is implicated in the control of microfilament dynamics. Cell 87 (2), 227–239.

Ghiglione, C., Jouandin, P., Cérézo, D., Noselli, S., 2018. The *Drosophila* insulin pathway controls *Profilin* expression and dynamic actin-rich protrusions during collective cell migration. Development 145 (14), dev161117.
Ghosh, A., Mazarakos, K., Zhou, H.-X., 2019. Three archetypical classes of macromolecular regulators of protein liquid–liquid phase separation. Proc. Natl. Acad. Sci. 116 (39), 19474–19483.
Giansanti, M.G., Fuller, M.T., 2012. What *Drosophila* spermatocytes tell us about the mechanisms underlying cytokinesis. Cytoskeleton 69 (11), 869–881.
Giansanti, M.G., Bonaccorsi, S., Williams, B., Williams, E.V., Santolamazza, C., Goldberg, M.L., Gatti, M., 1998. Cooperative interactions between the central spindle and the contractile ring during *Drosophila* cytokinesis. Genes Dev. 12 (3), 396–410.
Gieselmann, R., Kwiatkowski, D.J., Janmey, P.A., Witke, W., 2008. Distinct biochemical characteristics of the two human profilin isoforms. Eur. J. Biol. 229 (3), 621–628.
Goetz, S.C., Anderson, K.V., 2010. The primary cilium: a signaling centre during vertebrate development. Nat. Rev. Genet. 11 (5), 331–344.
Goldschmidt-Clermont, P.J., Furman, M.I., Wachsstock, D., Safer, D., Nachmias, V.T., Pollard, T.D., 1992. The control of actin nucleotide exchange by thymosin beta-4 and profilin. A potential regulatory mechanism for actin polymerization in cells. Mol. Biol. Cell 3 (9), 1015–1024.
Golemis, E.A., Scheet, P., Beck, T.N., Scolnick, E.M., Hunter, D.J., Hawk, E., Hopkins, N., 2018. Molecular mechanisms of the preventable causes of cancer in the United States. Genes Dev. 32 (13–14), 868–902.
Gordon, J.L., Sibley, L.D., 2005. Comparative genome analysis reveals a conserved family of actin-like proteins in apicomplexan parasites. BMC Genomics 6, 179.
Gordon, D.J., Boyer, J.L., Korn, E.D., 1977. Comparative biochemistry of non-muscle actins. J. Biol. Chem. 252 (22), 8300–8309.
Grossman, R.L., Heath, A.P., Ferretti, V., Varmus, H.E., Lowy, D.R., Kibbe, W.A., Staudt, L.M., 2016. Toward a shared vision for cancer genomic data. N. Engl. J. Med. 375 (12), 1109–1112.
Haarer, B.K., Brown, S.S., 1990. Structure and function of profilin. Cell Motil. Cytoskeleton 17 (2), 71–74.
Hall, A., 2009. The cytoskeleton and cancer. Cancer Metast. Rev. 28 (1–2), 5–14.
Hancock, W.O., 2014. Mitotic kinesins: a reason to delve into Kinesin-12. Curr. Biol. 24 (19), R968–R970.
Hansson, A., Skoglund, G., Lassing, I., Lindberg, U., Ingelman-Sundberg, M., 1988. Protein kinase C-dependent phosphorylation of profilin is specifically stimulated by phosphatidylinositol bisphosphate (PIP2). Biochem. Biophys. Res. Commun. 150 (2), 526–531.
Harper, J.D.I., McCurdy, D.W., Sanders, M.A., Salisbury, J.L., John, P.C.L., 1992. Actin dynamics during the cell cycle in *Chlamydomonas reinhardtii*. Cell Motil. Cytoskeleton 22 (2), 117–126.
Harris, E.H., 2001. *Chlamydomonas* as a model organism. Annu. Rev. Plant. Physiol. Plant. Mol. Biol. 52, 363–406.
Hauser, M., Roulias, A., Ferreira, F., Egger, M., 2010. Panallergens and their impact on the allergic patient. Allergy Asthma Clin. Immunol. 6 (1), 1.
Henty-Ridilla, J.L., Goode, B.L., 2015. Global resource distribution: allocation of actin building blocks by profilin. Dev. Cell 32 (1), 5–6.
Henty-Ridilla, J.L., Shimono, M., Li, J., Chang, J.H., Day, B., Staiger, C.J., 2013. The plant actin cytoskeleton responds to signals from microbe-associated molecular patterns. PLoS Pathog. 9 (4), e1003290.
Henty-Ridilla, J.L., Rankova, A., Eskin, J.A., Kenny, K., Goode, B.L., 2016. Accelerated actin filament polymerization from microtubule plus ends. Science 352 (6288), 1004.

Henty-Ridilla, J.L., Juanes, M.A., Goode, B.L., 2017. Profilin directly promotes microtubule growth through residues mutated in amyotrophic lateral sclerosis. Curr. Biol. 27 (22), 3535–3543.e4.
Herman, I.M., Pollard, T.D., 1979. Comparison of purified anti-actin and fluorescent-heavy meromyosin staining patterns in dividing cells. J. Cell Biol. 80 (3), 509–520.
Hernandez, P., Tirnauer, J.S., 2010. Tumor suppressor interactions with microtubules: keeping cell polarity and cell division on track. Dis. Model Mech. 3 (5–6), 304–315.
Hetrick, B., Han, M.S., Helgeson, L.A., Nolen, B.J., 2013. Small molecules CK-666 and CK-869 inhibit actin-related protein 2/3 complex by blocking an activating conformational change. Chem. Biol. 20 (5), 701–712.
Higgins, M., Obaidi, I., McMorrow, T., 2019. Primary cilia and their role in cancer. Oncol. Lett. 17 (3), 3041–3047.
Higgs, H.N., Pollard, T.D., 1999. Regulation of actin polymerization by Arp2/3 complex and WASp/scar proteins. J. Biol. Chem. 274 (46), 32531–32534.
Holzinger, A., Valenta, R., Lütz-Meindl, U., 2000. Profilin is localized in the nucleus-associated microtubule and actin system and is evenly distributed in the cytoplasm of the green algaMicrasterias denticulata. Protoplasma 212 (3), 197–205.
Horan, B.G., Zerze, G.H., Kim, Y.C., Vavylonis, D., Mittal, J., 2018. Computational modeling highlights the role of the disordered formin homology 1 domain in profilin-actin transfer. FEBS Lett. 592 (11), 1804–1816.
Horrevoets, A.J.G., 2007. Profilin-1: an unexpected molecule linking vascular inflammation to the actin cytoskeleton. Circ. Res. 101 (4), 328–330.
Huang, B., Wang, W., Bates, M., Zhuang, X., 2008. Three-dimensional super-resolution imaging by stochastic optical reconstruction microscopy. Science 319 (5864), 810–813.
Hurst, D.R., Welch, D.R., 2011. Metastasis suppressor genes at the interface between the environment and tumor cell growth. Int. Rev. Cell Mol. Biol. 286, 107–180.
Inoue, D., Obino, D., Pineau, J., Farina, F., Gaillard, J., Guerin, C., Blanchoin, L., Lennon-Duménil, A.-M., Théry, M., 2019. Actin filaments regulate microtubule growth at the centrosome. EMBO J. 38 (11), e99630.
Ishizaki, T., Morishima, Y., Okamoto, M., Furuyashiki, T., Kato, T., Narumiya, S., 2001. Coordination of microtubules and the actin cytoskeleton by the rho effector mDia1. Nat. Cell Biol. 3 (1), 8–14.
Isogai, T., van der Kammen, R., Innocenti, M., 2015. SMIFH2 has effects on Formins and p53 that perturb the cell cytoskeleton. Sci. Rep. 5, 9802.
Jack, B., Mueller, D.M., Fee, A.C., Tetlow, A.L., Avasthi, P., 2019. Partially redundant actin genes in *Chlamydomonas* control transition zone organization and flagellum-directed traffic. Cell Rep. 27 (8), 2459–2467.e3.
Janke, J., Schlüter, K., Jandrig, B., Theile, M., Kölble, K., Arnold, W., Grinstein, E., Schwartz, A., Estevéz-Schwarz, L., Schlag, P.M., Jockusch, B.M., Scherneck, S., 2000. Suppression of tumorigenicity in breast cancer cells by the microfilament protein profilin 1. J. Exp. Med. 191 (10), 1675–1686.
Jégou, A., Niedermayer, T., Orbán, J., Didry, D., Lipowsky, R., Carlier, M.-F., Romet-Lemonne, G., 2011. Individual actin filaments in a microfluidic flow reveal the mechanism of ATP hydrolysis and give insight into the properties of profilin. PLoS Biol. 9 (9), e1001161.
Jiang, H., Wang, S., Huang, Y., He, X., Cui, H., Zhu, X., Zheng, Y., 2015. Phase transition of spindle-associated protein regulate spindle apparatus assembly. Cell 163 (1), 108–122.
Jiang, C., Ding, Z., Joy, M., Chakraborty, S., Kim, S.H., Bottcher, R., Condeelis, J., Singh, S., Roy, P., 2017. A balanced level of profilin-1 promotes stemness and tumor-initiating potential of breast cancer cells. Cell Cycle 16 (24), 2366–2373.
Job, D., Valiron, O., Oakley, B., 2003. Microtubule nucleation. Curr. Opin. Cell Biol. 15 (1), 111–117.

Joy, M.E., Vollmer, L.L., Hulkower, K., Stern, A.M., Peterson, C.K., Boltz, R.C.D., Roy, P., Vogt, A., 2014. A high-content, multiplexed screen in human breast cancer cells identifies profilin-1 inducers with anti-migratory activities. PLoS One 9 (2), e88350.

Juanes, M.A., Bouguenina, H., Eskin, J.A., Jaiswal, R., Badache, A., Goode, B.L., 2017. Adenomatous polyposis coli nucleates actin assembly to drive cell migration and microtubule-induced focal adhesion turnover. J. Cell Biol. 216 (9), 2859–2875.

Juanes, M.A., Isnardon, D., Badache, A., Brasselet, S., Mavrakis, M., Goode, B.L., 2019. The role of APC-mediated actin assembly in microtubule capture and focal adhesion turnover. J. Cell Biol. 218 (10), 3415–3435.

Kaiser, D.A., Vinson, V.K., Murphy, D.B., Pollard, T.D., 1999. Profilin is predominantly associated with monomeric actin in *Acanthamoeba*. J. Cell Sci. 112, 3779–3790.

Kamagata, K., Kanbayashi, S., Honda, M., Itoh, Y., Takahashi, H., Kameda, T., Nagatsugi, F., Takahashi, S., 2020. Liquid-like droplet formation by tumor suppressor p53 induced by multivalent electrostatic interactions between two disordered domains. Sci. Rep. 10 (1), 580.

Kandasamy, M.K., McKinney, E.C., Meagher, R.B., 2002. Plant profilin isovariants are distinctly regulated in vegetative and reproductive tissues. Cell Motil. Cytoskeleton 52 (1), 22–32.

Kaverina, I., Straube, A., 2011. Regulation of cell migration by dynamic microtubules. Semin. Cell Dev. Biol. 22 (9), 968–974.

Kaverina, I., Rottner, K., Small, J.V., 1998. Targeting, capture, and stabilization of microtubules at early focal adhesions. J. Cell Biol. 142 (1), 181–190.

Kim, J.-G., Moon, M.-Y., Kim, H.-J., Li, Y., Song, D.-K., Kim, J.-S., Lee, J.-Y., Kim, J., Kim, S.-C., Park, J.-B., 2012. Ras-related GTPases Rap1 and RhoA collectively induce the phagocytosis of serum-opsonized zymosan particles in macrophages. J. Biol. Chem. 287 (7), 5145–5155.

Kim, M.-J., Lee, Y.-S., Han, G.-Y., Lee, H.-N., Ahn, C., Kim, C.-W., 2015. Profilin 2 promotes migration, invasion, and stemness of HT29 human colorectal cancer stem cells. Biosci. Biotechnol. Biochem. 79 (9), 1438–1446.

Kita, A.M., Swider, Z.T., Erofeev, I., Halloran, M.C., Goryachev, A.B., Bement, W.M., 2019. Spindle-F-actin interactions in mitotic spindles in an intact vertebrate epithelium. Mol. Biol. Cell 30 (14), 1645–1654.

Kollman, J.M., Polka, J.K., Zelter, A., Davis, T.N., Agard, D.A., 2010. Microtubule nucleating gamma-TuSC assembles structures with 13-fold microtubule-like symmetry. Nature 466 (7308), 879–882.

Kooij, V., Viswanathan, M.C., Lee, D.I., Rainer, P.P., Schmidt, W., Kronert, W.A., Harding, S.E., Kass, D.A., Bernstein, S.I., Van Eyk, J.E., Cammarato, A., 2016. Profilin modulates sarcomeric organization and mediates cardiomyocyte hypertrophy. Cardiovasc. Res. 110 (2), 238–248.

Korupolu, R.V., Achary, M.S., Aneesa, F., Sathish, K., Wasia, R., Sairam, M., Nagarajaram, H.A., Singh, S.S., 2009. Profilin oligomerization and its effect on poly(L-)proline binding and phosphorylation. Int. J. Biol. Macromol. 45 (3), 265–273.

Kotila, T., Kogan, K., Enkavi, G., Guo, S., Vattulainen, I., Goode, B.L., Lappalainen, P., 2018. Structural basis of actin monomer re-charging by cyclase-associated protein. Nat. Commun. 9 (1), 1892.

Kovar, D.R., Drøbak, B.K., Staiger, C.J., 2000. Maize profilin isoforms are functionally distinct. Plant Cell 12 (4), 583–598.

Kovar, D.R., Yang, P., Sale, W.S., Drobak, B.K., Staiger, C.J., 2001. Chlamydomonas reinhardtii produces a profilin with unusual biochemical properties. J. Cell Sci. 114 (23), 4293–4305.

Kovar, D.R., Kuhn, J.R., Tichy, A.L., Pollard, T.D., 2003. The fission yeast cytokinesis formin Cdc12p is a barbed end actin filament capping protein gated by profilin. J. Cell Biol. 161 (5), 875–887.

Kovar, D.R., Harris, E.S., Mahaffy, R., Higgs, H.N., Pollard, T.D., 2006. Control of the assembly of ATP- and ADP-actin by formins and profilin. Cell 124 (2), 423–435.

Krause, M., Dent, E.W., Bear, J.E., Loureiro, J.J., Gertler, F.B., 2003. Ena/VASP proteins: regulators of the actin cytoskeleton and cell migration. Annu. Rev. Cell Dev. Biol. 19, 541–564.

Krishnan, K., Moens, P.D.J., 2009. Structure and functions of profilins. Biophys. Rev. 1 (2), 71–81.

Kucera, K., Koblansky, A.A., Saunders, L.P., Frederick, K.B., De La Cruz, E.M., Ghosh, S., Modis, Y., 2010. Structure-based analysis of *Toxoplasma gondii* profilin: a parasite-specific motif is required for recognition by toll-like receptor 11. J. Mol. Biol. 403 (4), 616–629.

Kursula, P., Kursula, I., Massimi, M., Song, Y.-H., Downer, J., Stanley, W.A., Witke, W., Wilmanns, M., 2008. High-resolution structural analysis of mammalian profilin 2a complex formation with two physiological ligands: the formin homology 1 domain of mDia1 and the proline-rich domain of VASP. J. Mol. Biol. 375 (1), 270–290.

Kwak, J., Park, O.K., Jung, Y.J., Hwang, B.J., Kwon, S.-H., Kee, Y., 2013. Live image profiling of neural crest lineages in zebrafish transgenic lines. Mol. Cells 35 (3), 255–260.

Lai, S.-L., Chan, T.-H., Lin, M.-J., Huang, W.-P., Lou, S.-W., Lee, S.-J., 2008. Diaphanous-related formin 2 and profilin I are required for gastrulation cell movements. PLoS One 3 (10), e3439.

Lambrechts, A., Verschelde, J.L., Jonckheere, V., Goethals, M., Vandekerckhove, J., Ampe, C., 1997. The mammalian profilin isoforms display complementary affinities for PIP2 and proline-rich sequences. EMBO J. 16 (3), 484–494.

Lambrechts, A., Jonckheere, V., Peleman, C., Polet, D., De Vos, W., Vandekerckhove, J., Ampe, C., 2006. Profilin-I-ligand interactions influence various aspects of neuronal differentiation. J. Cell Sci. 119 (8), 1570–1578.

Lämmermann, T., Sixt, M., 2009. Mechanical modes of "amoeboid" cell migration. Curr. Opin. Cell Biol. 21 (5), 636–644.

Lassing, I., Lindberg, U., 1985. Specific interaction between phosphatidylinositol 4,5-bisphosphate and profilactin. Nature 314 (6010), 472–474.

LeCorgne, H., Tudosie, A.M., Lavik, K., Su, R., Becker, K.N., Moore, S., Walia, Y., Wisner, A., Koehler, D., Alberts, A.S., Williams, F.E., Eisenmann, K.M., 2018. Differential toxicity of mDia formin-directed functional agonists and antagonists in developing zebrafish. Front. Pharmacol. 9, 340.

Lederer, M., Jockusch, B.M., Rothkegel, M., 2005. Profilin regulates the activity of p42POP, a novel Myb-related transcription factor. J. Cell Sci. 118 (2), 331–341.

Lee, Y.-J., Mazzatti, D.J., Yun, Z., Keng, P.C., 2005. Inhibition of invasiveness of human lung cancer cell line H1299 by over-expression of cofilin. Cell Biol. Int. 29 (11), 877–883.

Lewkowicz, E., Herit, F., Le Clainche, C., Bourdoncle, P., Perez, F., Niedergang, F., 2008. The microtubule-binding protein CLIP-170 coordinates mDia1 and actin reorganization during CR3-mediated phagocytosis. J. Cell Biol. 183 (7), 1287–1298.

Li, R., Gundersen, G.G., 2008. Beyond polymer polarity: how the cytoskeleton builds a polarized cell. Nat. Rev. Mol. Cell Biol. 9 (11), 860–873.

Lila, T., Drubin, D.G., 1997. Evidence for physical and functional interactions among two *Saccharomyces cerevisiae* SH3 domain proteins, an adenylyl cyclase-associated protein and the actin cytoskeleton. Mol. Biol. Cell 8 (2), 367–385.

Lodish, H., Berk, A., Zipursky, S.L., et al., 2000. The actin cytoskeleton. In: Freeman, W.H. (Ed.), Molecular Cell Biology. Macmillan Education, New York, NY, USA.

Loisel, T.P., Boujemaa, R., Pantaloni, D., Carlier, M.F., 1999. Reconstitution of actin-based motility of *Listeria* and *Shigella* using pure proteins. Nature 401 (6753), 613–616.

Löwe, J., Li, H., Downing, K.H., Nogales, E., 2001. Refined structure of alpha beta-tubulin at 3.5 Å resolution. J. Mol. Biol. 313 (5), 1045–1057.

Lu, J., Pollard, T.D., 2001. Profilin binding to poly-L-proline and actin monomers along with ability to catalyze actin nucleotide exchange is required for viability of fission yeast. Mol. Biol. Cell 12 (4), 1161–1175.

Machesky, L.M., Pollard, T.D., 1993. Profilin as a potential mediator of membrane-cytoskeleton communication. Trends Cell Biol. 3 (11), 381–385.

Machesky, L.M., Cole, N.B., Moss, B., Pollard, T.D., 1994. *Vaccinia* virus expresses a novel profilin with a higher affinity for polyphosphoinositides than actin. Biochemistry 33 (35), 10815–10824.

Machesky, L.M., Mullins, R.D., Higgs, H.N., Kaiser, D.A., Blanchoin, L., May, R.C., Hall, M.E., Pollard, T.D., 1999. Scar, a WASp-related protein, activates nucleation of actin filaments by the Arp2/3 complex. Proc. Natl. Acad. Sci. 96 (7), 3739–3744.

Majesky, M.W., 2007. Developmental basis of vascular smooth muscle diversity. Arterioscler. Thromb. Vasc. Biol. 27 (6), 1248–1258.

Mammoto, A., Sasaki, T., Asakura, T., Hotta, I., Imamura, H., Takahashi, K., Matsuura, Y., Shirao, T., Takai, Y., 1998. Interactions of drebrin and gephyrin with profilin. Biochem. Biophys. Res. Commun. 243 (1), 86–89.

Manandhar, A., Kang, M., Chakraborty, K., Loverde, S.M., 2018. Effect of nucleotide state on the protofilament conformation of tubulin octamers. J. Phys. Chem. 122 (23), 6164–6178.

Mares-Mejía, I., Martínez-Caballero, S., Garay-Canales, C., Cano-Sánchez, P., Torres-Larios, A., Lara-González, S., Ortega, E., Rodríguez-Romero, A., 2016. Structural insights into the IgE mediated responses induced by the allergens Hev b 8 and Zea m 12 in their dimeric forms. Sci. Rep. 6, 32552.

Margolis, R.L., 1981. Role of GTP hydrolysis in microtubule treadmilling and assembly. Proc. Natl. Acad. Sci. 78 (3), 1586–1590.

McGrath, J.L., Tardy, Y., Dewey, C.F., Meister, J.J., Hartwig, J.H., 1998. Simultaneous measurements of actin filament turnover, filament fraction, and monomer diffusion in endothelial cells. Biophys. J. 75 (4), 2070–2078.

McIntosh, J.R., Hays, T., 2016. A brief history of research on mitotic mechanisms. Biology 5 (4), 55.

Melak, M., Plessner, M., Grosse, R., 2017. Actin visualization at a glance. J. Cell Sci. 130 (3), 525–530.

Melamed, Z., López-Erauskin, J., Baughn, M.W., Zhang, O., Drenner, K., Sun, Y., Freyermuth, F., McMahon, M.A., Beccari, M.S., Artates, J.W., Ohkubo, T., Rodriguez, M., Lin, N., Wu, D., Bennett, C.F., Rigo, F., Da Cruz, S., Ravits, J., Lagier-Tourenne, C., Cleveland, D.W., 2019. Premature polyadenylation-mediated loss of stathmin-2 is a hallmark of TDP-43-dependent neurodegeneration. Nat. Neurosci. 22 (2), 180–190.

Merino, F., Pospich, S., Funk, J., Wagner, T., Küllmer, F., Arndt, H.-D., Bieling, P., Raunser, S., 2018. Structural transitions of F-actin upon ATP hydrolysis at near-atomic resolution revealed by cryo-EM. Nat. Struct. Mol. Biol. 25 (6), 528–537.

Meyers, J.R., 2018. Zebrafish: development of a vertebrate model organism. Curr. Protoc. Lab. Tech. 16 (1), e19.

Michaelsen, K., Murk, K., Zagrebelsky, M., Dreznjak, A., Jockusch, B.M., Rothkegel, M., Korte, M., 2010. Fine-tuning of neuronal architecture requires two profilin isoforms. Proc. Natl. Acad. Sci. 107 (36), 15780–15785.

Michelot, A., Drubin, D.G., 2011. Building distinct actin filament networks in a common cytoplasm. Curr. Biol. 21 (14), R560–R569.

Miki, H., Suetsugu, S., Takenawa, T., 1998. WAVE, a novel WASP-family protein involved in actin reorganization induced by Rac. EMBO J. 17 (23), 6932–6941.

Milic, B., Chakraborty, A., Han, K., Bassik, M.C., Block, S.M., 2018. KIF15 nanomechanics and kinesin inhibitors, with implications for cancer chemotherapeutics. Proc. Natl. Acad. Sci. 115 (20), E4613–E4622.

Miller, A.L., 2011. The contractile ring. Curr. Biol. 21 (24), R976–R978.
Mitchison, T., 1993. Localization of an exchangeable GTP binding site at the plus end of microtubules. Science 261 (5124), 1044–1047.
Mitchison, T., Kirschner, M., 1984a. Microtubule assembly nucleated by isolated centrosomes. Nature 312 (5991), 232–237.
Mitchison, T., Kirschner, M., 1984b. Dynamic instability of microtubule growth. Nature 312 (5991), 237–242.
Mockrin, S.C., Korn, E.D., 1980. *Acanthamoeba* profilin interacts with G-actin to increase the rate of exchange of actin-bound adenosine 5′-triphosphate. Biochem. 19 (23), 5359–5362.
Mogilner, A., Oster, G., 1996. Cell motility driven by actin polymerization. Biophys. J. 71 (6), 3030–3045.
Molliex, A., Temirov, J., Lee, J., Coughlin, M., Kanagaraj, A.P., Kim, H.J., Mittag, T., Taylor, J.P., 2015. Phase separation by low complexity domains promotes stress granule assembly and drives pathological fibrillization. Cell 163 (1), 123–133.
Moreau, C.A., Bhargav, S.P., Kumar, H., Quadt, K.A., Piirainen, H., Strauss, L., Kehrer, J., Streichfuss, M., Spatz, J.P., Wade, R.C., Kursula, I., Frischknecht, F., 2017. A unique profilin-actin interface is important for malaria parasite motility. PLOS Pathog. 13 (5), e1006412.
Moreau, C.A., Quadt, K.A., Piirainen, H., Kumar, H., Bhargav, S.P., Strauss, L., Tolia, N.H., Wade, R.C., Spatz, J.P., Kursula, I., Frischknecht, F., 2020. A function of profilin in force generation during malaria parasite motility independent of actin binding. J. Cell Sci. 134 (5), jcs233775.
Moritz, M., Braunfeld, M.B., Guénebaut, V., Heuser, J., Agard, D.A., 2000. Structure of the gamma-tubulin ring complex: a template for microtubule nucleation. Nat. Cell Biol. 2 (6), 365–370.
Mostowy, S., Shenoy, A.R., 2015. The cytoskeleton in cell-autonomous immunity: structural determinants of host defense. Nat. Rev. Immunol. 15 (9), 559–573.
Mouneimne, G., Hansen, S.D., Selfors, L.M., Petrak, L., Hickey, M.M., Gallegos, L.L., Simpson, K.J., Lim, J., Gertler, F.B., Hartwig, J.H., Mullins, R.D., Brugge, J.S., 2012. Differential remodeling of actin cytoskeleton architecture by profilin isoforms leads to distinct effects on cell migration and invasion. Cancer Cell 22 (5), 615–630.
Moustakas, A., Heldin, C.-H., 2008. Dynamic control of TGF-β signaling and its links to the cytoskeleton. FEBS Lett. 582 (14), 2051–2065.
Mukhtar, E., Adhami, V.M., Mukhtar, H., 2014. Targeting microtubules by natural agents for cancer therapy. Mol. Cancer Ther. 13 (2), 275–284.
Mullins, R.D., Pollard, T.D., 1999. Structure and function of the Arp2/3 complex. Curr. Opin. Struct. Biol. 9 (2), 244–249.
Mullins, R.D., Heuser, J.A., Pollard, T.D., 1998. The interaction of Arp2/3 complex with actin: nucleation, high affinity pointed end capping, and formation of branching networks of filaments. Proc. Natl. Acad. Sci. 95 (11), 6181–6186.
Murk, K., Buchmeier, S., Jockusch, B.M., Rothkegel, M., 2009. In birds, profilin-2a is ubiquitously expressed and contributes to actin-based motility. J. Cell Sci. 122 (7), 957–964.
Müssar, K.J., Kandasamy, M.K., McKinney, E.C., Meagher, R.B., 2015. *Arabidopsis* plants deficient in constitutive class profilins reveal independent and quantitative genetic effects. BMC Plant Biol. 15, 177.
Neidt, E.M., Scott, B.J., Kovar, D.R., 2009. Formin differentially utilizes profilin isoforms to rapidly assemble actin filaments. J. Biol. Chem. 284 (1), 673–684.
Nejedla, M., Sadi, S., Sulimenko, V., de Almeida, F.N., Blom, H., Draber, P., Aspenström, P., Karlsson, R., 2016. Profilin connects actin assembly with microtubule dynamics. Mol. Biol. Cell 27 (15), 2381–2393.
Nejedla, M., Li, Z., Masser, A.E., Biancospino, M., Spiess, M., Mackowiak, S.D., Friedländer, M.R., Karlsson, R., 2017. A fluorophore fusion construct of human

profilin-I with non-compromised poly(L-)proline binding capacity suitable for imaging. J. Mol. Biol. 429 (7), 964–976.
Nodelman, I.M., Bowman, G.D., Lindberg, U., Schutt, C.E., 1999. X-ray structure determination of human profilin II: a comparative structural analysis of human profilins. J. Mol. Biol. 294 (5), 1271–1285.
Nogales, E., 2001. Structural insight into microtubule function. Annu. Rev. Biophys. Biomol. Struct. 30, 397–420.
Nogales, E., Wolf, S.G., Downing, K.H., 1998. Structure of the alpha beta tubulin dimer by electron crystallography. Nature 391 (6663), 199–203.
Nolen, B.J., Tomasevic, N., Russell, A., Pierce, D.W., Jia, Z., McCormick, C.D., Hartman, J., Sakowicz, R., Pollard, T.D., 2009. Characterization of two classes of small molecule inhibitors of Arp2/3 complex. Nature 460 (7258), 1031–1034.
Oakley, B.R., Paolillo, V., Zheng, Y., 2015. γ-Tubulin complexes in microtubule nucleation and beyond. Mol. Biol. Cell 26 (17), 2957–2962.
Oda, T., Aihara, T., Wakabayashi, K., 2016. Early nucleation events in the polymerization of actin, probed by time-resolved small-angle x-ray scattering. Sci. Rep. 6, 34539.
Onishi, M., Pringle, J.R., Cross, F.R., 2016. Evidence that an unconventional actin can provide essential F-actin function and that a surveillance system monitors F-actin integrity in *Chlamydomonas*. Genetics 202 (3), 977–996.
Ono, S., 2013. The role of cyclase-associated protein in regulating actin filament dynamics - more than a monomer-sequestration factor. J. Cell Sci. 126 (15), 3249–3258.
Ostrander, D.B., Ernst, E.G., Lavoie, T.B., Gorman, J.A., 1999. Polyproline binding is an essential function of human profilin in yeast. Eur. J. Biochem. 262 (1), 26–35.
Palucka, A.K., Coussens, L.M., 2016. The basis of oncoimmunology. Cell 164 (6), 1233–1247.
Pan, L.-N., Zhang, Y., Zhu, C.-J., Dong, Z.-X., 2017. Kinesin KIF4A is associated with chemotherapeutic drug resistance by regulating intracellular trafficking of lung resistance-related protein. J. Zhejiang Univ. Sci. B 18 (12), 1046–1054.
Pantaloni, D., Carlier, M.F., 1993. How profilin promotes actin filament assembly in the presence of Thymosin Beta 4. Cell 75 (5), 1007–1014.
Pantaloni, D., Le Clainche, C., Carlier, M.F., 2001. Mechanism of actin-based motility. Science 292 (5521), 1502–1506.
Panzica, M.T., Marin, H.C., Reymann, A.-C., McNally, F.J., 2017. F-actin prevents interaction between sperm DNA and the oocyte meiotic spindle in *C. elegans*. J. Cell Biol. 216 (8), 2273–2282.
Paplomata, E., O'Regan, R., 2014. The PI3K/AKT/mTOR pathway in breast cancer: targets, trials and biomarkers. Ther. Adv. Med. Oncol. 6 (4), 154–166.
Paul, A., Pollard, T., 2008. The role of the FH1 domain and profilin in formin-mediated actin-filament elongation and nucleation. Curr. Biol. 18 (1), 9–19.
Pazour, G.J., Witman, G.B., 2009. The Chlamydomonas flagellum as a model for human ciliary disease. In: Chlamydomonas Sourcebook. Academy Press, pp. 445–478.
Pelham, R.J., Chang, F., 2002. Actin dynamics in the contractile ring during cytokinesis in fission yeast. Nature 419 (6902), 82–86.
Percipalle, P., Vartiainen, M., 2019. Cytoskeletal proteins in the cell nucleus: a special nuclear actin perspective. Mol. Biol. Cell 30 (15), 1781–1785.
Perelroizen, I., Marchand, J.-B., Blanchoin, L., Didry, D., Carlier, M.-F., 1994. Interaction of profilin with G-actin and poly(L-)proline. Biochemistry 33 (28), 8472–8478.
Pernier, J., Shekhar, S., Jegou, A., Guichard, B., Carlier, M.-F., 2016. Profilin interaction with actin filament barbed end controls dynamic instability, capping, branching, and motility. Dev. Cell 36 (2), 201–214.
Peterson, J.R., Bickford, L.C., Morgan, D., Kim, A.S., Ouerfelli, O., Kirschner, M.W., Rosen, M.K., 2004. Chemical inhibition of N-WASP by stabilization of a native autoinhibited conformation. Nat. Struct. Mol. Biol. 11 (8), 747–755.

Petrella, E.C., Machesky, L.M., Kaiser, D.A., Pollard, T.D., 1996. Structural requirements and thermodynamics of the interaction of proline peptides with profilin. Biochemistry 35 (51), 16535–16543.

Pfajfer, L., Mair, N.K., Jiménez-Heredia, R., Genel, F., Gulez, N., Ardeniz, Ö., Hoeger, B., Bal, S.K., Madritsch, C., Kalinichenko, A., Chandra Ardy, R., Gerçeker, B., Rey-Barroso, J., Ijspeert, H., Tangye, S.G., Simonitsch-Klupp, I., Huppa, J.B., van der Burg, M., Dupré, L., Boztug, K., 2018. Mutations affecting the actin regulator WD repeat-containing protein 1 lead to aberrant lymphoid immunity. J. Allergy Clin. Immunol. Pract. 142 (5), 1589–1604.e11.

Pinto-Costa, R., Sousa, M.M., 2019. Profilin as a dual regulator of actin and microtubule dynamics. Cytoskeleton 77 (3–4), 76–83.

Piperno, G., Luck, D.J., 1979. Axonemal adenosine triphosphatases from flagella of *Chlamydomonas reinhardtii*. Purification of two dyneins. J. Biol. Chem. 254 (8), 3084–3090.

Plastino, J., Blanchoin, L., 2018. Dynamic stability of the actin ecosystem. J. Cell Sci. 132 (4).

Plattner, F., Yarovinsky, F., Romero, S., Didry, D., Carlier, M.-F., Sher, A., Soldati-Favre, D., 2008. *Toxoplasma* profilin is essential for host cell invasion and TLR11-dependent induction of an Interleukin-12 response. Cell Host Microbe 3 (2), 77–87.

Plessner, M., Knerr, J., Grosse, R., 2019. Centrosomal actin assembly is required for proper mitotic spindle formation and chromosome congression. iScience 15, 274–281.

Polet, D., Lambrechts, A., Ono, K., Mah, A., Peelman, F., Vandekerckhove, J., Baillie, D.L., Ampe, C., Ono, S., 2006. Caenorhabditis elegans expresses three functional profilins in a tissue-specific manner. Cell Motil. Cytoskeleton 63 (1), 14–28.

Pollard, T.D., 1986. Rate constants for the reactions of ATP- and ADP-actin with the ends of actin filaments. J. Cell Biol. 103 (6), 2747–2754.

Pollard, T.D., 2007. Regulation of actin filament assembly by Arp2/3 complex and formins. Annu. Rev. Biophys. Biomol. Struct. 36, 451–477.

Pollard, T.D., 2016. Actin and actin-binding proteins. Cold Spring Harb. Perspect. Biol. 8 (8), a018226.

Pollard, T.D., Borisy, G.G., 2003. Cellular motility driven by assembly and disassembly of actin filaments. Cell 112 (4), 453–465.

Pollard, T.D., O'Shaughnessy, B., 2019. Molecular mechanism of cytokinesis. Annu. Rev. Biochem. 88, 661–689.

Pollard, T.D., Wu, J.-Q., 2010. Understanding cytokinesis: lessons from fission yeast. Nat. Rev. Mol. Cell Biol. 11 (2), 149–155.

Pollard, T.D., Blanchoin, L., Mullins, R.D., 2000. Molecular mechanisms controlling actin filament dynamics in nonmuscle cells. Annu. Rev. Biophys. Biomol. Struct. 29, 545–576.

Popov, A.V., Severin, F., Karsenti, E., 2002. XMAP215 is required for the microtubule-nucleating activity of centrosomes. Curr. Biol. 12 (15), 1326–1330.

Posey, A.E., Ruff, K.M., Harmon, T.S., Crick, S.L., Li, A., Diamond, M.I., Pappu, R.V., 2018. Profilin reduces aggregation and phase separation of huntingtin N-terminal fragments by preferentially binding to soluble monomers and oligomers. J. Biol. Chem. 293 (10), 3734–3746.

Prezel, E., Elie, A., Delaroche, J., Stoppin-Mellet, V., Bosc, C., Serre, L., Fourest-Lieuvin, A., Andrieux, A., Vantard, M., Arnal, I., 2018. Tau can switch microtubule network organizations: from random networks to dynamic and stable bundles. Mol. Biol. Cell 29 (2), 154–165.

Prokop, A., Beaven, R., Qu, Y., Sánchez-Soriano, N., 2013. Using fly genetics to dissect the cytoskeletal machinery of neurons during axonal growth and maintenance. J. Cell Sci. 126 (11), 2331–2341.

Pruyne, D., Evangelista, M., Yang, C., Bi, E., Zigmond, S., Bretscher, A., Boone, C., 2002. Role of formins in actin assembly: nucleation and barbed-end association. Science 297 (5581), 612–615.

Qiao, Z., Sun, H., Ng, J.T.Y., Ma, Q., Koh, S.H., Mu, Y., Miao, Y., Gao, Y.-G., 2019. Structural and computational examination of the *Arabidopsis* profilin–poly-p complex reveals mechanistic details in profilin-regulated actin assembly. J. Biol. Chem. 294 (49), 18650–18661.

Quinlan, M.E., Heuser, J.E., Kerkhoff, E., Mullins, R.D., 2005. *Drosophila* spire is an actin nucleation factor. Nature 433 (7024), 382–388.

Rebowski, G., Boczkowska, M., Drazic, A., Ree, R., Goris, M., Arnesen, T., Dominguez, R., 2020. Mechanism of actin N-terminal acetylation. Sci. Adv. 6 (15), eaay8793.

Reeve, S.P., Bassetto, L., Genova, G.K., Kleyner, Y., Leyssen, M., Jackson, F.R., Hassan, B.A., 2005. The *Drosophila* fragile X mental retardation protein controls actin dynamics by directly regulating profilin in the brain. Curr. Biol. 15 (12), 1156–1163.

Reinhard, M., Giehl, K., Abel, K., Haffner, C., Jarchau, T., Hoppe, V., Jockusch, B.M., Walter, U., 1995. The proline-rich focal adhesion and microfilament protein VASP is a ligand for profilins. EMBO J. 14 (8), 1583–1589.

Ricketts, S.N., Francis, M.L., Farhadi, L., Rust, M.J., Das, M., Ross, J.L., Robertson-Anderson, R.M., 2019. Varying crosslinking motifs drive the mesoscale mechanics of actin-microtubule composites. Sci. Rep. 9 (1), 12831.

Rizvi, S.A., Neidt, E.M., Cui, J., Feiger, Z., Skau, C.T., Gardel, M.L., Kozmin, S.A., Kovar, D.R., 2009. Identification and characterization of a small molecule inhibitor of formin-mediated actin assembly. Chem. Biol. 16 (11), 1158–1168.

Rizwani, W., Fasim, A., Sharma, D., Reddy, D.J., Bin Omar, N.A.M., Singh, S.S., 2014. S137 phosphorylation of profilin 1 is an important signaling event in breast cancer progression. PLoS One 9 (8), e103868.

Rodal, A.A., Manning, A.L., Goode, B.L., Drubin, D.G., 2003. Negative regulation of yeast WASp by two SH3 domain-containing proteins. Curr. Biol. 13 (12), 1000–1008.

Rodionov, V.I., Hope, A.J., Svitkina, T.M., Borisy, G.G., 1998. Functional coordination of microtubule-based and actin-based motility in melanophores. Curr. Biol. 8 (3), 165–169.

Rodriguez, O.C., Schaefer, A.W., Mandato, C.A., Forscher, P., Bement, W.M., Waterman-Storer, C.M., 2003. Conserved microtubule-actin interactions in cell movement and morphogenesis. Nat. Cell Biol. 5 (7), 599–609.

Roeles, J., Tsiavaliaris, G., 2019. Actin-microtubule interplay coordinates spindle assembly in human oocytes. Nat. Commun. 10 (1), 4651.

Romero, S., Le Clainche, C., Didry, D., Egile, C., Pantaloni, D., Carlier, M.-F., 2004. Formin is a processive motor that requires profilin to accelerate actin assembly and associated ATP hydrolysis. Cell 119 (3), 419–429.

Roostalu, J., Surrey, T., 2017. Microtubule nucleation: beyond the template. Nat. Rev. Mol. Cell Biol. 18 (11), 702–710.

Rosenblatt, J., Cramer, L.P., Baum, B., McGee, K.M., 2004. Myosin II-dependent cortical movement is required for centrosome separation and positioning during mitotic spindle assembly. Cell 117 (3), 361–372.

Roth, L.W., Bormann, P., Bonnet, A., Reinhard, E., 1999. Beta-thymosin is required for axonal tract formation in developing zebrafish brain. Development 126 (7), 1365–1374.

Roth-Johnson, E.A., Vizcarra, C.L., Bois, J.S., Quinlan, M.E., 2014. Interaction between microtubules and the *Drosophila* formin cappuccino and its effect on actin assembly. J. Biol. Chem. 289 (7), 4395–4404.

Rotty, J.D., Wu, C., Haynes, E.M., Suarez, C., Winkelman, J.D., Johnson, H.E., Haugh, J.M., Kovar, D.R., Bear, J.E., 2015. Profilin-1 serves as a gatekeeper for actin assembly by Arp2/3-dependent and -independent pathways. Dev. Cell 32 (1), 54–67.

Rouiller, I., Xu, X.-P., Amann, K.J., Egile, C., Nickell, S., Nicastro, D., Li, R., Pollard, T.D., Volkmann, N., Hanein, D., 2008. The structural basis of actin filament branching by the Arp2/3 complex. J. Cell Biol. 180 (5), 887–895.

Roy, P., Jacobson, K., 2004. Overexpression of profilin reduces the migration of invasive breast cancer cells. Cell Motil. Cytoskeleton 57 (2), 84–95.
Rust, M.J., Bates, M., Zhuang, X., 2006. Sub-diffraction-limit imaging by stochastic optical reconstruction microscopy (STORM). Nat. Methods 3 (10), 793–796.
Safer, D., Elzinga, M., Nachmias, V.T., 1991. Thymosin-beta-4 and Fx, an actin-sequestering peptide, are indistinguishable. J. Biol. Chem. 266 (7), 4029–4032.
Sagot, I., Rodal, A.A., Moseley, J., Goode, B.L., Pellman, D., 2002. An actin nucleation mechanism mediated by Bni1 and profilin. Nat. Cell Biol. 4 (8), 626–631.
Salmon, W.C., Adams, M.C., Waterman-Storer, C.M., 2002. Dual-wavelength fluorescent speckle microscopy reveals coupling of microtubule and actin movements in migrating cells. J. Cell Biol. 158 (1), 31–37.
Sanger, J.W., 1975. Changing patterns of actin localization during cell division. Proc. Natl. Acad. Sci. 72 (5), 1913–1916.
Santos, A., Van Ree, R., 2011. Profilins: mimickers of allergy or relevant allergens. Int. Arch. Allergy Immunol. 155 (3), 191–204.
Sathish, K., Padma, B., Munugalavadla, V., Bhargavi, V., Radhika, K.V.N., Wasia, R., Sairam, M., Singh, S.S., 2004. Phosphorylation of profilin regulates its interaction with actin and poly(L)-proline. Cell. Signal. 16 (5), 589–596.
Schlett, K., 2017. More than a mere supply of monomers: G-actin pools regulate actin dynamics in dendritic spines. J. Cell Biol. 216 (8), 2255–2257.
Schlüter, K., Jockusch, B.M., Rothkegel, M., 1997. Profilins as regulators of actin dynamics. Biochim. Biophys. Acta 1359 (2), 97–109.
Schoppmeyer, R., Zhao, R., Cheng, H., Hamed, M., Liu, C., Zhou, X., Schwarz, E.C., Zhou, Y., Knörck, A., Schwär, G., Ji, S., Liu, L., Long, J., Helms, V., Hoth, M., Yu, X., Qu, B., 2017. Human profilin-1 is a negative regulator of CTL mediated cell-killing and migration. Eur. J. Immunol. 47 (9), 1562–1572.
Schutt, C.E., Myslik, J.C., Rozycki, M.D., Goonesekere, N.C.W., Lindber, U., 1993. The structure of crystalline profilin-β-actin. Nature 365, 810–816.
Sellers, J.R., Shi, S., Nishimura, Y., Zhang, F., Liu, R., Takagi, Y., Viasnoff, V., Bershadsky, A.D., 2020. The Formin inhibitor, SMIFH2, inhibits members of the myosin superfamily. Biophys. J. 118 (3), 125a.
Sept, D., McCammon, J.A., 2001. Thermodynamics and kinetics of actin filament nucleation. Biophys. J. 81 (2), 667–674.
Severson, A.F., Baillie, D.L., Bowerman, B., 2002. A Formin homology protein and a profilin are required for cytokinesis and Arp2/3-independent assembly of cortical microfilaments in *C. elegans*. Curr. Biol. 12 (24), 2066–2075.
Shao, J., Diamond, M.I., 2012. Protein Phosphatase-1 dephosphorylates Profilin-1 at ser-137. PLoS One 7 (3), e32802.
Shao, J., Welch, W.J., Diprospero, N.A., Diamond, M.I., 2008. Phosphorylation of profilin by ROCK1 regulates polyglutamine aggregation. Mol. Cell. Biol. 28 (17), 5196–5208.
Sherer, L.A., Zweifel, M.E., Courtemanche, N., 2018. Dissection of two parallel pathways for formin-mediated actin filament elongation. J. Biol. Chem. 293 (46), 17917–17928.
Shields, A.R., Spence, A.C., Yamashita, Y.M., Davies, E.L., Fuller, M.T., 2014. The actin-binding protein profilin is required for germline stem cell maintenance and germ cell enclosure by somatic cyst cells. Development 141 (1), 73–82.
Shirey, C.M., Scott, J.L., Stahelin, R.V., 2017. Notes and tips for improving quality of lipid-protein overlay assays. Anal. Biochem. 516, 9–12.
Simons, M., Schwartz, M.A., 2012. Profilin phosphorylation as a VEGFR effector in angiogenesis. Nat. Cell Biol. 14 (10), 985–987.
Singh, S.S., Chauhan, A., Murakami, N., Chauhan, V.P., 1996. Profilin and gelsolin stimulate phosphatidylinositol 3-kinase activity. Biochem. 35 (51), 16544–16549.

Skillman, K.M., Daher, W., Ma, C.I., Soldati-Favre, D., Sibley, L.D., 2012. Toxoplasma gondii profilin acts primarily to sequester G-actin while formins efficiently nucleate actin filament formation in vitro. Biochem. 51 (12), 2486–2495.

Skruber, K., Read, T.-A., Vitriol, E.A., 2018. Reconsidering an active role for G-actin in cytoskeletal regulation. J. Cell Sci. 131 (1).

Skruber, K., Warp, P.V., Shklyarov, R., Thomas, J.D., Swanson, M.S., Henty-Ridilla, J.L., Read, T.-A., Vitriol, E.A., 2020. Arp2/3 and Mena/VASP require Profilin 1 for actin network assembly at the leading edge. Curr. Biol. 30, 1–14.

Slater, P.G., Hayrapetian, L., Lowery, L.A., 2017. *Xenopus laevis* as a model system to study cytoskeletal dynamics during axon pathfinding. Genesis 55 (1–2), e22994.

Small, J.V., 1988. The actin cytoskeleton. Electron Microsc. Rev. 1 (1), 155–174.

Small, J.V., Celis, J.E., 1978. Filament arrangements in negatively stained cultured cells: the organization of actin. Cytobiologie 16 (2), 308–325.

Smith, T.E., Hong, W., Zachariah, M.M., Harper, M.K., Matainaho, T.K., Van Wagoner, R.M., Ireland, C.M., Vershinin, M., 2013. Single-molecule inhibition of human kinesin by adociasulfate-13 and -14 from the sponge *Cladocroce aculeata*. Proc. Natl. Acad. Sci. 110 (47), 18880–18885.

Söderberg, E., Hessle, V., von Euler, A., Visa, N., 2012. Profilin is associated with transcriptionally active genes. Nucleus 3 (3), 290–299.

Sohn, R.H., Chen, J., Koblan, K.S., Bray, P.F., Goldschmidt-Clermont, P.J., 1995. Localization of a binding site for phosphatidylinositol 4,5-bisphosphate on human profilin. J. Biol. Chem. 270 (36), 21114–21120.

Staiger, C.J., Sheahan, M.B., Khurana, P., Wang, X., McCurdy, D.W., Blanchoin, L., 2009. Actin filament dynamics are dominated by rapid growth and severing activity in the *Arabidopsis* cortical array. J. Cell Biol. 184 (2), 269–280.

Stehn, J.R., Haass, N.K., Bonello, T., Desouza, M., Kottyan, G., Treutlein, H., Zeng, J., Nascimento, P.R.B.B., Sequeira, V.B., Butler, T.L., Allanson, M., Fath, T., Hill, T.A., McCluskey, A., Schevzov, G., Palmer, S.J., Hardeman, E.C., Winlaw, D., Reeve, V.E., Dixon, I., Weninger, W., Cripe, T.P., Gunning, P.W., 2013. A novel class of anticancer compounds targets the actin cytoskeleton in tumor cells. Cancer Res. 73 (16), 5169–5182.

Sturgill, E.G., Norris, S.R., Guo, Y., Ohi, R., 2016. Kinesin-5 inhibitor resistance is driven by kinesin-12. J. Cell Biol. 213 (2), 213–227.

Stüven, T., Hartmann, E., Görlich, D., 2003. Exportin 6: a novel nuclear export receptor that is specific for profilin-actin complexes. EMBO J. 22 (21), 5928–5940.

Suarez, C., Kovar, D.R., 2016. Internetwork competition for monomers governs actin cytoskeleton organization. Nat. Rev. Mol. Cell Biol. 17 (12), 799–810.

Suarez, C., Carroll, R.T., Burke, T.A., Christensen, J.R., Bestul, A.J., Sees, J.A., James, M.L., Sirotkin, V., Kovar, D.R., 2015. Profilin regulates F-actin network homeostasis by favoring formin over Arp2/3 complex. Dev. Cell 32 (1), 43–53.

Suetsugu, S., Miki, H., Takenawa, T., 1998. The essential role of profilin in the assembly of actin for microspike formation. EMBO J. 17 (22), 6516–6526.

Sun, H., Qiao, Z., Chua, K.P., Tursic, A., Liu, X., Gao, Y.-G., Mu, Y., Hou, X., Miao, Y., 2018. Profilin negatively regulates Formin-mediated actin assembly to modulate PAMP-triggered plant immunity. Curr. Biol. 28 (12), 1882–1895.e7.

Svitkina, T.M., 2018. Ultrastructure of the actin cytoskeleton. Curr. Opin. Cell Biol. 54, 1–8.

Svitkina, T.M., Borisy, G.G., 1999. Arp2/3 complex and actin depolymerizing factor/cofilin in dendritic organization and treadmilling of actin filament array in lamellipodia. J. Cell Biol. 145 (5), 1009–1026.

Svitkina, T.M., Verkhovsky, A.B., McQuade, K.M., Borisy, G.G., 1997. Analysis of the actin-myosin II system in fish epidermal keratocytes: mechanism of cell body translocation. J. Cell Biol. 139 (2), 397–415.

Svitkina, T.M., Bulanova, E.A., Chaga, O.Y., Vignjevic, D.M., Kojima, S., Vasiliev, J.M., Borisy, G.G., 2003. Mechanism of filopodia initiation by reorganization of a dendritic network. J. Cell Biol. 160 (3), 409–421.
Symons, M.H., Mitchison, T.J., 1991. Control of actin polymerization in live and permeabilized fibroblasts. J. Cell Biol. 114 (3), 503–513.
Szikora, S., Földi, I., Tóth, K., Migh, E., Vig, A., Bugyi, B., Maléth, J., Hegyi, P., Kaltenecker, P., Sanchez-Soriano, N., Mihály, J., 2017. The formin DAAM is required for coordination of the actin and microtubule cytoskeleton in axonal growth cones. J. Cell Sci. 130, 2506–2519.
Tai, A.W., Chuang, J.Z., Bode, C., Wolfrum, U., Sung, C.H., 1999. Rhodopsin's carboxy-terminal cytoplasmic tail acts as a membrane receptor for cytoplasmic dynein by binding to the dynein light chain Tctex-1. Cell 97 (7), 877–887.
Tang, Y.-N., Ding, W.-Q., Guo, X.-J., Yuan, X.-W., Wang, D.-M., Song, J.-G., 2015. Epigenetic regulation of Smad2 and Smad3 by profilin-2 promotes lung cancer growth and metastasis. Nat. Commun. 6, 8230.
Tas, R.P., Chazeau, A., Cloin, B.M.C., Lambers, M.L.A., Hoogenraad, C.C., Kapitein, L.C., 2017. Differentiation between oppositely oriented microtubules controls polarized neuronal transport. Neuron 96 (6), 1264–1271.e5.
Thawani, A., Kadzik, R.S., Petry, S., 2018. XMAP215 is a microtubule nucleation factor that functions synergistically with the γ-tubulin ring complex. Nat. Cell Biol. 20 (5), 575–585.
Theriot, J.A., Mitchison, T.J., Tilney, L.G., Portnoy, D.A., 1992. The rate of actin-based motility of intracellular listeria monocytogenes equals the rate of actin polymerization. Nature 357 (6375), 257–260.
Theriot, J.A., Rosenblatt, J., Portnoy, D.A., Goldschmidt-Clermont, P.J., Mitchison, T.J., 1994. Involvement of profilin in the actin-based motility of *L. monocytogenes* in cells and in cell-free extracts. Cell 76 (3), 505–517.
Théry, M., Racine, V., Pépin, A., Piel, M., Chen, Y., Sibarita, J.-B., Bornens, M., 2005. The extracellular matrix guides the orientation of the cell division axis. Nat. Cell Biol. 7 (10), 947–953.
Thorn, K.S., Christensen, H.E., Shigeta, R., Huddler, D., Shalaby, L., Lindberg, U., Chua, N.H., Schutt, C.E., 1997. The crystal structure of a major allergen from plants. Structure 5 (1), 19–32.
Tilney, L.G., 1976. The polymerization of actin. Aggregates of nonfilamentous actin and its associated proteins: a storage form of actin. J. Cell Biol. 69 (1), 73–89.
Tilney, L.G., DeRosier, D.J., Weber, A., Tilney, M.S., 1992. How *Listeria* exploits host cell actin to form its own cytoskeleton. Nucleation, actin filament polarity, filament assembly, and evidence for a pointed end capper. J. Cell Biol. 118 (1), 83–93.
Toyoshima, F., Nishida, E., 2007. Spindle orientation in animal cell mitosis: roles of integrin in the control of spindle axis. J. Cell. Physiol. 213 (2), 407–411.
Ubersax, J.A., Woodbury, E.L., Quang, P.N., Paraz, M., Blethrow, J.D., Shah, K., Shokat, K.M., Morgan, D.O., 2003. Targets of the cyclin-dependent kinase Cdk1. Nature 425 (6960), 859–864.
Vargas, P., Barbier, L., Sáez, P.J., Piel, M., 2017. Mechanisms for fast cell migration in complex environments. Curr. Opin. Cell Biol. 48, 72–78.
Verheyen, E.M., Cooley, L., 1994. Profilin mutations disrupt multiple actin-dependent processes during *Drosophila* development. Development 120 (4), 717–728.
Vidali, L., Augustine, R.C., Kleinman, K.P., Bezanilla, M., 2007. Profilin is essential for tip growth in the moss Physcomitrella patens. Plant Cell 19 (11), 3705–3722.
Vignaud, T., Copos, C., Leterrier, C., Tseng, Q., Blanchoin, L., Mogilner, A., Théry, M., Kurzawa, L., 2020. Stress fibers are embedded in a contractile cortical network. bioRxiv https://doi.org/10.1101/2020.02.11.944579. Preprint.
Vinson, V., Archer, S., Lattman, E., Pollard, T., Torchia, D., 1993. Three-dimensional solution structure of *Acanthamoeba* profilin-I. J. Cell Biol. 122 (6), 1277–1283.

Vinson, V.K., De La Cruz, E.M., Higgs, H.N., Pollard, T.D., 1998. Interactions of *Acanthamoeba* profilin with actin and nucleotides bound to actin. Biochem. 37 (31), 10871–10880.

Vitriol, E.A., McMillen, L.M., Kapustina, M., Gomez, S.M., Vavylonis, D., Zheng, J.Q., 2015. Two functionally distinct sources of actin monomers supply the leading edge of lamellipodia. Cell Rep. 11 (3), 433–445.

Voter, W.A., Erickson, H.P., 1984. The kinetics of microtubule assembly. Evidence for a two-stage nucleation mechanism. J. Biol. Chem. 259 (16), 10430–10438.

Wade, R.H., 2007. Microtubules: an overview. *Meth.* Mol. Med. 137, 1–16.

Walter, L.M., Franz, P., Lindner, R., Tsiavaliaris, G., Hensel, N., Claus, P., 2020. Profilin2a-phosphorylation as a regulatory mechanism for actin dynamics. FASEB J. 34 (2), 2147–2160.

Wang, Y.L., 1984. Reorganization of actin filament bundles in living fibroblasts. J. Cell Biol. 99 (4), 1478–1485.

Wase, N., Tu, B., Rasineni, G.K., Cerny, R., Grove, R., Adamec, J., Black, P.N., DiRusso, C.C., 2019. Remodeling of *Chlamydomonas* metabolism using synthetic inducers results in lipid storage during growth. Plant Physiol. 181 (3), 1029–1049.

Watanabe, N., Madaule, P., Reid, T., Ishizaki, T., Watanabe, G., Kakizuka, A., Saito, Y., Nakao, K., Jockusch, B.M., Narumiya, S., 1997. p140mDia, a mammalian homolog of *Drosophila* diaphanous, is a target protein for rho small GTPase and is a ligand for profilin. EMBO J. 16 (11), 3044–3056.

Watanabe, S., Ando, Y., Yasuda, S., Hosoya, H., Watanabe, N., Ishizaki, T., Narumiya, S., 2008. mDia2 induces the actin scaffold for the contractile ring and stabilizes its position during cytokinesis in NIH 3T3 cells. Mol. Biol. Cell 19 (5), 2328–2338.

Waterman-Storer, C.M., Salmon, E.D., 1998. How microtubules get fluorescent speckles. Biophys. J. 75 (4), 2059–2069.

Weaver, B.A., 2014. How Taxol/paclitaxel kills cancer cells. Mol. Biol. Cell 25 (18), 2677–2681.

Wickramarachchi, D.C., Theofilopoulos, A.N., Kono, D.H., 2010. Immune pathology associated with altered actin cytoskeleton regulation. J. Autoimmun. 43 (1), 64–75.

Wieczorek, M., Bechstedt, S., Chaaban, S., Brouhard, G.J., 2015. Microtubule-associated proteins control the kinetics of microtubule nucleation. Nat. Cell Biol. 17 (7), 907–916.

Wiesner, S., Helfer, E., Didry, D., Ducouret, G., Lafuma, F., Carlier, M.-F., Pantaloni, D., 2003. A biomimetic motility assay provides insight into the mechanism of actin-based motility. J. Cell Biol. 160 (3), 387–398.

Wilson, N.F., Foglesong, M.J., Snell, W.J., 1997. The *Chlamydomonas* mating type plus fertilization tubule, a prototypic cell fusion organelle: isolation, characterization, and in vitro adhesion to mating type minus gametes. J. Cell Biol. 137 (7), 1537–1553.

Witke, W., 2004. The role of profilin complexes in cell motility and other cellular processes. Trends Cell Biol. 14, 461–469.

Witke, W., Podtelejnikov, A.V., Di Nardo, A., Sutherland, J.D., Gurniak, C.B., Dotti, C., Mann, M., 1998. In mouse brain profilin I and profilin II associate with regulators of the endocytic pathway and actin assembly. EMBO J. 17 (4), 967–976.

Witke, W., Sutherland, J.D., Sharpe, A., Arai, M., Kwiatkowski, D.J., 2001. Profilin I is essential for cell survival and cell division in early mouse development. Proc. Natl. Acad. Sci. 98 (7), 3832–3836.

Wittenmayer, N., Rothkegel, M., Jockusch, B.M., Schlüter, K., 2000. Functional characterization of green fluorescent protein-profilin fusion proteins: GFP-profilin fusion proteins. Eur. J. Biochem. 267 (16), 5247–5256.

Wittenmayer, N., Jandrig, B., Rothkegel, M., Schlüter, K., Arnold, W., Haensch, W., Scherneck, S., Jockusch, B.M., 2004. Tumor suppressor activity of profilin requires a functional actin binding site. Mol. Biol. Cell 15 (4), 1600–1608.

Wittmann, T., Waterman-Storer, C.M., 2001. Cell motility: Can Rho GTPases and microtubules point the way? J. Cell Sci. 114 (21), 3795–3803.

Wolven, A.K., Belmont, L.D., Mahoney, N.M., Almo, S.C., Drubin, D.G., 2000. In vivo importance of actin nucleotide exchange catalyzed by profilin. J. Cell Biol. 150 (4), 895–904.
Wu, C.-H., Fallini, C., Ticozzi, N., Keagle, P.J., Sapp, P.C., Piotrowska, K., Lowe, P., Koppers, M., McKenna-Yasek, D., Baron, D.M., Kost, J.E., Gonzalez-Perez, P., Fox, A.D., Adams, J., Taroni, F., Tiloca, C., Leclerc, A.L., Chafe, S.C., Mangroo, D., … Landers, J.E., 2012. Mutations in the profilin 1 gene cause familial amyotrophic lateral sclerosis. Nature 488 (7412), 499–503.
Yamashiro, S., Yamakita, Y., Hosoya, H., Matsumura, F., 1991. Phosphorylation of non-muscle caldesmon by p34cdc2 kinase during mitosis. Nature 349 (6305), 169–172.
Yang, D., Wang, Y., Jiang, M., Deng, X., Pei, Z., Li, F., Xia, K., Zhu, L., Yang, T., Chen, M., 2017. Downregulation of Profilin-1 expression attenuates cardiomyocytes hypertrophy and apoptosis induced by advanced glycation end products in H9c2 cells. Biomed. Res. Int. 2017, 9716087.
Yao, W., Ji, S., Qin, Y., Yang, J., Xu, J., Zhang, B., Xu, W., Liu, J., Shi, S., Liu, L., Liu, C., Long, J., Ni, Q., Li, M., Yu, X., 2014. Profilin-1 suppresses tumorigenicity in pancreatic cancer through regulation of the SIRT3-HIF1α axis. Mol. Cancer 13, 187.
Yarmola, E.G., Parikh, S., Bubb, M.R., 2001. Formation and implications of a ternary complex of profilin, thymosin beta 4, and actin. J. Biol. Chem. 276 (49), 45555–45563.
Yarovinsky, F., Zhang, D., Andersen, J.F., Bannenberg, G.L., Serhan, C.N., Hayden, M.S., Hieny, S., Sutterwala, F.S., Flavell, R.A., Ghosh, S., Sher, A., 2005. TLR11 activation of dendritic cells by a protozoan profilin-like protein. Science 308 (5728), 1626–1629.
Yuan, X., Song, M., Devine, P., Bruneau, B.G., Scott, I.C., Wilson, M.D., 2018. Heart enhancers with deeply conserved regulatory activity are established early in zebrafish development. Nat. Commun. 9 (1), 4977.
Zaremba-Niedzwiedzka, K., Caceres, E.F., Saw, J.H., Bäckström, D., Juzokaite, L., Vancaester, E., Seitz, K.W., Anantharaman, K., Starnawski, P., Kjeldsen, K.U., Stott, M.B., Nunoura, T., Banfield, J.F., Schramm, A., Baker, B.J., Spang, A., Ettema, T.J.G., 2017. Asgard archaea illuminate the origin of eukaryotic cellular complexity. Nature 541 (7637), 353–358.
Zhang, R., Alushin, G.M., Brown, A., Nogales, E., 2015. Mechanistic origin of microtubule dynamic instability and its modulation by EB proteins. Cell 162 (4), 849–859.
Zhang, H., Yang, W., Yan, J., Zhou, K., Wan, B., Shi, P., Chen, Y., He, S., Li, D., 2018. Loss of profilin 2 contributes to enhanced epithelial-mesenchymal transition and metastasis of colorectal cancer. Int. J. Oncol. 53 (3), 1118–1128.
Zhou, K., Chen, J., Wu, J., Xu, Y., Wu, Q., Yue, J., Song, Y., Li, S., Zhou, P., Tu, W., Yang, G., Jiang, S., 2019. Profilin 2 promotes proliferation and metastasis of head and neck cancer cells by regulating PI3K/AKT/β-catenin signaling pathway. Oncol. Res. 27 (9), 1079–1088.
Zoidakis, J., Makridakis, M., Zerefos, P.G., Bitsika, V., Esteban, S., Frantzi, M., Stravodimos, K., Anagnou, N.P., Roubelakis, M.G., Sanchez-Carbayo, M., Vlahou, A., 2012. Profilin 1 is a potential biomarker for bladder cancer aggressiveness. Mol. Cell. Proteomics 11 (4). M111.009449.
Zou, L., Jaramillo, M., Whaley, D., Wells, A., Panchapakesa, V., Das, T., Roy, P., 2007. Profilin-1 is a negative regulator of mammary carcinoma aggressiveness. Brit. J. Cancer 97 (10), 1361–1371.
Zou, L., Ding, Z., Roy, P., 2010. Profilin-1 overexpression inhibits proliferation of MDA-MB-231 breast cancer cells partly through p27kip1 upregulation. J. Cell. Physiol. 223 (3), 623–629.
Zweifel, M.E., Courtemanche, N., 2020. Competition for delivery of profilin-actin to barbed ends limits the rate of formin-mediated actin filament elongation. J. Biol. Chem. 295, 4513–4525.

CHAPTER SIX

Mechanics of actin filaments in cancer onset and progress

Mohammad Tafazzoli-Shadpour*, Ehsan Mohammadi, Elham Torkashvand
Faculty of Biomedical Engineering, Amirkabir University of Technology, Tehran, Iran
*Corresponding author: e-mail address: tafazoli@aut.ac.ir

Contents

Abstract

Cancer, as a major cause of mortality, is highly related to alterations in the structure and behavior of cells of cancerous tissues. The invasive nature of cancer cells is correlated with their increased traction force, high deformability and altered cell adhesion. These changes are directly attributed to the remodeling of cell cytoskeleton mostly in actin structure. While microtubules and intermediate filaments are mostly involved in mechanical properties of cytoskeleton, actin fibers actively contribute to not only mechanical properties, but also other aspects. Hence study of actin mechanics assists in a deeper understanding of cancer related events. Here, with a biomechanical perspective, we describe the cytoskeleton changes in cancer onset and progress in fiber and protein levels, with focus on actin structure in terms of content and arrangement. Cytoskeleton remodeling and particularly alterations in the content and arrangement of actin structure, highly influence cell mechanical properties, force generation and adhesion potentials.

International Review of Cell and Molecular Biology, Volume 355
ISSN 1937-6448
https://doi.org/10.1016/bs.ircmb.2020.05.002

1. Introduction

Cancer, as a major source of world mortality, is related to changes in cell behavior including accelerated proliferation and tumor formation in a tissue, enhanced motility and invasion to other tissues leading to disturbances in the function of organs and state of hemostasis. Cancerous cells are less sensitive to hypoxia and survive well during initial buildup of the tumor (Vaupel, 2010). By the growth of the primary tumor, the core of the tumor becomes necrotic and a metastatic stage is triggered by which, cells tend to migrate to other tissues through biological processes of intravasation and extravasation (Gatenby et al., 2006; Welch and Hurst, 2019).

It has been stated that the metastatic phase is related to changes in the cell behavior and phenotype through epithelial-mesenchymal transition (EMT), to enable cells to detach from the initial tumor and invade to other tissues (Radisky, 2005; Tomaskovic-Crook et al., 2009). Cancer related changes in the behavior of cells have also been hypothesized in the concept of cancer stem cells, which are considered as one of the initial cell sources that originates tumor formation (Bjerkvig et al., 2005), and due to their un-differentiated nature are less sensitive to chemotherapy targeting (Phi et al., 2018).

Despite the uncertainties about the inherent mechanisms and diversity of cancer cells in different tissues, cancer initiation and progression is a multiphase process during which, there are genetic adjustments that affect cell behavior, enabling appropriate conditions to colonize and later to invade. Such process requires physical adjustments in the structure of cells through alterations in the cytoskeletal elements. After disconnection from the primary tumor, highly deformable and motile cells penetrate the extracellular matrix of the tissue and reach the neighboring circulatory vessels, pass through the endothelial intercellular junctions of small blood vessels or lymphatic system of the host tissue due to their higher deformability and enter circulation (Cross et al., 2007, 2008; Suresh, 2007). The mechanical adjustments in the cell structure assist circulatory cancer cells to withstand various hemodynamic forces depending on the flow conditions and enter other tissues when physical circumstances allow (Kumar and Weaver, 2009). Metastatic cells pass through the endothelial junction of the vessels of the new tissue provided by their mechanical properties and migrate till they find locations which are structurally suitable for further colony formation. The success of the metastasis phase is influenced by a variety of biochemical and biomechanical factors such as chemokine signals from the host tissue, flexibility of cells to interact with extra cellular matrix (ECM) and endothelial cells, interaction of cells with blood immune cells, shearing forces of blood stream and its flow rate, and physical properties of the initial and secondary tissues (Azevedo et al., 2015).

Depending on the tissue they reside, cells with the same phenotype, such as epithelial cells, have their own structure to fulfill their function. Aside from differences among cells, cancer cells undergo common events during initiation and progress of cancer, including alterations in cytoskeletal structure to survive and later invade. Alterations in the behavior of cancerous cells are related not only to the genetic transformations and chemical cues, but also to the mechanical properties of cells and their environment. These changes result in altered mechanical properties which assist tumor formation and in later stages, invasion to other tissues. This paper contributes to develop an understanding of common changes in the cytoskeleton of cancer cells with emphasis on actin fibers, from healthy to cancerous cells and the way they change with cancer progress until reaching metastasis. We discuss how cancer initiation and progress is associated with the changes in the mechanics of cells, and how these changes are related to the cell cytoskeleton and more specifically to the content and organization of actin microfilaments. In addition to direct contribution of actin structure to the physical stability and mechanical properties, actin fibers are actively involved in generation of contractile forces and cell adhesion. Hence study of actin mechanics can give useful insights in analysis of the behavior and function of cancer cells.

2. Cancer and cell structure

The invasive nature of cancerous cells necessitates continuous alterations in the cytoskeletal properties of cell body that cause higher deformability of cells, generation of extra contractile forces and improved motility, and altered cell adhesion (Fife et al., 2014; Hall, 2009; Olson and Sahai, 2009). In this regard, physical properties of malignant cells change to meet the requirements of metastasis (Suresh, 2007), most notably through cytoskeletal reorganization, and remodeling of cell-cell and cell-ECM adhesion cascades (Radisky, 2005; Takeichi, 1993). Changes in physical properties of cells have been widely studied and have been suggested to be considered as biomarkers of cancerous cells to serve as diagnostic tools (Alibert et al., 2017).

It has been indicated that alterations in cell structural properties are triggered by genetic signals; on the other hand, gene and protein markers are in turn sensitive to the chemical environmental stimuli, as well as mechanical stimuli such as external forces and ECM mechanical properties that define the mechanical micro-environment (Artemenko et al., 2016). It has been shown that both chemical and physical microenvironments are involved in changes in gene and protein expressions responsible for cancer initiation and progress (Walker et al., 2018).

A growing body of research has focused on protein-based alterations in cancer onset and progress, through recognition of mutations in different domains of proteins of functional cells (Vogelstein et al., 2013). A variety of structural mutations in metabolic and chemical cascades of protein networks, from cell membrane to the cell nucleus have been recognized which are responsible for cancer related changes in cell behavior among different tissues (Pavlova and Thompson, 2016; Seyfried and Shelton, 2010), including proliferation and migration through up/down regulation of specific markers and changes in the signaling pathways (Sever and Brugge, 2015). A variety of proteins in cell membrane, cytoplasm, and nucleus are involved in many cancer types such as growth factor receptors (EGFR) and G-protein receptors (GPCR) in cell membrane (Bar-Shavit et al., 2016; Hynes and Lane, 2005). Alterations in the function of membrane proteins cause changes in the downstream proteins including MAPK and PI3K, and further down cause changes in mTOR activity (De Luca et al., 2012; Kampen, 2011) Such alterations extend to the cell nucleus through different cascades which eventually affect DNA.

Despite advancements in recognition of changes induced by cancer in a variety of protein cascades, there have been less scientific attention to the cancer related structural changes of cellular components which are essential for cell function/behavior, but not directly related to the metabolic state of cells. As an example, cell cytoskeleton which provides structural integrity of the cell body, defines its physical properties, and contributes to the motility and force generation of cells is closely related to the cancer initiation and progress (Fife et al., 2014; Hall, 2009). To invade, cancer cells need to deform easily, i.e. they are softer than normal cells and have lower elastic modulus (Wirtz et al., 2012; Yilmaz and Christofori, 2009). Hence cell cytoskeleton, and specifically actin structure, both in terms of arrangement and content, are determinants of major behaviors which are crucial in cancer invasion (Fife et al., 2014; Nürnberg et al., 2011).

Among cytoskeletal elements, actin fibers are believed to play pivotal role in physical state of the cell body (Fife et al., 2014; Zhang et al., 2017). Due to the critical role of actin structure in cytoskeleton, changes in cell motility (Kedrin et al., 2007; Olson and Sahai, 2009), generation of contractile forces (Hall, 2009), and cell deformability (Suresh, 2007) are closely interrelated. For instance, cancer related alterations in cell adhesion were correlated with changes in cell elastic properties (Azadi et al., 2016); or cell contractile forces which are due to actin-myosin interaction were correlated with actin content and arrangement which in turn determines cell mechanical properties (Cai et al., 2006). There are few cancer drugs which directly target cytoskeleton, among which targeting microtubule binding proteins are more practical (Mukhtar et al., 2014; Zhou and Giannakakou, 2005). There are also treatment strategies targeting actin binding and regulatory proteins, as well as actin remodeling mechanisms (Choi et al., 2016; Jurmeister et al., 2012; Rao and Li, 2004). Aside from direct targeting cytoskeleton structure in cancer treatment, an interesting concept is to consider effects of cancer drugs that target chemical cascades on cytoskeletal remodeling during treatment. Although biochemical pathways are distinct from physical pathways, cancer development affects both cascades simultaneously. Hence it is of great interest to investigate the correlation and synergy between these two cascades and the possible relations between them, that might enhance the current treatment strategies. Many cancer treatments that block specific chemical signals, also affect the cytoskeleton and consequently cell behaviors that are related to cytoskeletal features (Peterson and Mitchison, 2002). As an example, in vitro application of cancer treatment anti-EGFR Antibody agents on cultured breast cancer cell lines led to cytoskeleton remodeling and the subsequent cell stiffening and increased cell-cell adhesion (Azadi et al., 2016). The usage of these drugs which are designed for suppressing specific chemical signals, synergistically elevated cell-cell adhesion and caused cell stiffening, two mechanisms that oppose invasion.

3. Cell cytoskeleton

Cell cytoskeleton which frames biological organelles is covered by the membrane and provides structural integrity for cells and defines their mechanical deformability (Fletcher and Mullins, 2010); it also supports cell organelles and facilitates cell signaling both in terms of physical and chemical signals (Jamora and Fuchs, 2002; Janmey, 1998; Kim and Coulombe, 2010). The cell-cell and cell-ECM interactions are partly sensed and performed through cytoskeleton which are vital in tissue development, growth, healing and remodeling (Geiger, 1989; Jamora and Fuchs, 2002; Martins-Green and Bissell, 1995; Mège et al., 2006). While cytoplasmic organelles of cells are mostly responsible for biological and metabolic functions of cells, cytoskeleton synergistically provides physical characteristics which are necessary for cell function through providing structural and mechanical properties and cell signaling (Forgacs et al., 2004), although due to physical and chemical interactions with organelles, cytoskeleton indirectly influences biological function of cells (Blain, 2009).

The cytoskeletal fibers make a complex connective structure which spans from cell nucleus to the membrane in different spatial forms depending on cell phenotype and function, and surrounds and protects organelles. Within the inner part of the cell body, the cytoskeletal fibers generate a network like structure that provides mechanical integrity and support, and toward the cell periphery, the fibrous structure is dominated by actin fibers either in the form of stress fibers or other fibers that branch and form lamellipodium and filopodium at the leading edges of cells which are responsible for cell migration (Le Clainche and Carlier, 2008).

The variation of fibers from nucleus toward the cell periphery is in a way that meets predefined requirements. Close to the nucleus and within the cytoplasmic space, the mesh like structure of different types of cytoskeletal fibers are functionally more engaged in mechanical integrity of the cell body that allows organelles to interact and properly function, most notably in protein synthesis (Bershadsky and Vasiliev, 1988). Due to physical engagement, cytoskeletal fibers share delivery of synthesized molecules, specially different types of collagen, from organelles toward membrane (Manneville et al., 2003). Furthermore, they contribute to the cell signaling that transfers environmental stimuli from the membrane to the nucleus (Janmey, 1998). These signals are mostly mechanical forces, although cytoskeletal fibers are occasionally involved in transduction of chemical signals as well (Frieden and Gatenby, 2019). Cells, depending on the tissue they reside in, are exposed to physical stimuli commonly mechanical forces in the forms of tensile, compressive and shearing (Wang and Li, 2010), and somehow sense the positional information within the tissue which is necessary for the state of homeostasis and biological development (Frankel, 1992). Such forces induce mechanical stresses within the cell body which are sensed by the cytoskeleton and the adjacent ECM and are determinants in biological responses during morphogenesis, growth, remodeling and adaptation, disease development and healing (Janmey and Miller, 2011). Such biological responses are triggered by cells after receiving mechanical and chemical stimuli (Stamenović and Wang, 2011).

Cell cytoskeleton has a heterogeneous structure within the cell body and its structure is mainly composed of three major fibrous components, i.e. actin microfilaments, microtubules, and intermediate filaments which interactively form the responsive structure to the internal and external microenvironmental stimuli. Such structure is not entirely continuous and uniform (Aranda-Espinoza, 2015), and the content and arrangement of fibers and their connections change from nucleus toward cell periphery (Hafner and Rieger, 2016). The ratio of different components and their hierarchical fibrous structures, together with the arrangement of the fibrous skeleton differ within the cell body to meet the requirements that cytoskeleton satisfies in terms of mechanical properties, generation of contractile force, and migration (Szymanski et al., 2015).

Within the cytoplasmic domain and in the vicinity of the nucleus, microtubules start from the centrosome and form a dense fibrous network and are extended toward periphery of the cell. Intermediate filaments are highly abundant in this zone among microtubule fibers. Such composite structure of fibers defines mechanical solidity of the cell body (Alberts, 2008). Closer to the cell membrane, actin fibers are richer and dominate cytoskeleton with a variety of composite structures such as stress fibers, and cell edge structures which are actively involved in motility related behavior of cells, as well as other cell-cell and cell-ECM interactions (Svitkina, 2018). The thick bundles of actin fibers reach the focal adhesion sites and transfer generated contractile forces to cell surface to move cell body through polymerization (addition of new monomers and lengthening of the fibers) and depolymerization at adhesion sites resulting in a stepwise crawling of the cell (Davies, 2013; Parsons et al., 2010).

Microtubules are the largest fibers, similar to hollow tubes which are made from the subunit tubulin. The long structure is formed by end-to-end polymerization of alpha and beta tubulin dimmer which allows further extension (Lodish et al., 2000). Among different fibers, they provide mechanical support of the cell body the most. While cable like actin fibers undergo extension, they cannot stand compression (Mehrbod and Mofrad, 2011). However, microtubules have higher bending stiffness and persistence length and can resist compression and buckling, hence they play a central role in mechanical integrity of the cell body (Holy and Leibler, 1994).

In addition to the physical tasks, microtubules contribute to some other cell functions such as migration, mitosis and intracellular trafficking (Hammond et al., 2008).

Intermediate filaments, smaller than microtubules and larger than actin microfibers, are abundant in the cytoskeleton with six sub-categories which share the structural units of alpha helix proteins (Bloemendal and Pieper, 1989; Fuchs and Weber, 1994). Similar to microtubules, they are extended from cell nucleus toward cell periphery within the cytoplasmic domain with higher concentrations closer to the nucleus. Due to their abundance, they highly contribute to cross linking of cytoskeletal structure as well as maturation of adhesion points (Cooper, 2000). The latter is related to vimentin, as one of intermediate filaments (Cooper, 2000). Mechanically, intermediate filaments are highly stretchable due to their hierarchical structure (Charrier and Janmey, 2016; Hu et al., 2019). Because of their abundance, they cover and protect biological organelles from external overloading, hence they contribute to physical stability of the cell body, however, unlike two other major cytoskeletal components they do not play a key role in cell motility (Cooper, 2000). It has been shown that in cell-cell adherence their extensions meet the adjacent cell, a phenomenon that facilitates transforming physical stimuli, specially shear forces, within the cell body as a physical signaling cascade (Sanghvi-Shah and Weber, 2017).

Unlike microtubules and intermediate filaments which are mostly located within the inner part of the cell body, actin fibrous structure is the most abundant fiber in the periphery of the cell. Actin microfilaments are physically the most important components of cell cytoskeleton, since they play key roles in all aspects of cell deformability, cell motility and generation of contractile forces (Ananthakrishnan and Ehrlicher, 2007; Dong and Lei, 2000). While microtubules and intermediate fibers passively contribute to the mechanical stability of cell body, actin microfilaments are actively involved in the mechanical processes of cell function that generate contractile force through movement of myosin on actin microfilaments as well as cell migration predominantly by actin polymerization and depolymerization at focal adhesion sites (Pollard, 2016; Pollard and Borisy, 2003; Pollard and Cooper, 1986; Svitkina, 2018). Compared to the other two cytoskeletal fibers, actin microfilaments are the smallest components and can be diffused to the nucleus. Hence, they exist virtually in all parts of the cell body (Williams et al., 2019).

The subunit of actin is the globular actin (G-actin) and within the form of fibers they are assembled to form filamentous actin (F-actin) through single stranded helix structure which are polarized and can be used in rapid assembly and disassembly during cell movement at adhesion sites (Cooper, 2000). Actin microfilaments are composed of linear polymers and during cell activities undergo tensile stresses, while structurally they cannot stand compressive forces (Lodish et al., 2008). The movement of myosin proteins along the actin filaments generates contractile forces and is transferred to the environment mostly along the stress fibers and results in cell-ECM force interaction which is responsible for cell movement (Hotulainen and Lappalainen, 2006). Depending on the type of cells, actin is categorized as alpha, beta, and gamma, as alpha actin is related to muscular cells with the highest contractile force generation (Kashina, 2020; Small, 1985).

Actin filaments are formed in different structural arrangements depending on the location and function, such as stress fiber, focal adhesion, cortical actin, filodopodium, microvilli and lamellipodium (Rottner et al., 2017; Svitkina, 2018). Generally, actin fibers are assembled in two forms, mesh or network like structures which are dominant within the inner part of cells and are responsible for mechanical support, and bundles of fibers which are within the outer part and as the thick stress fibers, they reach the focal adhesion sites (Cooper, 2000). The whole connective assembly of actin structure is among the signaling routes for transduction of both physical and chemical signals (Humphrey et al., 2014). Hence, in addition to direct mechanical function, actin structure is considered as one of major cell signaling cascades (Frieden and Gatenby, 2019; Kustermans et al., 2008). Mutually, the assembly and disassembly of actin filaments are highly influenced by cell signaling, which is an important remodeling algorithm that allows cells to adapt to their microenvironment and external stimuli (Wear et al., 2000).

Different models have attempted to relate physical function of cells to their structure, which among them tensegrity models are distinct due to their multiscale nature. A spatial architecture of cables and rods that represents actin and microtubule fibers respectively, simulates cell cytoskeleton (Chen et al., 2010). Under external forces, cables undergo extension, while they do not stand compression. The initial pre-stress within the cables indicates the contractile forces. The mechanical properties of the whole structure are determined by the mechanical properties of the components and their arrangement (Pugh, 1976). Any changes in the architecture of the sub-structural components result in altered mechanical properties of the whole cell body and vice versa. To some extent, this model can be applied in defining alterations in the cytoskeletal fiber properties due to alterations in cell body mechanical properties caused by cancer (Tadeo et al., 2014).

4. Cancer and cell mechanical properties

Cancer initiation and progress is accompanied by changes in the expression of intracellular and extracellular proteins of the cells (mostly epithelial cells) of the diseased tissue (Salama et al., 2008; Schmitt et al., 2007). Simultaneously, some proteins undergo mutations by which the structure and sequence of a small portion of molecular domains change (Yang et al., 2015). Despite small deviation from the original structure, the mutated structure shows marked alterations in its physical properties including force-displacement relation (Omidvar et al., 2014). Such protein based alterations, either in the form of expression or mutation, as well as possible secondary level changes in proteins in the form of post-translational modifications, induce changes in the function of cells that define the state of cancer (Leśniak, 2011). These events mostly occur in chemical cascades along nucleus to cell membrane including adhesive proteins (Azadi et al., 2016). concurrently, cell cytoskeleton, as the determinant of cell body structural properties, undergoes evident changes both in terms of fiber content and arrangement induced by cancer onset and progress (Ben-Ze'ev, 1985; Gowing et al., 1984; Lambrechts et al., 2004; Pachenari et al., 2014; Rao and Cohen, 1991; Yamazaki et al., 2005; Yilmaz and Christofori, 2009). These chances lead to alterations in elastic and viscoelastic properties of cells which are the subject of this section. Alterations in cytoskeleton that result in changes in cell traction forces and cell adhesion are subjects of the subsequent sections. Fig. 1 summarizes the alterations in cell actin structure during cancer formation in fiber and protein levels by a schematic representation.

It has been shown that cancer affects all three major fibers, i.e., microtubules, intermediate fibers and actin microfilaments (Ben-Ze'ev, 1985) mostly in the form of alterations in spatial arrangement and prestress values (Pachenari et al., 2014), although changes in actin content and arrangement are more prominent (Olson and Sahai, 2009; Yamaguchi and Condeelis, 2007; Yamazaki et al., 2005). Since cytoskeletal fibrous content and arrangement describe elastic and viscoelastic properties of cell body, cancer causes defined changes in the cell mechanical properties depending on the state of disease.

While actin fibers undergo the most intense changes caused by cancer, the other two main components, i.e. microtubules and intermediate filaments also show cancer related alterations. Since microtubules frame and support cytoplasmic organelles, changes to the content and/or arrangement of these fibers might affect the movement and interaction of organelles

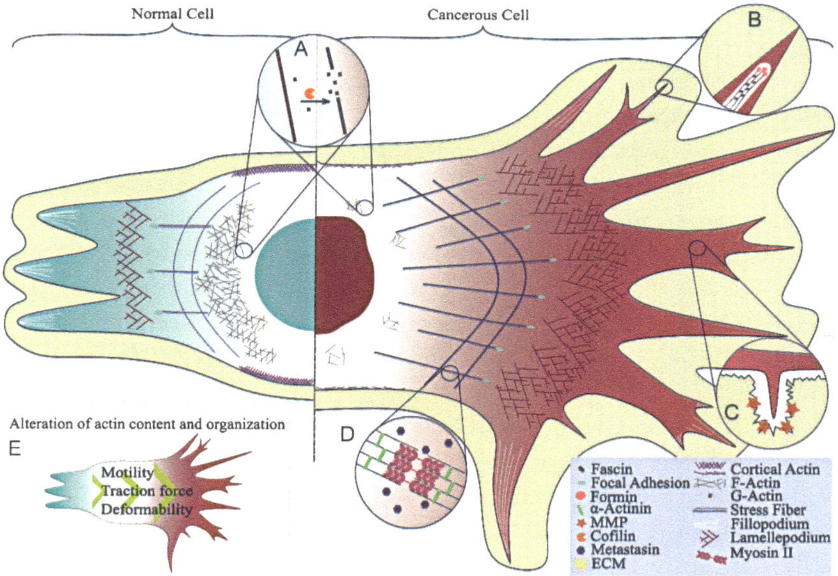

Fig. 1 A schematic representation of alterations in actin content and organization caused by cancer. Figure also describes actin based and binding proteins. (A) Cofilin as an actin-binding protein severs actin filaments and is overexpressed in cancerous cells. Cofilin depolymerizes actin filaments and decreases actin content. (B) Filopodium at the leading edge of the cell is the structure of parallel actin fibers connected by actin-bundling protein, fascin. Formin is located at the tip of actin filament, which polymerizes actin filaments and forms protrusive edges. (C) Invadopodium penetrates to ECM and secrets MMP family proteins to degrade ECM for invasion (side view cross-section of a typical cancerous cell). (D) Stress fibers are upregulated with the multidirectional trend in cancerous cells. Stress fiber is formed due to the connection of actin filaments mostly by myosin II, and α-actinin. Metastasin protein, which is associated with metastasis, is also shown. (E) Alterations in actin content and reorganization lead to an increase in motility, traction force, and deformability of cancer cells.

which in turn, influence cell behaviors (Kulić et al., 2008). Cancer affects expression of tubulin and microtubule associated proteins (MAPs) (Parker et al., 2014). In some chemotherapy treatments, to prevent the high proliferation of cancer cells, some agents target tubulins (Jordan et al., 1998). Cancer has been shown to be correlated with the overexpression of proteins which interact microtubules with actin filaments at focal adhesion sites (Schroeder et al., 2019). Hence some chemotherapy agents target dynamics of microtubules to suppress cancer, since limitation of microtubule activities leads to limitations of actin filaments specially within the lamellipodium which in turn suppress migration ability of cancer cells (Honore et al., 2005).

Similarly, cancer induces changes among intermediate filaments. For instance, the expression of Keratin as one of types of intermediate filament molecules of epithelial cells, undergo alterations with cancer (Hendrix et al., 1996). Hence this expression is known to be a biomarker for diagnosis of cancer among epithelial cells (Karantza, 2011). The overexpression of other intermediate filament related proteins, such as Vimentin and Nestin is also another sign of cancer progress (Liu et al., 2015). Another role of intermediate filaments which might be affected by cancer is related to their connectivity between cell membrane (at Desmosome junctions) with cytoplasmic content which affect many cell behaviors such as proliferation, apoptosis and invasion through cell-cell adhesion (Brooke et al., 2012; Holthöfer et al., 2007).

The initiation and progress of cancer is highly correlated with changes in the content and organization of actin filaments in all architectural forms available in cell body including cell cortex, stress fibers, filopodium and lamellipodium (Chugh and Paluch, 2018; Izdebska et al., 2018), as previously said "form follows function" (Fletcher and Mullins, 2010). Among architectural features of actin structure, generation of bundled stress fibers and cross linking of actin fibers to produce a fibrous network, and interaction with cytoskeletal microfilaments and other proteins such as Myosin II are prominent features in changes of cell mechanical properties (Fletcher and Mullins, 2010).

Altogether, a majority of studies in this field agrees that alterations in cytoskeleton induced by cancer cause decrease in cell body stiffness (Lekka et al., 1999). Cancer treatments, based on chemotherapy, have shown that in addition to adjustment of cell behaviors such as proliferation and migration, they result in cell stiffening (Suresh, 2007; Lam et al., 2007). This shows the key role of cell softening in cancer progress.

A growing body of research has examined cancer cells, either primary or cell lines, for different aspects of mechanical properties using various experimental techniques and compared the results to those of healthy cells (Cross et al., 2007; Hayashi and Iwata, 2015). These techniques include micropipette aspiration (MA), atomic force microscopy (AFM), magnetic twisting cytometry (MTC), particle tracking microrheology (PTM), parallel plate rheometry, cell monolayer rheology, and optical tweezers (Wu et al., 2018), among which the first two are the most common. Results describe a general tendency of cancer induced softening of malignant cells of all tissues both in suspended and adhered states. The suspended state of cells is related to the passive structural state in which cells do not perform major biological behaviors such as cell adherence and migration. In MA technique, cells are pulled into a needle, based on the negative aspiration pressure, and the aspirated length versus pressure is recorded. Using mathematical models such as Hertz model, the effective Young's modulus of elasticity of the cell body is calculated (Hochmuth, 2000). If the applied pressure remains constant, the cell body will relax with time due to viscoelastic property and the aspirated length versus time describes a stress relaxation model and the relevant viscoelastic parameters are calculated accordingly (Sato et al., 1990).

The elastic and viscoelastic moduli of different locations of cell body can also be obtained, using indentation technique of AFM. In this technique cells are seeded and adhered on the substrates and a localized nano-indentation is performed on different parts of cell surface, and the compressive force versus indented length are recorded and the Young's modulus and/or viscoelastic parameters are calculated for each indented point. Both techniques, along other techniques to obtain load-displacement data on cell body, have been vastly applied on cancer cells and results confirm softening of cell body induced by cancer compared to healthy cells with the same phenotype. The cell softening is more prominent among invasive cells compared to non-invasive cancer cells (Lekka et al., 1999). Table 1 shows typical data of the results of changes in cell mechanical properties induced by cancer among non-invasive and invasive cell lines. Due to employing different methods on mechanical characterization and depending on the state of cells (suspended vs. adhered, and whole body vs. localized zones), the resultant quantities obtained from different techniques are different (Wu et al., 2018), however, the trend of cell softening by cancer persists among different techniques.

In addition to elastic properties of cells, a growing body of research has shown alterations in viscoelastic properties of cells induced by cancer. The viscoelastic nature of cell body is due to the presence of cytoplasmic fluid as well as the engagement of cytoskeletal fibers that in response to external forces undergo viscous energy dissipation in addition to elastic deformation (Khani et al., 2015). The viscoelastic properties of cells have been evaluated mostly by AFM and MA techniques through time dependent experiments (Jannatbabaei et al., 2019; Khani et al., 2014). Based on the experimental results, different viscoelastic models have been fitted to the data, among them three and two element models have been the most applicable. The standard three element model utilizes a spring parallel to a series of a spring and a dashpot, and Kelvin model is represented by a spring and dashpot in series, and both have been used in evaluation of viscoelastic parameters of cells.

Table 1 Elastic properties of cells in course of cancer and invasion using different techniques.

Method	Tissue/cell	Cell line	Young's modulus (kPa)	References
AFM	Bladder	Hu609 (*non-cancerous*)	9.7 ± (3.6), 12.9 ± (4.8)	Lekka et al. (1999)
		SV-HUC-U1 (*non-cancerous*)	27.57	Canetta et al. (2014)
		HCV29 (*non-cancerous*)	7.5 ± (3.6), 10 ± (4.6)	Lekka et al. (1999)
			16 ± (0.9), 33 ± (2)	Ramos et al. (2014)
			10.3 ± (3.4)	Lekka et al. (2001)
		Hu456 (*non-invasive*)	0.3 ± (0.2), 0.4 ± (0.3)	Lekka et al. (1999)
		T24 (*invasive*)	0.8 ± (0.4), 1 ± (0.5)	Lekka et al. (1999)‘
			2.9 ± (0.5), 6,2 ± (0.6)	Ramos et al. (2014)
			1.8 ± (0.6)	Lekka et al. (2001)
		HT-1376 (*invasive*)	5.2 ± (0.1)	Ramos et al. (2014)
		HTB-9 (*invasive*)	3.0 ± (0.1), 6.4 ± (0.5)	Ramos et al. (2014)
		MGH-U1 (*invasive*)	2.46	Canetta et al. (2014)
	Prostate	PZHPV-7 (*non-cancerous*)	3.09 ± (0.84)	Lekka et al. (2012)
		LNCaP (*invasive*)	0.287 ± (0.05)	Faria et al. (2008)
			0.45 ± (0.21)	Lekka et al. (2012)
		PC-3 (*invasive*)	1.4 ± (0.16)	Faria et al. (2008)
			1.95 ± (0.47)	Lekka et al. (2012)
		Du145 (*invasive*)	1.36 ± (0.42)	Lekka et al. (2012)
			0.6 ± (0.4)	Efremov et al. (2015)

Table 1 Elastic properties of cells in course of cancer and invasion using different techniques.—cont'd

Method	Tissue/cell	Cell line	Young's modulus (kPa)	References
	Breast	A184A1 (*non-cancerous*)	2.26 ± (0.56)	Lekka et al. (2012)
		T47-D (*slightly-invasive*)	1.2 ± (0.28)	Lekka et al. (2012)
			0.94 ± (0.25)	Omidvar et al. (2014)
		MCF-7 (*non-invasive*)	1.24 ± (0.46)	Lekka et al. (2012)
			1.04 ± (0.27)	Omidvar et al. (2014)
		MDA-MB-231 (*highly-invasive*)	0.62 ± (0.21)	Omidvar et al. (2014)
	Ovarian	IOSE (*non-cancerous*)	2.47 ± (2.04)	Xu et al. (2012)
		OVCAR4 (*highly-invasive*)	1.12 ± (0.86)	Xu et al. (2012)
		HEY (*low-invasive*)	0.88 ± (0.52)	Xu et al. (2012)
		OVCAR3 (*low-invasive*)	0.57 ± (0.23)	Xu et al. (2012)
		HEYA8 (*highly-invasive*)	0.49 ± (0.22)	Xu et al. (2012)
	Fibroblast	BALB 3T3 (*non-cancerous*)	1.01 ± (0.4)	Park et al. (2005)
		H-ras (*invasive*)	0.42 ± (0.35)	Park et al. (2005)
		SV-T2 (*invasive*)	0.48 ± (0.51)	Park et al. (2005)
Micropipette aspiration	Breast	MCF-10A (*non-cancerous*)	0.21 ± (0.018)	Tabatabaei et al. (2019)
		T47-D (*slightly-invasive*)	0.06 ± (0.012)	Tabatabaei et al. (2019)

Continued

Table 1 Elastic properties of cells in course of cancer and invasion using different techniques.—cont'd

Method	Tissue/cell	Cell line	Optical deformability (OD) (%)	References
Optimal stretcher	Breast	MCF10A (*non-cancerous*)	10.5 ± (0.8)	Guck et al. (2005)
		MCF-7 (*non-invasive*)	21.4 ± (1.1)	Guck et al. (2005)
		MDA-MB-231 (*highly-invasive*)	33.7 ± (1.4)	Guck et al. (2005)
	Fibroblast	BALB 3T3 (*non-cancerous*)	8.4 ± (1.0)	Guck et al. (2005)
		SV-T2 (*invasive*)	11.7 ± (1.1)	Guck et al. (2005)
Method	**Tissue**	**Cell line**	**Stiffness (Dyne/cm^2)**	**References**
Magnetic twisting cytometry	–	Adherent endothelial cells (*non-cancerous*)	30–90	Wang et al. (1993)
Particle tracking microrheology	Kidney	COS7 (*non-cancerous*)	330 ± 250 (perinuclear region)	Yamada et al. (2000)
			820 ± 520 (lamellar region)	Yamada et al. (2000)

Similar to cancer related softening of cells, in viscoelastic analysis, cancer cells generally show lower viscoelastic parameters and this trend is intensified by invasion. Using three parameter viscoelastic models to fit data of viscoelastic tests on cell bodies, results have confirmed viscoelastic behavior of malignant cells with lower constants of the model (Bu et al., 2019; Carmichael et al., 2015). Table 2 reviews typical data regarding alterations in viscoelastic parameters of cancer cell lines. These changes are mostly attributed to the altered cytoskeletal fiber organization and content, most notably among actin fibers (Tabatabaei et al., 2019).

The reported results of mechanical properties of cells differ when different techniques are employed. As an example, the values reported by AFM

Table 2 Viscoelastic properties of cancerous and non-cancerous cells obtained from different experimental methods.

Method	Tissue	Model	Cell line	Viscoelastic parameters	References
AFM	Gastric	$J(t) = \frac{1}{E}\left[1 + c\left(1 - e^{\frac{-t}{\tau}}\right)\right]$	GES-1 (*non-cancerous*)	$c = 1.606$ $\tau = 0.9042 \pm 0.0149$ s	Bu et al. (2019b)
			NCI-N87 (*invasive*)	$c = 2.624$ $\tau = 2.054 \pm 0.025$ s	
			SGC7901 (*invasive*)	$c = 2.092$ $\tau = 1.676 \pm 0.023$ s	
	Liver		HepG2 (*medium-invasive*)	$c = 2.35$ $\tau = 2.085 \pm 0.027$ s	
	Human articular chondrocytes	$\delta(t) = \frac{3F}{8RE_1}\left[-\left(\frac{E_1}{E_2}\right)\exp\left(\frac{-E_2 t}{3\mu}\right) + 1 + \frac{E_1}{E_2}\right]$	Sample extraction	$E_1 = 0.63 \pm 0.51$ kPa $E_\infty = 0.24 \pm 0.11$ kPa $\mu = 3 \pm 1.8$ kPa. s	Koay et al. (2003)
	Breast	Power-law rheology $E_{(t)} = E_0\left(1 + \frac{t}{t'}\right)^{-\alpha}$	MCF-7 (*non-invasive*)	$E_0 = 1.29$ kPa $\alpha = 0.219$	Efremov et al. (2017)
			MDA-MB-231 (*highly-invasive*)	$E_0 = 1.32$ kPa $\alpha = 0.166$	
	Fibroblast		NIH3T3 (*non-cancerous*)	$E_0 = 2.8$ kPa $\alpha = 0.089$	
	Mouse		NMuMG	$E_0 = 2.9$ kPa $\alpha = 0.078$	

Continued

Table 2 Viscoelastic properties of cancerous and non-cancerous cells obtained from different experimental methods.—cont'd

Method	Tissue	Model	Cell line	Viscoelastic parameters	References
Micropipette aspiration	Breast	Linear viscoelastic solid model $J(t) = \frac{e^{\frac{-t}{\tau}}}{k_1 + k_2} + \frac{1}{k_1}\left(1 - e^{\frac{-t}{\tau}}\right)$	MCF10A (*non-cancerous*)	$k_1 = 0.14 \pm 0.01$ kPa $k_2 = 6.53 \pm 0.097$ kPa $\tau = 0.81 \pm 0.009$ Pa. s	Tabatabaei et al. (2019)
			T47D (*slightly-invasive*)	$k_1 = 0.08 \pm 0.01$ kPa $k_2 = 2.18 \pm 0.07$ kPa $\tau = 0.01 \pm 0.008$ Pa. s	
	Human peripheral blood	Standard viscoelastic solid model $J(t) = \frac{k_2}{k_1(k_1 + k_2)} \exp\left[-\frac{k_1 k_2}{\mu(k_1 + k_2)} t\right]$	Sample extraction	$k_1 = 12.12$ Pa $k_2 = 29.3$ Pa $\tau = 22.38$ Pa. s	Sung et al. (1988)
				$k_1 = 0.75$ Pa $k_2 = 23.8$ Pa $\tau = 33$ Pa. s	
Optical tweezers	Fibroblast	$G(f) = G_0\left(\frac{f}{f_0}\right)^{\gamma}(1 + i\eta)\zeta(1 - \gamma)\cos\left(\frac{\pi\gamma}{2}\right) + 2\pi i\mu f$	NIH3T3 (*non-cancerous*)	$G_0 = 55 \pm 5$ Pa $\mu = 0.15 \pm 0.05$ Pa. s	Ayala et al. (2016)

are almost one order of magnitude higher than those obtained by MA. This is mostly due to test conditions as AFM indent very locally with compressive force in nano-scale, and can probe very localized portions of cell membrane, while cells are adhered to the surface and cytoskeletal fibers are actively involved. On the other hand, MA applies tensional pressure on cell membrane while cells are suspended and cytoskeletal fibers are not activated. Hence the aim of most reported results is not to present absolute values of mechanical properties of cancerous vs. normal cells, however to show the trend of alterations in those properties due to initiation of cancer and later disease progression to a more invasive step.

Interestingly, cancerous tissues are stiffer than healthy tissues, despite the fact that cancer cells are softer than normal cells. Application of nano-indentation on slices of cancerous tissues has revealed that the cellular part of cancerous tissues is softer than those of normal tissues, however, the non-cellular parts are significantly stiffer (Ansardamavandi et al., 2016; Lekka et al., 2012). The stiffer nature of ECM among cancerous tissues is due to different parameters most notably intense synthesis of ECM and elevation of interstitial fluid pressure (Mohammadi and Sahai, 2018). Cancerous tissues not only contain cancerous cells mostly with epithelial origin, but also they include cancer associated fibroblasts (CAFs) and other minor cell types. These cells and the cross talk between them considerably activate ECM synthesis, specially collagen deposition by CAFs, and secretion of growth factors and other enzymes such as MMPs leading to regulation of ECM through degradation and remodeling, followed by accelerated buildup of thick bundled collagen fibers which are oriented and guide cancer cell in their migration path, while cause stiffening of the cancerous tissue (Mohammadi and Sahai, 2018; Stylianopoulos et al., 2018).

5. Cancer and actin dependent cell traction forces

The high deformability of cancerous cells during invasion is necessary but not sufficient. Cancerous cells are very motile. To be so, they generate high contractile forces (Kraning-Rush et al., 2012), transfer them to the cell surface and produce conditions of cell-cell and cell-ECM adherence for improved motility (Gkretsi and Stylianopoulos, 2018; Jiang et al., 2009; Sahai and Marshall, 2002; Van Roy and Mareel, 1992). There is a direct correlation between invasion of cancer cells and their traction forces. The traction force at the cell-substrate interface is due to contractile forces along the bundled actin fibers, called stress fibers which are extended toward cell membrane and generate force through actomyosin bridges driven by ATP hydrolysis causing cell contraction, and transfer them to the ECM at focal adhesion sites (Balaban et al., 2001; Kolega et al., 1991; Sanger et al., 1983). These forces are enhanced by actin polymerization at focal adhesion sites that triggers cell migration process (Bereiter-Hahn, 2005). Such polymerization leads to lengthening of actin fibers and together with exertion of traction forces cause movement of cell edges, and by support of microtubules and depolymerization of actin fibers at the opposite points, cells move forward (Elosegui-Artola et al., 2016; Owen et al., 2017). By the depolymerization of actin fibers at the opposite site and release of focal adhesions, stress fibers pull the body in the direction of migration. The major parameter of cell motility is based on generation of traction forces (Ananthakrishnan and Ehrlicher, 2007), and is enhanced in cancer related behavior by alterations in the content and arrangement of cytoskeleton (Yamazaki et al., 2005), expression of myosin II, enrichment of stress fibers (Ouderkirk and Krendel, 2014; Tavares et al., 2017) and alterations in the expression of cell-cell (Cavallaro and Christofori, 2004; Frixen et al., 1991) and cell-ECM (Jinka et al., 2012) adhesion proteins.

The elevation of traction forces due to rearrangement of actin structure, generation of more stress fibers, and enhanced ATP hydrolysis enables cancer cells for improved penetration to the ECM through thin and long filopodial protrusions (Svitkina, 2018). Cell traction force as the local force which is exerted by cell on the unit area of the substrate, was first quantified based on the shapes and number of wrinkles originated due to application of contractile forces on soft and thin substrates (Harris et al., 1980). Using more accurate methods of strain calculations, traction force microscopy (TFM) has evolved to a general application mostly using confocal microscopes and fluorescent beads to show how the soft substrates deform under cell traction forces (Butler et al., 2002). This requires a measurable deformation by substrates which is carried out by use of substrates with low elastic modulus. Using substrate elastic modulus and displacement of beads, cell traction forces are calculated either directly through constitutive equations or other methods (Hur et al., 2020; Polacheck and Chen, 2016). Various studies using traction force microscopy have concluded that cancerous cells have higher contractility than normal cells, and that compared to less invasive cancer cells, highly invasive cancer cells produce significantly higher traction forces in a variety of cancerous tissues or cancer cell lines (Koch et al., 2012; Kraning-Rush et al., 2012; Li et al., 2017). Table 3 provides information on the values of traction forces generated by cancer cells of different types and invasive potentials.

Table 3 Traction forces generated by cells with different degrees of invasiveness.

Tissue	Cell line	Status	Traction force (nN)	References
Breast	MCF-10A	Non-cancerous	70–270	Kraning-Rush et al. (2012)
	MDA-MB-231	Highly invasive	100–380	
Bladder	T24	Invasive	17.4	Peschetola et al. (2013)
	RT112	Invasive	22.8	Peschetola et al. (2013)
Prostate	PrEC	Non-cancerous	80–330	Kraning-Rush et al. (2012)
	PC3	Invasive	100–410	
Lung	BEAS-2B	Non-cancerous	40–100	
	A549	Invasive	50–220	
Fibroblast	3T3	Non-cancerous	70	Maskarinec et al., (2009)
Keratocytes	Molly fish Poecillia	Non-cancerous	1–2	Lee et al. (1994)

In protein level, the generation of intracellular contraction is due to coupling of myosin and actin. Myosin is a family of protein motors that are actin dependent to create force through ATP hydrolysis (Lodish et al., 2000). The subcategory of myosin family involves a high number of proteins which are involved in force related various functions. In general, myosins are either muscle myosins of muscle cells, or non-muscle myosins of other cells such as epithelial cells. Hence, the non-muscle myosins are involved with different actin dependent cytoskeletal structures in the cell body such as stress fibers and focal adhesions, filopodia, microvilli, etc. (Betapudi et al., 2006; Lodish et al., 2008; Makowska et al., 2015). While some myosins are not closely related to motor activity and perform other functions (such as being an anchor protein), others are directly involved in motor activity among which myosin II subfamily is of particular interest with a higher impact on force generation compared to other myosin family members (Cai et al., 2006). In addition to active role of force generation, due to their interaction with actin microfilaments, myosins passively contribute to bundling of cancer fibers.

It is of great importance to understand the role of myosins in cancer related events. While cancer cells generate higher contractile forces, paradoxically they exhibit lower content of actin microfilaments. There are several possible scenarios to overcome this dilemma, such as the likelihood of changes in actin structure and arrangement in cancer cells to compensate for the lower content of actin; or the possibility of increased number of myosin-actin bridges despite decrease of actin content most likely in enriched stress fibers. In general, several studies have shown a correlation between cancer initiation/progress with alterations in myosin expression/content (Ouderkirk and Krendel, 2014). It has been shown that during tumor progression, different types of myosin family enhance or prohibit tumor progress (Betapudi et al., 2006; Makowska et al., 2015; Ouderkirk and Krendel, 2014).

Myosin IIA is limited to stress fibers and is directly involved in regulation of cell contractility (Maupin et al., 1994). Although some studies have shown the role of myosin II in tumor suppression in specific cancer cells, others have indicated enhancing role in cancer progress (Ouderkirk and Krendel, 2014). Higher expressions of non-muscle myosin IIA in human gastric cancer samples have been shown (Liu et al., 2012). The expression of myosin light chain 9 (MYL9) as the activator of non-muscle myosin II has been shown to be significantly higher in malignant esophageal squamous cells than normal cells (Wang et al., 2017). Same trend was also reported among colorectal cancer samples and the promotional effects of this marker on metastasis was confirmed. Paradoxically, the expressions of MYL9 and MYLK were both lower in lung cancer cells than those of normal cells while higher in lymph metastasis (Tan and Chen, 2014). However, another study reported significant relation between expression levels of myosin light chain kinase and metastasis among lung cancer cells (Ishihara et al., 2013; Minamiya et al., 2005).

Myosin IIB is less involved with stress fibers and is more abundant in cell edge structures which are fundamental in motility and consequently polarity of cells, and its alterations has been reported during EMT, as well as its upregulation together with downregulation of myosin IIC with cancer progress (Beach et al., 2011). The expression of myosin IIB is high in invasive cancer cells, however in less invasive cancer cells the expression of myosin IIC is reported to be high (Ouderkirk and Krendel, 2014).

While there are diverse views on the role of myosin superfamily during cancer onset and progress, an interesting concept has emerged that correlates cancer invasion to myosin activity. Metastasin (S100A4), which is associated with invasion and metastasis, is a binding protein to non-muscle myosin in cancerous cells which increases cell spreading and subsequently enhances the motile ability of cancer cells (Kiss et al., 2012; Yu et al., 2013). Hence the high presence of this marker in tumorous cells intensifies the migration of those cells, regardless of level of expression of myosin II.

Since non-muscle myosin II is a binding partner of metastasin, by further spread of cells, more myosin II A and B are employed, especially at cell edge structures which optimizes the generated traction force for cell movement (Betapudi et al., 2006).

6. Cancer and actin dependent cell adhesion

The ability of cell adhesion to the ECM or other cells, is among essential characteristics that most of cells cannot survive without them. Cells, as units of life, are required to transmit signals with the environment to be actively involved with the functions and purposes of the structure they reside in. Cell adhesion which provides the cell-microenvironment interface is necessary to meet these requirements. Moreover, cell-ECM adhesion is closely related to the cell motility since traction forces are exerted at adhesion sites and the interactive force by the ECM is perceived by adhesive proteins, mostly members of the integrin family at the membrane, and contributes to the movement of the cells through activation of a cascades of proteins that connects adhesion sites to the cytoskeleton such as talin, vinculin, and filamin (Critchley, 2009; Das et al., 2012; Loo et al., 1998; Ziegler et al., 2008). Interestingly, in addition to cell-ECM adhesion, cell-cell adhesion also plays role in cell motility related behaviors. This is of importance when cells migrate in sheet like forms instead of individual migration (Friedl et al., 2004). In this form of migration, Tight and Desmosomes adherence junctions provide cell-cell adhesion through proteins such as Claudin, E-cadherin and Desmoglein which are also connected to the actin filaments via intermediate proteins.

It has been shown that cancer onset and progress is significantly correlated with cell-cell and cell-ECM adhesions which are mostly mediated by E-cadherin and integrin adhesion membrane proteins respectively (Azadi et al., 2016; Birchmeier and Behrens, 1994; Frixen et al., 1991). Tumor cells sense and experience events from microenvironment through adhesive and surface proteins such as integrins, cadherins, cell surface proteoglycans, and tetraspanins either in contact with other cells or ECM, which in turn cause regulation of cell-cell and cell-ECM adhesions (Canel et al., 2013). The transformation activities through adhesion receptors and coupling with extracellular ligands of microenvironment provide appropriate conditions for enhanced migration, invasion and proliferation of cancer cells. The cell-ECM adherence is mostly provided through physical biding of integrin surface proteins with ECM ligands such as laminin, collagen, and fibronectin (Jinka et al., 2012). The mechanical attachment to ECM that integrins provide, is additionally considered as a signal transduction pathway (Aksorn and Chanvorachote, 2019). Such pathway links integrins to the cytoskeleton actin via connective proteins including talin, kindlins, and vinculin (Canel et al., 2013).

The overall alterations in cell adhesion properties induced by cancer include weakening of cell-cell adhesion and change in cell-ECM adhesion in a way that enhances cell motility. In this regard suppression of adhesion between cells (Abduljauwad, 2019) and enhancement of part of integrin family (Desgrosellier and Cheresh, 2010), and alterations in expression of other integrins (Cooper and Giancotti, 2019; Seguin et al., 2015) have been reported.

Cell-ECM adherence is defined by integrin family multifunctional receptors which undergo transformation during tumor progress to provide suitable conditions for migration and invasion (Jinka et al., 2012). While part of integrins (both in α and β subcategories) show higher expressions during cancer initiation (Taherian et al., 2011) and progression toward invasion (Rathinam and Alahari, 2010), some other types of integrin undergo structural alterations with cancer (Kren et al., 2007). Due to significant role of integrins in cancer, several studies have proposed inhibition and suppression of certain integrins as a treatment method in cancer that downturn migration and invasion abilities which consequently induce apoptosis and blocks angiogenesis (Aksorn and Chanvorachote, 2019).

In addition to changes in cell-ECM adhesion with cancer progress, cell-cell adhesion also changes to maintain metastatic trend. While healthy epithelial cells show strong adherence among themselves within a healthy tissue, the tough adhesion bonds become disrupted during cancer progress and metastasis (Takeichi, 1993). There is an inverse correlation between cancer invasion and cell-cell adhesion, to allow cancer cells to detach from the primary tumor and metastasize (Suresh, 2007). It has been shown that EMT is associated with lower expression of epithelial cadherin (E-cadherin), responsible for a firm adhesion of epithelial cells, while mesenchymal markers such as non-epithelial cadherin and vimentin are highly expressed (Omidvar et al., 2014; Radisky, 2005; Takeichi, 1993; Tomaskovic-Crook et al., 2009). E-cadherin molecules provide high binding forces with each other, however, non-epithelial cadherin molecules are adhered noticeably weaker (Chu et al., 2004; Panorchan et al., 2006). Different mechanisms of loss of E-cadherin function induced by cancer have been distinguished (Jeanes et al., 2008; Wong and Gumbiner, 2003). The dysfunction of this protein is related to structural mutations in the protein or encoding gene CDH1 (Berx et al., 1998; Shargh et al., 2014), abnormality in some proteins such as α and β catenin (Jeanes et al., 2008), or down regulation of this protein which is suggested to be the most common mechanism among different types of cancer (Wendt et al., 2011). Some studies have aimed in up regulation of E-cadherin as a cancer treatment strategy by modifying relevant signaling pathways (Song et al., 2019a,b). The downregulation of E-cadherin induced in different types of cancer is usually accompanied by a switching mechanism of abnormally high expression of non-epithelial cadherin which is an important determinant of metastasis (Gravdal et al., 2007; Hui et al., 2013; Loh et al., 2019; Nakajima et al., 2004). E-cadherin is interactively engaged with actin microfilaments, hence cell adhesion due to E-cadherin is correlated with actin structure (Chu et al., 2004; Hazan and Norton, 1998). This shows an interesting subject of mutual effects of cell adhesion-elastic properties to further explore.

7. Cancer and actin structure: Content vs arrangement

Cancer events are accompanied by changes in the cytoskeleton fibrous network. Microtubules and intermediate filaments have direct physical contribution to the cell mechanical integrity and less to other cell behaviors. Of particular interest is altered actin structure that enhances not only deformability of cell body, but also generation of higher traction forces and alterations in cell adhesion which are vital during invasion (Fig. 1). In fiber level, such effects require alterations both in the content and structural arrangement of actin fibers. Cancer induced changes in content/arrangement of actin fibers have been vastly reported, although the proposed degree and trends of such changes are not similar among published data.

In most studies cancer has been shown to be correlated with reduction of actin content, however, in some cell lines an opposite trend has been shown (Ali et al., 2016). Early researches have reported significant reduction in actin content brought by cancer (Stark et al., 1982; Varani et al., 1983). A pioneer study compared tumorigenic and non- tumorigenic cells of different lines and indicated altered actin organization among two groups as well as significant reduction in actin concentration among tumorigenic cells either per cell or per protein (Gowing et al., 1984). The tumorigenic cell lines showed 35% decrease in actin content compared to non-tumorigenic ones, regardless of cell density.

A recent study that compared normal and invasive breast cancer cell lines in terms of mechanical properties and actin structure in suspended and spread conditions, has shown a significant reduction in cell actin index, as a measure of cell cortex in suspended cancerous cells (Tabatabaei et al., 2019). This might be due to the content reduction and/or concentration of actin fibers close to the cell membrane in cancerous cell. Results also indicated a decrease in actin intensity for spread cancer cells describing lower actin content for them, while the intensity of microtubules was not significantly different between two cell lines. Previously, the decrease in the actin cortex was also confirmed in other cancer cell lines, and it was shown that the ratio of actin to microtubule content decreased by invasion (Pachenari et al., 2014). Another recent study has further shown reduction of actin content in invasive lung cells compared to normal cells, mostly within the cytoplasmic domain compared to membrane content where actin fibers are required to contribute in cancer related motility (Ali et al., 2016). Alterations in actin content are due to changes in protein and signaling pathways that are directly or indirectly connected to actin binding proteins, and affect actin polymerization and depolymerization, actin bundling and branching, and generation of actin network (Smith et al., 2015). Some main examples include Coflin-1 which contributes to actin fiber depolymerization and prevents G-actin polymerization, Profilin-1 which improves actin fiber polymerization through transforming ADP-actin monomer to ATP-actin, and Septin-9 which binds with actin microfilaments and reduces the effects of Cofilin and Myosin during depolymerization (Smith et al., 2015). It has been shown that suppression of NCKK1 (the Sodium Potassium Chloride co-transporter) during cancer cell migration lowers invasion and causes alteration in actin content and decrease in cell actin layer (Schiapparelli et al., 2017).

An important concept is the study of actin structure and arrangement, in addition to its content. While most studies confirm lower content of cytoskeletal actin induced by cancer initiation and progress, it is expected to observe poor cell behaviors which are related to the decrease in actin content such as motility and generation of contractile forces. In opposition, cancer cells are very motile and generate high contractile forces. Such paradoxical behavior might be seen from two viewpoints: changes in *fibrous arrangement and orientation* of actin structure which enhances cell motility despite lower actin content, as well as the role of *actin binding proteins*. The cytoskeletal remodeling due to cancer results in fewer actin microfilaments in cytoplasmic domain which causes cell softening, however actin structure and arrangement of actin fibers change, especially close to membrane and cell edges which contributes to higher migration. The lower actin content is accompanied by significant changes in the distribution of F-actins and their anisotropy with higher concentration of actin fibers in the cell periphery leading to thinner cortex in suspended condition and rich stress fibers at focal adhesion sites in spread conditions (Tabatabaei et al., 2019). Various studies have shown that cancer causes remodeling of cytoskeletal structure, particularly actin bundling in different parts of cell body such as cell cortex, stress fibers, and cell edge structures such as filopodium, microvillus, and invadopodium (Stevenson et al., 2012). The actin structure remodeling occurs both during initiation and invasion of cancer. It has been suggested that the remodeling of actin structure is a determinant of EMT (Shankar and Nabi, 2015).

An early research, using quantified data, indicated loss of actin microfilament organization among tumorigenic cells compared to non-tumorigenic cells, although tumorigenic cells had more stress fibers (Gowing et al., 1984). The fluorescent intensity of stained cells indicated that the total actin organization was disrupted among tumorigenic cells. Another study on cultured human colonic epithelial cells showed loss of actin arrangement from benign to malignant tumor cells, as well as in vivo colonic epithelial cells (Friedman et al., 1984). It was concluded that loss of actin organization is an indicator of transition of cancerous cells to invasive and metastatic cells. Furthermore, actin cytoskeleton rearrangement has been reported to be part of both aging and malignant transformation (Rao and Cohen, 1991).

Recent studies have attributed the cancer driven cell softening, not only to the reduced actin content, but also to the altered actin arrangement. While normal cells have mesh like actin structure, invasive cancer cells show disruption of actin network within the cytoplasmic domain, together with formation of stress fibers (Tabatabaei et al., 2019). The stress fibers were oriented parallel in normal cells, however, in invasive cells, bundles of stress fibers were multi-directional to reach focal adhesions. The diverse focal adhesion sites in cancerous cells might be partly due to higher spreading caused by myosin binding metastasin protein (Kiss et al., 2012). Recently, application of cancer treatment drugs through EGFR blockade on invasive and non-invasive cancer breast cell lines has shown more uni-directional parallel arrangement of stress fibers among treated cells (Azadi et al., 2019). Similar results on unidirectional orientation of actin fibers were observed after treatment of Hela cells (Yang et al., 2014).

Interestingly, in addition to altered structure of actin fibers, microtubules also change with smaller and less fibrous forms among cancer cells, although with lower grade compared to actin fibers (Tabatabaei et al., 2019). This might be another reason for cancer induced cell softening, since microtubules resist compression and deformation.

In addition to changes in the conventional actin based cell structures, cancer invasion is also associated with formation of cancer specific actin rich structures such as invadopodia as a cell edge structure. They have been observed in vivo and contain actin rich chemosensing protrusions that highly contribute to localized directional ECM degradation by secretion of MMPs to allow cell invasion (Weaver, 2006; Williams et al., 2019). Invadopodia has been considered as a target for blocking cancer invasion and metastasis (Meirson and Gil-Henn, 2018). The invasion specific formation of invadopodia is associated with altered expression of a variety of proteins among different cancer cell lines. For instance, an overexpression of SNX27 (involved in protein trafficking toward membrane) triggers MMP1 secretion (Sharma et al., 2020), or an up-regulation of Cx43 (a gap junction associated protein) is correlated with higher MMP2 level (Chepied et al., 2020), both enhancing function of invadopodia in ECM degradation.

Alterations in actin structure of cancer cells can be interpreted in protein level among different actin partner proteins. An example is α-Actinin, a major protein that crosslinks thin actin filaments into bundles and exists in almost all actin based structures such as stress fibers and cell edge structures. Upregulation and mutation among two family members of non-muscle cells, α-Actinin1 and α-Actinin4 are vastly reported, while the latter shows higher impact due to both structural bundling and mechanosensory role in force generation (Azatov et al., 2016).

A variety of proteins which are involved in actin bundling undergo changes during cancer progress. Most of these proteins are located in cell protrusions, and vastly contribute to the cell migration. Fascin, a monomer with two actin binding sites, crosslinks actins into bundles and is mostly localized in cell protrusions. It is highly an indicator of cancer among different tissues (Machesky and Lia, 2010), and is among top upregulated actin related proteins in metastatic tumors. Its expression is correlated with invasion among different types of cancer (Vignjevic et al., 2007; Yoder et al., 2005), and its inhibition blocks cancer metastasis (McGuire et al., 2019). T-Plastin is also involved in crosslinking of F-actin to form bundles and localizes in lamellipodia and microvilli, and is highly expressed in different types of cancer (Breitsprecher and Goode, 2013; Hisano et al., 1996). Formins act as tools of actin polymerization and are responsible for rearrangement of actin structure. Alterations in Formin family have been recognized with cancer progress (DeWard et al., 2010). Profilin I is another actin binding protein at the leading edge of cells during migration which plays a key role in the control of actin polymerization and is down regulated in different types of cancer (Lorente et al., 2014). Due to the presence of many actin-binding proteins during cancer initiation and metastasis, it is not likely to mention a particular protein as the major cancer and metastasis driven protein. Similar to transductive cascades in biological cells, a network of signaling pathways are responsible for alterations in the structure and function of actin structure during cancer invasion (Stevenson et al., 2012).

8. Conclusion

In this chapter book we discussed the correlation of cancer invasion with alterations in cytoskeletal fibers, with actin fibers in the focus. It was shown that the invasive nature of cancerous cells in terms of physical behaviors such as motility, adhesion and mechanical properties is highly related to the cytoskeleton remodeling, dominantly by alterations in the content and arrangement of actin structure. Such phenomenon is particularly of interest, since it shows the effects of cell structural properties on its biological behavior. Hence many cancer treatment strategies on blocking metabolic activities of cancerous cells by targeting chemical cascades, can be combined with targeting physical and mechanotransductive pathways that are related to cytoskeleton structure and cause cell softening, enhanced motility and altered cell adhesion. Such synergy can enhance the efficiency of treatment plans. Furthermore, it can be concluded that even with different chemical and mechanical cascades in cells, this phenomenon describes an inherent relationship between chemical and physical pathways. The mechanism of this relation, either in terms of relevant intervening proteins or other possible incentives remains an interesting field of study to further explore.

References

Abduljauwad, S.N., 2019. Enhancing cancer cell adhesion with clay nanoparticles for countering metastasis. Sci. Rep. 9 (1), 1–12.
Aksorn, N., Chanvorachote, P., 2019. Integrin as a molecular target for anti-cancer approaches in lung cancer. Anticancer Res 39 (2), 541–548.
Alberts, B., 2008. Molecular Biology of the Cell. Garland Science.
Ali, M., Heyob, K., Jacob, N.K., Rogers, L.K., 2016. Alterative expression and localization of profilin 1/VASPpS157 and cofilin 1/VASPpS239 regulates metastatic growth and is modified by DHA supplementation. Mol. Cancer Ther. 15 (9), 2220–2231.
Alibert, C., Goud, B., Manneville, J.B., 2017. Are cancer cells really softer than normal cells? Biol. Cell 109 (5), 167–189.

Ananthakrishnan, R., Ehrlicher, A., 2007. The forces behind cell movement. Int. J. Biol. Sci. 3 (5), 303–317.
Ansardamavandi, A., Tafazzoli-Shadpour, M., Omidvar, R., Jahanzad, I., 2016. Quantification of effects of cancer on elastic properties of breast tissue by atomic force microscopy. J. Mech. Behav. Biomed. Mater. 60, 234–242.
Aranda-Espinoza, H., 2015. Mechanobiology of the Endothelium. CRC Press.
Artemenko, Y., Axiotakis, L., Borleis, J., Iglesias, P.A., Devreotes, P.N., 2016. Chemical and mechanical stimuli act on common signal transduction and cytoskeletal networks. Proc. Natl. Acad. Sci. U. S. A. 113 (47), E7500–E7509.
Ayala, Y.A., et al., 2016. Rheological properties of cells measured by optical tweezers. BMC Biophys. 9 (1), 5.
Azadi, S., Tafazzoli-Shadpour, M., Omidvar, R., Moradi, L., Habibi-Anbouhi, M., 2016. Epidermal growth factor receptor targeting alters gene expression and restores the adhesion function of cancerous cells as measured by single cell force spectroscopy. Mol. Cell. Biochem. 423 (1–2), 129–139.
Azadi, S., Tafazzoli-Shadpour, M., Soleimani, M., Warkiani, M.E., 2019. Modulating cancer cell mechanics and actin cytoskeleton structure by chemical and mechanical stimulations. J. Biomed. Mater. Res. - Part A 107 (8), 1569–1581.
Azatov, M., Goicoechea, S.M., Otey, C.A., Upadhyaya, A., 2016. The actin crosslinking protein palladin modulates force generation and mechanosensitivity of tumor associated fibroblasts. Sci. Rep. 6, 1–12.
Azevedo, A.S., Follain, G., Patthabhiraman, S., Harlepp, S., Goetz, J.G., 2015. Metastasis of circulating tumor cells: favorable soil or suitable biomechanics, or both? Cell Adh. Migr. 9 (5), 345–356.
Balaban, N.Q., et al., 2001. Force and focal adhesion assembly: a close relationship studied using elastic micropatterened substrates. Nat. Cell Biol. 3, 466–472.
Bar-Shavit, R., et al., 2016. G protein-coupled receptors in cancer. Int. J. Mol. Sci. 17 (8), 1–16.
Beach, J.R., et al., 2011. Myosin II isoform switching mediates invasiveness after TGF-β–induced epithelial–mesenchymal transition. Proc. Natl. Acad. Sci. 108 (44), 17991–17996.
Ben-Ze'ev, A., 1985. The cytoskeleton in cancer cells. BBA-Rev. Cancer 780 (3), 197–212.
Bereiter-Hahn, J., 2005. Mechanics of crawling cells. Med. Eng. Phys. 27 (9), 743–753.
Bershadsky, A.D., Vasiliev, J.M., 1988. Regulation of synthesis of cytoskeletal proteins. In: Cytoskeleton. Springer, pp. 203–216.
Berx, G., Becker, K., Höfler, H., Van Roy, F., 1998. Mutations of the human E-cadherin (CDH1) gene. Hum. Mutat. 12 (4), 226–237.
Betapudi, V., Licate, L.S., Egelhoff, T.T., 2006. Distinct roles of nonmuscle myosin II isoforms in the regulation of MDA-MB-231 breast cancer cell spreading and migration. Cancer Res. 66 (9), 4725–4733.
Birchmeier, W., Behrens, J., 1994. Cadherin expression in carcinomas: role in the formation of cell junctions and the prevention of invasiveness. BBA-Rev. Cancer 1198 (1), 11–26.
Bjerkvig, R., Tysnes, B.B., Aboody, K.S., Najbauer, J., Terzis, A.J.A., 2005. The origin of the cancer stem cell: current controversies and new insights. Nat. Rev. Cancer 5 (11), 899–904.
Blain, E.J., 2009. Involvement of the cytoskeletal elements in articular cartilage homeostasis and pathology. Int. J. Exp. Pathol. 90 (1), 1–15.
Bloemendal, H., Pieper, F.R., 1989. Intermediate filaments: known structure, unknown function. BBA-Gene Struct. Expr. 1007 (3), 245–253.
Breitsprecher, D., Goode, B.L., 2013. Formins at a glance. J. Cell Sci. 126 (1), 1–7.
Brooke, M.A., Nitoiu, D., Kelsell, D.P., 2012. Cell-cell connectivity: desmosomes and disease. J. Pathol. 226 (2), 158–171.

Bu, Y., Li, L., Yang, C., Li, R., Wang, J., 2019. Measuring viscoelastic properties of living cells. Acta Mech. Solida Sin. 32 (5), 599–610.
Butler, J.P., Tolic-Nørrelykke, I.M., Fabry, B., Fredberg, J.J., 2002. Traction fields, moments, and strain energy that cells exert on their surroundings. Am. J. Physiol. Physiol. 282 (3), C595–C605.
Cai, Y., et al., 2006. Nonmuscle myosin IIA-dependent force inhibits cell spreading and drives F-actin flow. Biophys. J. 91 (10), 3907–3920.
Canel, M., Serrels, A., Frame, M.C., Brunton, V.G., 2013. E-cadherin-integrin crosstalk in cancer invasion and metastasis. J. Cell Sci. 126 (2), 393–401.
Canetta, E., Riches, A., Borger, E., Herrington, S., Dholakia, K., Adya, A.K., 2014. Discrimination of bladder cancer cells from normal urothelial cells with high specificity and sensitivity: combined application of atomic force microscopy and modulated Raman spectroscopy. Acta Biomater. 10 (5), 2043–2055.
Carmichael, B., Babahosseini, H., Mahmoodi, S.N., Agah, M., 2015. The fractional viscoelastic response of human breast tissue cells. Phys. Biol. 12 (4). https://doi.org/10.1088/1478-3975/12/4/046001.
Cavallaro, U., Christofori, G., 2004. Cell adhesion and signalling by cadherins and Ig-CAMs in cancer. Nat. Rev. Cancer 4 (2), 118–132.
Charrier, E.E., Janmey, P.A., 2016. Mechanical Properties of Intermediate Filament Proteins, first ed. vol. 568. Elsevier Inc.
Chen, T.-J., Wu, C.-C., Tang, M.-J., Huang, J.-S., Su, F.-C., 2010. Complexity of the tensegrity structure for dynamic energy and force distribution of cytoskeleton during cell spreading. PLoS One 5 (12). https://doi.org/10.1371/journal.pone.0014392.
Chepied, A., Daoud-omar, Z., Laird, D.W., Mesnil, M., Defamie, N., 2020. Involvement of the gap junction protein, connexin43, in the formation and function of invadopodia in the human U251 glioblastoma cell line. Cells 43, 117.
Choi, J.A., Jung, Y.S., Kim, J.Y., Kim, H.M., Lim, I.K., 2016. Inhibition of breast cancer invasion by TIS21/BTG2/Pc3-Akt1-Sp1-Nox4 pathway targeting actin nucleators, mDia genes. Oncogene 35 (1), 83–93.
Chu, Y.S., et al., 2004. Force measurements in E-cadherin-mediated cell doublets reveal rapid adhesion strengthened by actin cytoskeleton remodeling through Rac and Cdc42. J. Cell Biol. 167 (6), 1183–1194.
Chugh, P., Paluch, E.K., 2018. The actin cortex at a glance. J. Cell Sci. 131 (14), 1–9.
Cooper, G., 2000. The Cell: A Molecular Approach. Intermediate Filaments. Sinauer Associates, Sunderland, MA.
Cooper, J., Giancotti, F.G., 2019. Integrin signaling in cancer: mechanotransduction, stemness, epithelial plasticity, and therapeutic resistance. Cancer Cell 35 (3), 347–367.
Critchley, D.R., 2009. Biochemical and structural properties of the integrin-associated cytoskeletal protein Talin. Annu. Rev. Biophys. 38 (1), 235–254.
Cross, S.E., Jin, Y.S., Rao, J., Gimzewski, J.K., 2007. Nanomechanical analysis of cells from cancer patients. Nat. Nanotechnol. 2 (12), 780–783.
Cross, S.E., Jin, Y.S., Tondre, J., Wong, R., Rao, J.Y., Gimzewski, J.K., 2008. AFM-based analysis of human metastatic cancer cells. Nanotechnology 19 (38).
Das, M., Ithychanda, S., Bledzka, K., Qin, J., Plow, E.F., 2012. Competition Between Filamin and Kindlin-2 Regulates Integrin Activation. American Heart Association.
Davies, J., 2013. Mechanisms of Morphogenesis. Academic Press.
De Luca, A., Maiello, M.R., D'Alessio, A., Pergameno, M., Normanno, N., 2012. The RAS/RAF/MEK/ERK and the PI3K/AKT signalling pathways: role in cancer pathogenesis and implications for therapeutic approaches. Expert Opin. Ther. Targets 16 (SUPPL.2), 17–27.
Desgrosellier, J.S., Cheresh, D.A., 2010. Integrins in cancer: biological implications and therapeutic opportunities. Nat. Rev. Cancer 10 (1), 9–22.

DeWard, A.D., Eisenmann, K.M., Matheson, S.F., Alberts, A.S., 2010. The role of formins in human disease. Biochim. Biophys. Acta-Mol. Cell Res. 1803 (2), 226–233.
Dong, C., Lei, X.X., 2000. Biomechanics of cell rolling: shear flow, cell-surface adhesion, and cell deformability. J. Biomech. 33 (1), 35–43.
Efremov, Y.M., Dokrunova, A.A., Efremenko, A.V., Kirpichnikov, M.P., Shaitan, K.V., Sokolova, O.S., 2015. Distinct impact of targeted actin cytoskeleton reorganization on mechanical properties of normal and malignant cells. Biochim. Biophys. Acta (BBA)-Molecular Cell Res. 1853 (11), 3117–3125.
Efremov, Y.M., Wang, W.-H., Hardy, S.D., Geahlen, R.L., Raman, A., 2017. Measuring nanoscale viscoelastic parameters of cells directly from AFM force-displacement curves. Sci. Rep. 7 (1), 1–14.
Elosegui-Artola, A., et al., 2016. Mechanical regulation of a molecular clutch defines force transmission and transduction in response to matrix rigidity. Nat. Cell Biol. 18 (5), 540–548.
Faria, E.C., et al., 2008. Measurement of elastic properties of prostate cancer cells using AFM. Analyst 133 (11), 1498–1500.
Fife, C.M., McCarroll, J.A., Kavallaris, M., 2014. Movers and shakers: cell cytoskeleton in cancer metastasis. Br. J. Pharmacol. 171 (24), 5507–5523.
Fletcher, D.A., Mullins, R.D., 2010. Cell mechanics and the cytoskeleton. Nature 463 (7280), 485–492.
Forgacs, G., Yook, S.H., Janmey, P.A., Jeong, H., Burd, C.G., 2004. Role of the cytoskeleton in signaling networks. J. Cell Sci. 117 (13), 2769–2775.
Frankel, J., 1992. Positional information in cells and organisms. Trends Cell Biol. 2 (9), 256–260.
Frieden, B.R., Gatenby, R.A., 2019. Signal transmission through elements of the cytoskeleton form an optimized information network in eukaryotic cells. Sci. Rep. 9 (1), 1–10.
Friedl, P., Hegerfeldt, Y., Tusch, M., Friedl, P., 2004. Collective cell migration in morphogenesis and cancer. 449, 441–449.
Friedman, E., Winawer, S., Verderame, M., Pollack, R., 1984. Actin cytoskeletal organization loss in the benign-to-malignant tumor transition in cultured human colonic epithelial cells. Cancer Res. 44 (7), 3040–3050.
Frixen, U.H., et al., 1991. E-cadherin-mediated cell-cell adhesion prevents invasiveness of human carcinoma cells. J. Cell Biol. 113 (1), 173–185.
Fuchs, E., Weber, K., 1994. Intermediate filaments: structure, dynamics, function and disease. Annu. Rev. Biochem. 63 (1), 345–382.
Gatenby, R.A., Gawlinski, E.T., Gmitro, A.F., Kaylor, B., Gillies, R.J., 2006. Acid-mediated tumor invasion: a multidisciplinary study. Cancer Res. 66 (10), 5216–5223.
Geiger, B., 1989. Cytoskeleton-associated cell contacts. 1 (1), 103–109.
Gkretsi, V., Stylianopoulos, T., 2018. Cell adhesion and matrix stiffness: coordinating cancer cell invasion and metastasis. Front. Oncol. 8, 145.
Gowing, L.R., Tellam, R.L., Banyard, M.R., 1984. Microfilament organization and total actin content are decreased in hybrids derived from the fusion of HeLa cells with human fibroblasts. J. Cell Sci. 69 (1), 137–146.
Gravdal, K., Halvorsen, O.J., Haukaas, S.A., Akslen, L.A., 2007. A switch from E-cadherin to N-cadherin expression indicates epithelial to mesenchymal transition and is of strong and independent importance for the progress of prostate cancer. Clin. Cancer Res. 13 (23), 7003–7011.
Guck, J., et al., 2005. Optical deformability as an inherent cell marker for testing malignant transformation and metastatic competence. Biophys. J. 88 (5), 3689–3698.
Hafner, A.E., Rieger, H., 2016. Spatial organization of the cytoskeleton enhances cargo delivery to specific target areas on the plasma membrane of spherical cells. Phys. Biol. 13 (6).

Hall, A., 2009. The cytoskeleton and cancer. Cancer Metastasis Rev. 28 (1–2), 5–14.
Hammond, J.W., Cai, D., Verhey, K.J., 2008. Tubulin modifications and their cellular functions. Curr. Opin. Cell Biol. 20 (1), 71–76.
Harris, A.K., Wild, P., Stopak, D., 1980. Silicone rubber substrata: a new wrinkle in the study of cell locomotion. Science (80-.). 208 (4440), 177–179.
Hayashi, K., Iwata, M., 2015. Stiffness of cancer cells measured with an AFM indentation method. J. Mech. Behav. Biomed. Mater. 49, 105–111.
Hazan, R.B., Norton, L., 1998. The epidermal growth factor receptor modulates the interaction of E-cadherin with the actin cytoskeleton. J. Biol. Chem. 273 (15), 9078–9084.
Hendrix, M.J.C., Seftor, E.A., Chu, Y.W., Trevor, K.T., Seftor, R.E.B., 1996. Role of intermediate filaments in migration, invasion and metastasis. Cancer Metastasis Rev. 15 (4), 507–525.
Hisano, T., Ono, M., Nakayama, M., Naito, S., Kuwano, M., Wada, M., 1996. Increased expression of T-plastin gene in cisplatin-resistant human cancer cells: identification by mRNA differential display. FEBS Lett. 397 (1), 101–107.
Hochmuth, R.M., 2000. Micropipette aspiration of living cells. J. Biomech. 33 (1), 15–22.
Holthöfer, B., Windoffer, R., Troyanovsky, S., Leube, R.E., 2007. Structure and function of desmosomes. Int. Rev. Cytol. 264 (07), 65–163.
Holy, T.E., Leibler, S., 1994. Dynamic instability of microtubules as an efficient way to search in space. Proc. Natl. Acad. Sci. U. S. A. 91 (12), 5682–5685.
Honore, S., Pasquier, E., Braguer, D., 2005. Understanding microtubule dynamics for improved cancer therapy. Cell. Mol. Life Sci. 62 (24), 3039–3056.
Hotulainen, P., Lappalainen, P., 2006. Stress fibers are generated by two distinct actin assembly mechanisms in motile cells. J. Cell Biol. 173 (3), 383–394.
Hu, J., et al., 2019. High stretchability, strength, and toughness of living cells enabled by hyperelastic vimentin intermediate filaments. Proc. Natl. Acad. Sci. U. S. A. 116 (35), 17175–17180.
Hui, L., Zhang, S., Dong, X., Tian, D., Cui, Z., Qiu, X., 2013. Prognostic significance of twist and N-cadherin expression in NSCLC. PLoS One 8 (4), 1–10.
Humphrey, J.D., Dufresne, E.R., Schwartz, M.A., 2014. Mechanotransduction and extracellular matrix homeostasis. Nat. Rev. Mol. Cell Biol. 15 (12), 802–812.
Hur, S.S., Jeong, J.H., Ban, M.J., Park, J.H., Yoon, J.K., Hwang, Y., 2020. Traction force microscopy for understanding cellular mechanotransduction. BMB Rep. 53 (2), 74.
Hynes, N.E., Lane, H.A., 2005. ERBB receptors and cancer: the complexity of targeted inhibitors. Nat. Rev. Cancer 5 (5), 341–354.
Ishihara, S., et al., 2013. Irradiation-tolerant lung cancer cells acquire invasive ability dependent on dephosphorylation of the myosin regulatory light chain. FEBS Lett. 587 (6), 732–736.
Izdebska, M., Zielińska, W., Grzanka, D., Gagat, M., 2018. The role of actin dynamics and actin-binding proteins expression in epithelial-to-mesenchymal transition and its association with cancer progression and evaluation of possible therapeutic targets. Biomed. Res. Int.
Jamora, C., Fuchs, E., 2002. Intercellular adhesion, signalling and the cytoskeleton. Nat. Cell Biol. 4 (4), 101–108.
Janmey, P.A., 1998. The cytoskeleton and cell signaling: component localization and mechanical coupling. Physiol. Rev. 78 (3), 763–781.
Janmey, P.A., Miller, R.T., 2011. Mechanisms of mechanical signaling in development and disease. J. Cell Sci. 124 (1), 9–18.
Jannatbabaei, A., Tafazzoli-Shadpour, M., Seyedjafari, E., Fatouraee, N., 2019. Cytoskeletal remodeling induced by substrate rigidity regulates rheological behaviors in endothelial cells. J. Biomed. Mater. Res. Part A 107 (1), 71–80.

Jeanes, A., Gottardi, C.J., Yap, A.S., 2008. Cadherins and cancer: how does cadherin dysfunction promote tumor progression? Oncogene 27 (55), 6920–6929.

Jiang, P., Enomoto, A., Takahashi, M., 2009. Cell biology of the movement of breast cancer cells: intracellular signalling and the actin cytoskeleton. Cancer Lett. 284 (2), 122–130.

Jinka, R., Kapoor, R., Sistla, P.G., Raj, T.A., Pande, G., 2012. Alterations in cell-extracellular matrix interactions during progression of cancers. Int. J. Cell Biol.

Jordan, A., Hadfield, J.A., Lawrence, N.J., Mcgown, A.T., 1998. Tubulin as a target for anticancer drugs: agents which interact with the mitotic spindle. Med. Res. Rev. 18 (4), 259–296.

Jurmeister, S., et al., 2012. MicroRNA-200c represses migration and invasion of breast cancer cells by targeting actin-regulatory proteins FHOD1 and PPM1F. Mol. Cell. Biol. 32 (3), 633–651.

Kampen, K.R., 2011. Membrane proteins: the key players of a cancer cell. J. Membr. Biol. 242 (2), 69–74.

Karantza, V., 2011. Keratins in health and cancer: more than mere epithelial cell markers. Oncogene 30 (2), 127–138.

Kashina, A.S., 2020. Regulation of actin isoforms in cellular and developmental processes. Semin. Cell Dev. Biol. 102, 113–121.

Kedrin, D., van Rheenen, J., Hernandez, L., Condeelis, J., Segall, J.E., 2007. Cell motility and cytoskeletal regulation in invasion and metastasis. J. Mammary Gland Biol. Neoplasia 12 (2–3), 143–152.

Khani, M.-M., Tafazzoli-Shadpour, M., Rostami, M., Peirovi, H., Janmaleki, M., 2014. Evaluation of mechanical properties of human mesenchymal stem cells during differentiation to smooth muscle cells. Ann. Biomed. Eng. 42 (7), 1373–1380.

Khani, M.-M., Tafazzoli-Shadpour, M., Goli-Malekabadi, Z., Haghighipour, N., 2015. Mechanical characterization of human mesenchymal stem cells subjected to cyclic uniaxial strain and TGF-β1. J. Mech. Behav. Biomed. Mater. 43, 18–25.

Kim, S., Coulombe, P.A., 2010. Emerging role for the cytoskeleton as an organizer and regulator of translation. Nat. Rev. Mol. Cell Biol. 11, 75–81.

Kiss, B., Duelli, A., Radnai, L., Kékesi, K.A., Katona, G., Nyitray, L., 2012. Crystal structure of the S100A4-nonmuscle myosin IIA tail fragment complex reveals an asymmetric target binding mechanism. Proc. Natl. Acad. Sci. U. S. A. 109 (16), 6048–6053.

Koay, E.J., Shieh, A.C., Athanasiou, K.A., 2003. Creep indentation of single cells. J. Biomech. Eng. 125 (3), 334–341.

Koch, T.M., Münster, S., Bonakdar, N., Butler, J.P., Fabry, B., 2012. 3D traction forces in cancer cell invasion. PLoS One 7 (3), e33476.

Kolega, J., Janson, L.W., Taylor, D.L., 1991. The role of solation-contraction coupling in regulating stress fiber dynamics in nonmuscle cells. J. Cell Biol. 114 (5), 993–1003.

Kraning-Rush, C.M., Califano, J.P., Reinhart-King, C.A., 2012. Cellular traction stresses increase with increasing metastatic potential. PLoS One 7 (2), e32572.

Kren, A., et al., 2007. Increased tumor cell dissemination and cellular senescence in the absence of β1-integrin function. EMBO J. 26 (12), 2832–2842.

Kulić, I.M., et al., 2008. The role of microtubule movement in bidirectional organelle transport. Proc. Natl. Acad. Sci. U. S. A. 105 (29), 10011–10016.

Kumar, S., Weaver, V.M., 2009. Mechanics, malignancy, and metastasis: the force journey of a tumor cell. Cancer Metastasis Rev. 28 (1–2), 113–127.

Kustermans, G., Piette, J., Legrand-Poels, S., 2008. Actin-targeting natural compounds as tools to study the role of actin cytoskeleton in signal transduction. Biochem. Pharmacol. 76 (11), 1310–1322.

Lam, W.A., Rosenbluth, M.J., Fletcher, D.A., 2007. Chemotherapy exposure increases leukemia cell stiffness. Blood 109 (8), 3505–3508.

Lambrechts, A., Van Troys, M., Ampe, C., 2004. The actin cytoskeleton in normal and pathological cell motility. Int. J. Biochem. Cell Biol. 36 (10), 1890–1909.

Le Clainche, C., Carlier, M.-F., 2008. Regulation of actin assembly associated with protrusion and adhesion in cell migration. Physiol. Rev. 88 (2), 489–513.

Lee, J., Leonard, M., Oliver, T., Ishihara, A., Jacobson, K., 1994. Traction forces generated by locomoting keratocytes. J. Cell Biol. 127 (6 II), 1957–1964.

Lekka, M., Laidler, P., Gil, D., Lekki, J., Stachura, Z., Hrynkiewicz, A.Z., 1999. Elasticity of normal and cancerous human bladder cells studied by scanning force microscopy. Eur. Biophys. J. 28 (4), 312–316.

Lekka, M., et al., 2001. The effect of chitosan on stiffness and glycolytic activity of human bladder cells. Biochim. Biophys. Acta (BBA)-Molecular Cell Res. 1540 (2), 127–136.

Lekka, M., et al., 2012. Cancer cell detection in tissue sections using AFM. Arch. Biochem. Biophys. 518 (2), 151–156.

Leśniak, W., 2011. Epigenetic regulation of S100 protein expression. Clin. Epigenetics 2 (2), 77–83.

Li, Z., et al., 2017. Cellular traction forces: a useful parameter in cancer research. Nanoscale 9 (48), 19039–19044.

Liu, D., et al., 2012. Clinicopathological significance of NMIIA overexpression in human gastric cancer. Int. J. Mol. Sci. 13 (11), 15291–15304.

Liu, C.-Y., Lin, H.-H., Tang, M.-J., Wang, Y.-K., 2015. Vimentin contributes to epithelial-mesenchymal transition cancer cell mechanics by mediating cytoskeletal organization and focal adhesion maturation. Oncotarget 6 (18), 15966.

Lodish, H., Berk, A., Zipursky, S.L., Matsudaira, P., Baltimore, D., Darnell, J., 2000. Microtubule dynamics and associated proteins. In: Molecular Cell Biology, fourth ed. WH Freeman.

Lodish, H., et al., 2008. Molecular Cell Biology. Macmillan.

Loh, C.-Y., et al., 2019. The E-cadherin and N-cadherin switch in epithelial-to-mesenchymal transition: signaling, therapeutic implications, and challenges. Cell 8 (10), 1118.

Loo, D.T., Kanner, S.B., Aruffo, A., 1998. Filamin binds to the cytoplasmic domain of the β1-integrin: identification of amino acids responsible for this interaction. J. Biol. Chem. 273 (36), 23304–23312.

Lorente, G., Syriani, E., Morales, M., 2014. Actin filaments at the leading edge of cancer cells are characterized by a high mobile fraction and turnover regulation by profilin I. PLoS One 9 (1), e85817.

Machesky, L.M., Lia, A., 2010. Fascin: invasive filopodia promoting metastasis. Commun. Integr. Biol. 3 (3), 263–270.

Makowska, K.A., Hughes, R.E., White, K.J., Wells, C.M., Peckham, M., 2015. Specific myosins control actin organization, cell morphology, and migration in prostate cancer cells. Cell Rep. 13 (10), 2118–2125.

Manneville, J.B., Etienne-Manneville, S., Skehel, P., Carter, T., Ogden, D., Ferenczi, M., 2003. Interaction of the actin cytoskeleton with microtubules regulates secretory organelle movement near the plasma membrane in human endothelial cells. J. Cell Sci. 116 (19), 3927–3938.

Martins-Green, M., Bissell, M.J., 1995. Cell-ECM interactions in development. Semin. Dev. Biol. 6 (2), 149–159.

Maskarinec, S.A., Franck, C., Tirrell, D.A., Ravichandran, G., 2009. Quantifying cellular traction forces in three dimensions. Proc. Natl. Acad. Sci. USA 106 (52), 22108–22113.

Maupin, P., Phillips, C.L., Adelstein, R.S., Pollard, T.D., 1994. Differential localization of myosin-II isozymes in human cultured cells and blood cells. J. Cell Sci. 107 (11), 3077–3090.

McGuire, S., et al., 2019. Inhibition of fascin in cancer and stromal cells blocks ovarian cancer metastasis. Gynecol. Oncol. 153 (2), 405–415.
Mège, R.-M., Gavard, J., Lambert, M., 2006. Regulation of cell–cell junctions by the cytoskeleton. Curr. Opin. Cell Biol. 18 (5), 541–548.
Mehrbod, M., Mofrad, M.R.K., 2011. On the significance of microtubule flexural behavior in cytoskeletal mechanics. PLoS One 6 (10), e25627.
Meirson, T., Gil-Henn, H., 2018. Targeting invadopodia for blocking breast cancer metastasis. Drug Resist. Updat. 39, 1–17.
Minamiya, Y., et al., 2005. Increased expression of myosin light chain kinase mRNA is related to metastasis in non-small cell lung cancer. Tumor Biol. 26 (3), 153–157.
Mohammadi, H., Sahai, E., 2018. Mechanisms and impact of altered tumour mechanics. Nat. Cell Biol. 20 (7), 766–774.
Mukhtar, E., Adhami, V.M., Mukhtar, H., 2014. Targeting microtubules by natural agents for cancer therapy. Mol. Cancer Ther. 13 (2), 275–284.
Nakajima, S., et al., 2004. N-cadherin expression and epithelial-mesenchymal transition in pancreatic carcinoma. Clin. Cancer Res. 10 (12 I), 4125–4133.
Nürnberg, A., Kitzing, T., Grosse, R., 2011. Nucleating actin for invasion. Nat. Rev. Cancer 11 (3), 117–187.
Olson, M.F., Sahai, E., 2009. The actin cytoskeleton in cancer cell motility. Clin. Exp. Metastasis 26 (4), 273–287.
Omidvar, R., Tafazzoli-shadpour, M., Shokrgozar, M.A., Rostami, M., 2014. Atomic force microscope-based single cell force spectroscopy of breast cancer cell lines: an approach for evaluating cellular invasion. J. Biomech. 47 (13), 3373–3379.
Ouderkirk, J.L., Krendel, M., 2014. Non-muscle myosins in tumor progression, cancer cell invasion, and metastasis. Cytoskeleton 71 (8), 447–463.
Owen, L.M., et al., 2017. A cytoskeletal clutch mediates cellular force transmission in a soft, three-dimensional extracellular matrix. Mol. Biol. Cell 28 (14), 1959–1974.
Pachenari, M., Seyedpour, S.M., Janmaleki, M., Shayan, S.B., Taranejoo, S., Hosseinkhani, H., 2014. Mechanical properties of cancer cytoskeleton depend on actin filaments to microtubules content: investigating different grades of colon cancer cell lines. J. Biomech. 47 (2), 373–379.
Panorchan, P., Thompson, M.S., Davis, K.J., Tseng, Y., Konstantopoulos, K., Wirtz, D., 2006. Single-molecule analysis of cadherin-mediated cell-cell adhesion. J. Cell Sci. 119 (1), 66–74.
Park, S., Koch, D., Cardenas, R., Käs, J., Shih, C.-K., 2005. Cell motility and local viscoelasticity of fibroblasts. Biophys. J. 89 (6), 4330–4342.
Parker, A.L., Kavallaris, M., McCarroll, J.A., 2014. Microtubules and their role in cellular stress in cancer. Front. Oncol. 4, 1–19.
Parsons, J.T., Horwitz, A.R., Schwartz, M.A., 2010. Cell adhesion: integrating cytoskeletal dynamics and cellular tension. Nat. Rev. Mol. Cell Biol. 11 (9), 633–643.
Pavlova, N.N., Thompson, C.B., 2016. The emerging hallmarks of cancer metabolism. Cell Metab. 23 (1), 27–47.
Peschetola, V., et al., 2013. Time-dependent traction force microscopy for cancer cells as a measure of invasiveness. Cytoskeleton 70 (4), 201–214.
Peterson, J.R., Mitchison, T.J., 2002. Small molecules, big impact: a history of chemical inhibitors and the cytoskeleton. Chem. Biol. 9 (12), 1275–1285.
Phi, L.T.H., et al., 2018. Cancer stem cells (CSCs) in drug resistance and their therapeutic implications in cancer treatment. Stem Cells Int. 2018, 1–16.
Polacheck, W.J., Chen, C.S., 2016. Measuring cell-generated forces: a guide to the available tools. Nat. Methods 13 (5), 415.
Pollard, T.D., 2016. Actin and actin-binding proteins. Cold Spring Harb. Perspect. Biol. 8, 1–17.

Pollard, T.D., Borisy, G.G., 2003. Cellular motility driven by assembly and disassembly of actin filaments. Cell 112 (4), 453–465.
Pollard, T.D., Cooper, J.A., 1986. Actin and actin-binding proteins. A critical evaluation of mechanisms and functions. Annu. Rev. Biochem. 55 (1), 987–1035.
Pugh, A., 1976. An Introduction to Tensegrity. Univ of California Press.
Radisky, D.C., 2005. Epithelial-mesenchymal transition. J. Cell Sci. 118 (19), 4325–4326.
Ramos, J.R., Pabijan, J., Garcia, R., Lekka, M., 2014. The softening of human bladder cancer cells happens at an early stage of the malignancy process. Beilstein J. Nanotechnol. 5 (1), 447–457.
Rao, K.M.K., Cohen, H.J., 1991. Actin cytoskeletal network in aging and cancer. Mutat. Res. DNAging 256 (2–6), 139–148.
Rao, J.Y., Li, N., 2004. Microfilament actin remodeling as a potential target for cancer drug development. Curr. Cancer Drug Targets 4 (4), 345–354.
Rathinam, R., Alahari, S.K., 2010. Important role of integrins in the cancer biology. Cancer Metastasis Rev. 29 (1), 223–237.
Rottner, K., Faix, J., Bogdan, S., Linder, S., Kerkhoff, E., 2017. Actin assembly mechanisms at a glance. J. Cell Sci. 130 (20), 3427–3435.
Sahai, E., Marshall, C.J., 2002. RHO–GTPases and cancer. Nat. Rev. Cancer 2 (2), 133–142.
Salama, I., Malone, P.S., Mihaimeed, F., Jones, J.L., 2008. A review of the S100 proteins in cancer. Eur. J. Surg. Oncol. 34 (4), 357–364.
Sanger, J.W., Sanger, J.M., Jockusch, B.M., 1983. Differences in the stress fibers between fibroblasts and epithelial cells. J. Cell Biol. 96 (4), 961–969.
Sanghvi-Shah, R., Weber, G.F., 2017. Intermediate filaments at the junction of mechanotransduction, migration, and development. Front. Cell Dev. Biol. 5, 1–19.
Sato, M., Theret, D.P., Wheeler, L.T., Ohshima, N., Nerem, R.M., 1990. Application of the micropipette technique to the measurement of cultured porcine aortic endothelial cell viscoelastic properties. J. Biomech. Eng. 112 (3), 263–268.
Schiapparelli, P., et al., 2017. NKCC1 regulates migration ability of glioblastoma cells by modulation of actin dynamics and interacting with cofilin. EBioMedicine 21 (2016), 94–103.
Schmitt, E., Gehrmann, M., Brunet, M., Multhoff, G., Garrido, C., 2007. Intracellular and extracellular functions of heat shock proteins: repercussions in cancer therapy. J. Leukoc. Biol. 81 (1), 15–27.
Schroeder, C., et al., 2019. Aberrant expression of the microtubule-associated protein tau is an independent prognostic feature in prostate cancer. BMC Cancer 19 (1), 193.
Seguin, L., Desgrosellier, J.S., Weis, S.M., Cheresh, D.A., 2015. Integrins and cancer: regulators of cancer stemness, metastasis, and drug resistance. Trends Cell Biol. 25 (4), 234–240.
Sever, R., Brugge, J.S., 2015. Signal transduction in cancer. Cold Spring Harb. Perspect. Med. 5 (4), a006098.
Seyfried, T.N., Shelton, L.M., 2010. Cancer as a metabolic disease. Nutr. Metab. (Lond.) 7, 1–22.
Shankar, J., Nabi, I.R., 2015. Actin cytoskeleton regulation of epithelial mesenchymal transition in metastatic cancer cells. PLoS One 10 (3), 1–12.
Shargh, S.A., et al., 2014. Downregulation of E-cadherin expression in breast cancer by promoter hypermethylation and its relation with progression and prognosis of tumor. Med. Oncol. 31 (11), 1–6.
Sharma, P., et al., 2020. SNX27–retromer assembly recycles MT1-MMP to invadopodia and promotes breast cancer metastasis Retromer recycles MT1-MMP to promote cancer invasion. J. Cell Biol. 219 (1). https://doi.org/10.1083/jcb.201812098.
Small, J.V., 1985. Geometry of actin-membrane attachments in the smooth muscle cell: the localisations of vinculin and alpha-actinin. EMBO J. 4 (1), 45–49.

Smith, C., Dolat, L., Angelis, D., Forgacs, E., Spiliotis, E.T., Galkin, V.E., 2015. Septin 9 exhibits polymorphic binding to F-actin and inhibits myosin and cofilin activity. J. Mol. Biol. 427 (20), 3273–3284.
Song, Y., Ye, M., Zhou, J., Wang, Z., Zhu, X., 2019a. Targeting E-cadherin expression with small molecules for digestive cancer treatment. Am. J. Transl. Res. 11 (7), 3932.
Song, Y., Ye, M., Zhou, J., Wang, Z.w., Zhu, X., 2019b. Restoring E-cadherin expression by natural compounds for anticancer therapies in genital and urinary cancers. Mol. Ther. Oncolytics 14 (109), 130–138.
Stamenović, D., Wang, N., 2011. Stress transmission within the cell. Compr. Physiol. 1 (1), 499–524.
Stark, R., Liebes, L.F., Nevrla, D., Conklyn, M., Silber, R., 1982. Decreased actin content of lymphocytes from patients with chronic lymphocytic leukemia. 59, 536–541.
Stevenson, R.P., Veltman, D., Machesky, L.M., 2012. Actin-bundling proteins in cancer progression at a glance. J. Cell Sci. 125 (5), 1073–1079.
Stylianopoulos, T., Munn, L.L., Jain, R.K., 2018. Reengineering the physical microenvironment of tumors to improve drug delivery and efficacy: from mathematical modeling to bench to bedside. Trends Cancer 4 (4), 292–319.
Sung, K.L., Dong, C., Schmid-SchOenbein, G.W., Chien, S., Skalak, R., 1988. Leukocyte relaxation properties. Biophys. J. 54 (2), 331–336.
Suresh, S., 2007. Biomechanics and biophysics of cancer cells. Acta Mater. 55 (12), 3989–4014.
Svitkina, T., 2018. The actin cytoskeleton and actin-based motility. Cold Spring Harb. Perspect. Biol. 10 (1), 1–21.
Szymanski, W.G., Zauber, H., Erban, A., Gorka, M., Wu, X.N., Schulze, W.X., 2015. Cytoskeletal components define protein location to membrane microdomains. Mol. Cell. Proteomics 14 (9), 2493–2509.
Tabatabaei, M., Tafazzoli-Shadpour, M., Khani, M.M., 2019. Correlation of the cell mechanical behavior and quantified cytoskeletal parameters in normal and cancerous breast cell lines. Biorheology 1, 1–13.
Tadeo, I., Berbegall, A.P., Escudero, L.M., Álvaro, T., Noguera, R., 2014. Biotensegrity of the extracellular matrix: physiology, dynamic mechanical balance, and implications in oncology and mechanotherapy. Front. Oncol. 4, 1–10.
Taherian, A., Li, X., Liu, Y., Haas, T.A., 2011. Differences in integrin expression and signaling within human breast cancer cells. BMC Cancer 11 (1), 293.
Takeichi, M., 1993. Cadherins in cancer: implications for invasion and metastasis. Curr. Opin. Cell Biol. 5 (5), 806–811.
Tan, X., Chen, M., 2014. MYLK and MYL9 expression in non-small cell lung cancer identified by bioinformatics analysis of public expression data. Tumor Biol. 35 (12), 12189–12200.
Tavares, S., et al., 2017. Actin stress fiber organization promotes cell stiffening and proliferation of pre-invasive breast cancer cells. Nat. Commun. 8, 1–18.
Tomaskovic-Crook, E., Thompson, E.W., Thiery, J.P., 2009. Epithelial to mesenchymal transition and breast cancer. Breast Cancer Res. 11 (6), 1–10.
Van Roy, F., Mareel, M., 1992. Tumour invasion: effects of cell adhesion and motility. Trends Cell Biol. 2 (6), 163–169.
Varani, J., Wass, J.A., Rao, K.M.K., 1983. Actin changes in normal human and rat leukocytes and in transformed human leukocytic cells. J. Natl. Cancer Inst. 70 (5), 805–809.
Vaupel, P., 2010. Metabolic microenvironment of tumor cells: a key factor in malignant progression. Exp. Oncol. 32 (3), 125–127.
Vignjevic, D., et al., 2007. Fascin, a novel target of β-catenin-TCF signaling, is expressed at the invasive front of human colon cancer. Cancer Res. 67 (14), 6844–6853.
Vogelstein, B., Papadopoulos, N., Velculescu, V.E., Zhou, S., Diaz, L.A., Kinzler, K.W., 2013. Cancer genome landscapes. Science (80-.). 339 (6127), 1546–1558.

Walker, C., Mojares, E., Del Río Hernández, A., 2018. Role of extracellular matrix in development and cancer progression. 19 (10), 1–31.
Wang, J.H.C., Li, B., 2010. Mechanics rules cell biology. Sports. Med. Arthrosc. Rehabil. Ther. Technol. 2 (1), 1–7.
Wang, N., Butler, J.P., Ingber, D.E., 1993. Mechanotransduction across the cell surface and through the cytoskeleton. Science (80-.). 260 (5111), 1124–1127.
Wang, J.-H., et al., 2017. Expression and prognostic significance of MYL9 in esophageal squamous cell carcinoma. PLoS One 12 (4), e0175280.
Wear, M.A., Schafer, D.A., Cooper, J.A., 2000. Actin dynamics: assembly and disassembly of actin networks. Curr. Biol. 10 (24), R891–R895.
Weaver, A.M., 2006. Invadopodia: specialized cell structures for cancer invasion. Clin. Exp. Metastasis 23 (2), 97–105.
Welch, D.R., Hurst, D.R., 2019. Defining the hallmarks of metastasis. Cancer Res. 79 (12), 3011–3027.
Wendt, M.K., Taylor, M.A., Schiemann, B.J., Schiemann, W.P., 2011. Down-regulation of epithelial cadherin is required to initiate metastatic outgrowth of breast cancer. Mol. Biol. Cell 22 (14), 2423–2435.
Williams, K.C., et al., 2019. Invadopodia are chemosensing protrusions that guide cancer cell extravasation to promote brain tropism in metastasis. Oncogene 38 (19), 3598–3615.
Wirtz, D., Konstantopoulos, K., Searson, P.C., 2012. The physics of cancer: the role of physical interactions and mechanical forces in metastasis. Nat. Rev. Cancer 11 (7), 512–522.
Wong, A.S.T., Gumbiner, B.M., 2003. Adhesion-independent mechanism for suppression of tumor cell invasion by E-cadherin. J. Cell Biol. 161 (6), 1191–1203.
Wu, P.H., et al., 2018. A comparison of methods to assess cell mechanical properties. Nat. Methods 15 (7), 491–498.
Xu, W., Mezencev, R., Kim, B., Wang, L., McDonald, J., Sulchek, T., 2012. Cell stiffness is a biomarker of the metastatic potential of ovarian cancer cells. PLoS One 7 (10), e46609.
Yamada, S., Wirtz, D., Kuo, S.C., 2000. Mechanics of living cells measured by laser tracking microrheology. Biophys. J. 78 (4), 1736–1747.
Yamaguchi, H., Condeelis, J., 2007. Regulation of the actin cytoskeleton in cancer cell migration and invasion. Biochim. Biophys. Acta-Mol. Cell Res. 1773 (5), 642–652.
Yamazaki, D., Kurisu, S., Takenawa, T., 2005. Regulation of cancer cell motility through actin reorganization. Cancer Sci. 96 (7), 379–386.
Yang, D.-H., Lee, J.-W., Lee, J., Moon, E.-Y., 2014. Dynamic rearrangement of F-actin is required to maintain the antitumor effect of trichostatin A. PLoS One 9 (5), e97352.
Yang, F., Petsalaki, E., Rolland, T., Hill, D.E., Vidal, M., Roth, F.P., 2015. Protein domain-level landscape of cancer-type-specific somatic mutations. PLoS Comput. Biol. 11 (3), 1–31.
Yilmaz, M., Christofori, G., 2009. EMT, the cytoskeleton, and cancer cell invasion. Cancer Metastasis Rev. 28 (1–2), 15–33.
Yoder, B.J., et al., 2005. The expression of fascin, an actin-bundling motility protein, correlates with hormone receptor-negative breast cancer and a more aggressive clinical course. Clin. Cancer Res. 11 (1), 186–192.
Yu, H.-J., et al., 2013. Tumor stiffness is unrelated to myosin light chain phosphorylation in cancer cells. PLoS One 8 (11), e79776.
Zhang, X., Pei, Z., Ji, C., Zhang, X., Xu, J., Wang, J., 2017. Novel insights into the role of the cytoskeleton in cancer. In: Cytoskeleton-Structure, Dynamics, Function and Disease. IntechOpen.
Zhou, J., Giannakakou, P., 2005. Targeting microtubules for cancer chemotherapy. Curr. Med. Chem. Anticancer Agents 5 (1), 65–71.
Ziegler, W.H., Gingras, A.R., Critchley, D.R., Emsley, J., 2008. Integrin connections to the cytoskeleton through Talin and vinculin. Biochem. Soc. Trans. 36 (2), 235–239.

Kolibri 360 GmbH
Ferdinand-Jühlke-Straße 7
99095 Erfurt, Deutschland
produktsicherheit@kolibri360.de